EQUINE
Artificial Insemination

This book is dedicated to:
Mum and Dad, as a thanks for all they have done for me;
and God, the creator of the subject of this book.

EQUINE
Artificial Insemination

MINA C.G. DAVIES MOREL

Welsh Institute of Rural Studies,
University of Wales
Aberystwyth
UK

CABI *Publishing*

CABI *Publishing* is a division of CAB *International*

CABI Publishing
CAB International
Wallingford
Oxon OX10 8DE
UK

CABI Publishing
10 E 40th Street
Suite 3203
New York, NY 10016
USA

Tel: +44 (0)1491 832111
Fax: +44 (0)1491 833508
Email: cabi@cabi.org

Tel: +1 212 481 7018
Fax: +1 212 686 7993
Email: cabi-nao@cabi.org

© CAB *International* 1999. All rights reserved. No part of this publication may be reproduced in any form or by any means, electronically, mechanically, by photocopying, recording or otherwise, without the prior permission of the copyright owners.

A catalogue record for this book is available from the British Library, London, UK.

Library of Congress Cataloging-in-Publication Data
Davies Morel, M. C. G. (Mina C. G.)
    Equine artificial insemination / M.C.G. Davies Morel.
        p.   cm.
    Includes bibliographical references (p.   ) and index.
    ISBN 0-85199-315-X (alk. paper)
    1. Horses--Artificial insemination.   2. Horses--Breeding.
I. Title.
SF291.D286 1999
636.1'08'245--dc21

98-33256
CIP

ISBN 0 85199 315 X

Typeset in Garamond by Columns Design Ltd, Reading
Printed and bound in the UK by Biddles Ltd, Guildford and King's Lynn

Contents

Acknowledgements		x
1. Introduction		**1**
1.1.	Introduction	1
1.2.	Advantages of Artificial Insemination	1
1.3.	Disadvantages of Artificial Insemination	5
1.4.	Conclusion	8
2. Historical Development and Present Use of Artificial Insemination		**9**
2.1.	Introduction	9
2.2.	Early History of the Technique	9
2.3.	Decline in the Horse Population and Equine Research	12
2.4.	Development of Techniques for the Collection of Semen	14
	2.4.1. Early methods of collection	14
	2.4.2. Development of the artificial vagina	15
2.5.	Development of Techniques for the Evaluation of Semen	17
2.6.	Development of Techniques for the Handling and Storage of Semen	18
2.7.	Development of Techniques for the Insemination of Semen	20
2.8.	Present Use of Equine AI	21
	2.8.1. Regulations governing the use of equine AI	24
2.9.	Current Reasons for Using Equine AI	27
	2.9.1. Removal of geographical restrictions	27
	2.9.2. Reduction in transfer of infection	28
	2.9.3. Improvement of native stock	29
	2.9.4. Development of gene banks	30
	2.9.5. Breeding from difficult mares	30
	2.9.6. Breeding from difficult stallions	33
	2.9.7. Reduction in labour costs	35
2.10.	Conclusion	35

3.	**Stallion Reproductive Anatomy and Control**		37
	3.1.	Introduction	37
	3.2.	Stallion Anatomy	37
		3.2.1. The penis	37
		3.2.2. Deposition of spermatozoa	41
		3.2.3. The accessory glands	46
		3.2.4. The vas deferens	50
		3.2.5. The epididymis	51
		3.2.6. The testis	53
	3.3.	Control of Stallion Reproduction	62
		3.3.1. Commencement of reproductive activity	62
		3.3.2. Seasonality	63
		3.3.3. The hypothalamic–pituitary–gonadal axis	67
		3.3.4. Hypothalamic and pituitary hormones	69
		3.3.5. Testicular hormones	72
		3.3.6. Behavioural control	75
		3.3.7. Hormonal regulation and control of spermatogenesis	75
	3.4.	Conclusion	77
4.	**Production of Spermatozoa**		78
	4.1.	Introduction	78
	4.2.	Seminal Plasma	82
		4.2.1. Protein	84
		4.2.2. Glucose	85
		4.2.3. Sorbitol	85
		4.2.4. Lactic acid	85
		4.2.5. Citric acid	86
		4.2.6. Inositol	86
		4.2.7. Ergothionine	86
		4.2.8. Glycerylphosphorylcholine	86
		4.2.9. Minerals	87
		4.2.10. Glycosidases	87
		4.2.11. Additional minor seminal plasma components	87
		4.2.12. Seminal plasma abnormalities	87
	4.3.	Spermatozoa	88
		4.3.1. Anatomy	88
		4.3.2. Spermatogenesis	104
		4.3.3. Metabolism and capacitation	139
		4.3.4. Function of the spermatozoon	145
		4.3.5. Movement of the spermatozoon	146
	4.4.	Conclusion	149
5.	**Semen Collection**		151
	5.1.	Introduction	151
	5.2.	The Artificial Vagina	151
		5.2.1. The Cambridge and Colorado models	151

	5.2.2. The Nishikawa model	155

        5.2.2. The Nishikawa model   155
        5.2.3. The Missouri model   156
        5.2.4. Alternative models   158
        5.2.5. Preparation and maintenance   160
  5.3. The Condom   166
  5.4. Preparation of the Stallion for Semen Collection   167
  5.5. The Collecting Area   170
  5.6. Sexual Stimulation of the Stallion and Encouragement to Ejaculate   170
        5.6.1. The jump mare and her preparation for use   172
        5.6.2. The dummy mare   174
  5.7. Alternative Methods of Semen Collection   178
        5.7.1. Manual stimulation   178
        5.7.2. Minimal restraint   178
        5.7.3. Collection without mounting   179
  5.8. Collection Procedure   181
  5.9. Frequency of Collection   185
  5.10. Training the Stallion for Collection   186
  5.11. Conclusion   189

6. Semen Evaluation   190
  6.1. Introduction   190
  6.2. Semen Evaluation as Part of an AI Programme   191
  6.3. General Semen Handling   191
  6.4. Gross Evaluation   193
        6.4.1. Volume   194
        6.4.2. Appearance   196
        6.4.3. Osmolarity   201
        6.4.4. Seminal fluid pH   201
  6.5. Microscopic Evaluation   202
        6.5.1. Motility   203
        6.5.2. Longevity   208
        6.5.3. Concentration   209
        6.5.4. Morphology   213
        6.5.5. Live:dead ratio   219
        6.5.6. Cytology   221
        6.5.7. Bacteriology   221
        6.5.8. Virology   225
        6.5.9. Mycoplasma   226
  6.6. Functional Tests   226
        6.6.1. Biochemical analysis   226
        6.6.2. Membrane integrity tests   227
        6.6.3. Flow cytometry   228
        6.6.4. Filtration assay   228
        6.6.5. Hypo-osmotic stress test   229
        6.6.6. Cervical mucus penetration test   230

		6.6.7.	Oviductal epithelial cell explant test	230

        6.6.7.   Oviductal epithelial cell explant test   230
        6.6.8.   Zona-free hamster ova penetration assay   231
        6.6.9.   Heterospermic insemination and competitive
                 fertilization   231
        6.6.10. Hemizona assay   232
        6.6.11. Chromatin analysis   232
  6.7. Sexing Spermatozoa   232
  6.8. Conclusion   233

7. **Semen Storage and Transportation**   234
  7.1. Introduction   234
  7.2. Extenders   235
        7.2.1.   Historical development   235
        7.2.2.   Present use of extenders   237
        7.2.3.   Milk and milk-product extenders   238
        7.2.4.   Cream–gel extenders   242
        7.2.5.   TRIS extenders   244
        7.2.6.   Egg yolk extenders   245
        7.2.7.   Other major components within extenders   247
        7.2.8.   Removal of seminal plasma   252
        7.2.9.   Osmolarity   258
        7.2.10. pH   258
  7.3. Fresh or Raw Semen   258
        7.3.1.   Extenders for use with fresh semen   259
        7.3.2.   Removal of seminal plasma   260
        7.3.3.   Dilution rates   260
        7.3.4.   Conception rates   260
        7.3.5.   Conclusion   261
  7.4. Chilled Semen   261
        7.4.1.   Cold shock   262
        7.4.2.   Cooling rates   266
        7.4.3.   Storage temperature   267
        7.4.4.   Length of storage time   268
        7.4.5.   Extenders for use with chilled semen   269
        7.4.6.   Removal of seminal plasma   271
        7.4.7.   Packaging for chilled semen   271
        7.4.8.   Methods of transportation   271
        7.4.9.   Dilution rates   275
        7.4.10. Conception rates   276
        7.4.11. Conclusion   276
  7.5. Cryopreservation of Semen   276
        7.5.1.   Principles   277
        7.5.2.   Extenders for use with frozen semen   280
        7.5.3.   Removal of seminal plasma   288
        7.5.4.   Packaging for frozen semen   289
        7.5.5.   Cooling rates   292

		7.5.6. Thawing rates and extenders	294
		7.5.7. Dilution rates	295
		7.5.8. Conception rates	296
		7.5.9. Factors affecting success rates of frozen semen	298
		7.5.10. Conclusion	300
	7.6.	Conclusion	301
8.	**Mare Insemination**		302
	8.1.	Introduction	302
	8.2.	Selection of Suitable Mares for Insemination	302
		8.2.1. History of the mare	303
		8.2.2. Temperament	303
		8.2.3. General condition	304
		8.2.4. Age	304
		8.2.5. Competence of the reproductive tract	304
	8.3.	Preparation of the Mare for Insemination	312
		8.3.1. Manipulation of the oestrous cycle	312
		8.3.2. Veterinary examination	322
		8.3.3. Physical preparation of the mare	322
	8.4.	Preparation of Semen for Insemination	322
		8.4.1. Thawing or warming semen	323
		8.4.2. Insemination dose	326
		8.4.3. Insemination volume	328
	8.5.	Methods of Insemination	329
		8.5.1. Insemination per vagina	329
		8.5.2. Insemination per rectum	330
	8.6.	Factors Affecting the Success of AI	333
		8.6.1. Timing and frequency of insemination	333
	8.7.	Conception Rates to AI	335
	8.8.	Conclusion	335
9.	**Conclusion**		337
References			342
Index			399

Acknowledgements

I am most grateful to many people for their help and support throughout the preparation of this book. In particular I would like to thank: Prof. Barry Ball (University of California, Davis), Dr Stuart Revell (Genus, UK) and Dr Joyce Parlevliet (University of Utrecht, The Netherlands) for their invaluable advice and guidance with the text of the book; Prof. Ian Gordon (University of Dublin, Republic of Ireland), Martin Boyle (Cambridge Stallion Services, UK) and Dr Desmond Leadon (Irish Equine Centre, Republic of Ireland) for their help and information provided; Mrs Shirley George, Ms Hilary Yeoman, Ms Alison Bramwell and Mr Alan Leather (University of Wales) for drawing, and help with labelling, the diagrams; Ms Julie Baumber (University of Wales and University of California, Davis), Dr Victor Medina (University of California, Davis), Hamilton Thorne Ltd (Beverly, USA) and Ms Jennifer James for providing many of the photographs; Mr John Saycell (University of Wales) for help with the collation of the photographs. Last, but not least, I would like to thank my family, Roger, Christopher and Andrew, for putting up with all the ups and downs that go with a project such as this.

1 Introduction

1.1. Introduction

Artificial insemination (AI) involves the collection of semen from a male, usually of superior genetic merit, followed by the transfer of that semen into a sexually receptive female at the time of ovulation, in order to result in fertilization. It is practised in numerous mammals, including humans, farm livestock and exotic zoological species. It has a long history, with the first reputed use being in the 14th century. However, significant development of the technique did not occur until the end of the 19th century. Research was largely carried out in horses and dogs, with the first commercial application being in horses in Russia at the end of the 19th century. Before its use in horses could be established, the advent of the combustion engine and the subsequent decline in horse numbers drove AI research towards its use in other farm livestock. Though some countries continued their interest in and use of equine AI, usually on a small scale, many countries concentrated their interests on bovine and then ovine and porcine AI, which had the greater earning potential. The recent upsurge in interest in equine AI during the last 15 years has been a reflection of the increase in horse numbers along with the developing leisure interest in horse riding and the realization of the economic advantages of AI.

1.2. Advantages of Artificial Insemination

There are many advantages in the use of equine AI, including the following.

Facilitation and acceleration of genetic improvement of stock

The use of AI results in an increase in the use of males with superior genetic merit and accelerates the introduction of new genetic material through importation of foreign semen. This will increase the incidence of genetically advantageous traits within the population and encourage the widespread use of outstanding stallions and hence worldwide dissemination of valuable genetic material. This can also be to the advantage of breeding programmes, allowing accelerated improvement and facilitating progeny testing. However, care must be taken in defining 'genetically superior', as this may only be a result of temporary fashion. Concentration on one trait will be to the exclusion of others, thus limiting the genetic base from which future 'fashionable' traits may be drawn.

Ensuring routine semen evaluation and monitoring

The collection of semen prior to AI allows evaluation of the sample. This, plus the associated routine evaluations, allows problems to be detected quickly so that immediate remedial action can be taken.

Improvement in the reproductive potential of sub-fertile stallions

AI may be used to increase the reproductive potential of stallions with poor semen quality. Treatment of semen, including concentration and filtration, may be used to produce a good quality sample from a number of poorer quality ejaculates. The insemination of this sample into the mare will enhance the chances of fertilization. Semen extenders can be designed specifically to contain supportive and protective factors for spermatozoa, which may then enhance pregnancy rates from sub-fertile stallions.

Allowing the covering of problem mares, precluded from natural service

AI may be used to overcome problems that may preclude a mare from natural service. Such problems include skeletal abnormalities or weaknesses, laminitis, navicular, or a highly strung or nervous temperament. It may also be used on mares that have undergone caslick surgery and removes the need for the normal episiotomy and resuturing at covering. However, in breeding from such animals, great care should be taken to ensure that the problem is not the result of an inherited trait, as there is little advantage in perpetuating such disadvantageous traits within the population.

Allowing mares with a heightened post-coital immunological response to be covered

AI may be used in mares that habitually demonstrate a heightened immunological response to spermatozoa *post coitum*. Such mares will benefit from covering with smaller doses of spermatozoa than would normally be ejaculated at natural service. Manipulation of oestrus and close monitoring of such mares prior to insemination should allow a single appropriately timed insemination to be successful rather than the normally practised several natural coverings per ovulation.

Permitting mares who suffer severe post-coital endometritis to be covered

AI may be used to decrease the severity of post-coital endometritis. All mares demonstrate some degree of post-coital endometritis, as a natural response to the immunological challenge presented by an ejaculate. Some mares are unable to respond appropriately to this challenge and habitually display acute endometritis *post coitum*. Such a severe inflammatory response may compromise the ability of the uterus to support a developing pregnancy. The insemination of semen extended in antibiotic extender reduces the severity of the challenge and the response in mares that are particularly susceptible.

Removal of geographical restrictions

One of the main economic advantages of AI is that it enables individuals that are geographically isolated to be mated without the need for transportation of animals over long, and often prohibitive, distances. Semen is now chilled and transported worldwide and, as the success of freezing semen improves, geographical barriers cease to exist. Such transportation allows the widespread distribution of an array of genetic traits, allowing their introduction into different populations and increasing the scope for genetic improvement.

Permitting the storage of semen for posterity

The possibility of freezing semen allows the storage of semen for posterity, with the potential for its use in the distant future after the death of the stallion. Along with this is the potential to develop a gene bank where samples from stallions of particularly high merit can be laid down for posterity, with the possibility of reintroducing these genes into the population long after that stallion has died.

Increasing the number of mares covered per stallion

AI allows a significant increase in the number of mares a stallion may cover during one season. Income in studs standing high-priced stallions is generated largely from covering fees or nominations. The greater the number of nominations, the greater is the income, which is of obvious benefit to stallion owners; but it may also lead to a reduction in the nomination fee, which is of advantage to mare owners and often the breed as whole, thus making valuable stallions accessible to the ordinary mare owner. The use of AI also reduces the risk of physical overuse of popular stallions while still allowing widespread access.

Allowing stallions to run concurrent careers as performance horses and breeding stallions

The use of AI allows stallions to carry out a full competitive career as well as covering a full book of mares. As part of the daily or weekly routine, semen may be collected at a specific time within that routine. It can be processed for storage and transportation, or even for immediate use on the yard later that day or as mares are brought in. If collection is carried out early in the morning and only on a couple of days per week, the stallion is then free for significant periods to compete, without interfering with his covering duties. The stallion's routine is standardized and the mares can be inseminated as soon as it is appropriate. Such a system also eases management by standardizing general yard routine.

Assistance in the preservation of rare breeds

The collection and storage of semen from rare breeds can be used to help to guarantee their perpetuation. Used with care, semen from such breeds can be reintroduced at a later date to widen the genetic pool within a restricted population. The limitation of the genetic pool is one of the major problems encountered by rare breed survival organizations.

Permitting the use of fixed-time AI and so potentially reducing labour costs

AI should ideally allow the use of 'fixed-time insemination'. Such a system aims to manipulate the mare's oestrus and ovulation to fix the time of ovulation and, therefore, insemination. This, in theory, would remove any need for teasing (with its accompanying risk of disease transfer or injury) and veterinary examination. Although this prediction of ovulation and use of fixed-time AI is very successful in cattle, as yet no means has been identified to time, or even predict precisely, the timing of ovulation in the mare: too much variation exists. At present, manipulation of the oestrous cycle along with the monitoring of ovarian activity via rectal palpation or ultrasonography will allow a form of fixed-time AI to be practised, removing the need at least for animal contact.

Allowing stock to be covered that are isolated on health grounds

AI allows mares and stallions isolated due to health restrictions (for example, influenza or strangles outbreaks) still to be covered at the planned time, as no direct contact between stock is required.

Aiding in the control of disease

The use of AI can aid in disease control as semen can be tested prior to insemination. The addition of antibiotics to semen extenders also reduces the chances of venereal transmission of bacterial diseases when compared with the risk with natural service.

Reduction in the risk of injury

The avoidance of direct contact between the mare and the stallion removes all risk of contact injury. This is of particular advantage to owners of valuable stock, as injury would at best result in a temporary withdrawal from the breeding programme, but may at worst result in permanent physical or psychological damage, and infertility.

Permitting the use of injured stallions

A stallion with injuries, especially to the hindquarters and pelvic and back regions, may not be able to mount a mare for natural service. The use of AI will prolong his breeding career. Care must be taken to ensure that any damage is the result of an injury, and not an inherited fault which could be passed on to subsequent generations.

Reinforcement of natural service

AI can be used to reinforce natural service in stallions that fail to cover naturally due to an inability to achieve full tumescence or penetration.

Encouragement of routine examination of the mare's reproductive tract

Invariably, prior to insemination, a mare's tract will be examined internally to ascertain the suitability of that mare for insemination. As a result, problems that may otherwise have gone unnoticed are identified and others are diagnosed earlier, thus facilitating more rapid and more successful treatment.

Extension of the breeding season

Semen may be collected and frozen during the non-breeding season and stored for use during the following breeding season. In essence this allows the breeding season to be extended and increases the total amount of semen available.

1.3. Disadvantages of Artificial Insemination

There are some disadvantages or concerns expressed regarding the use of AI in horses, including the following.

Reduction of the genetic pool

Fear has been expressed by many people that the use of AI will result in the domination of breeding lines by a few elite stallions, with a resultant reduction in the genetic pool. This is a possible danger with the uncontrolled use of AI. However, many countries and breed societies limit the number of mares that a stallion is allowed to cover per year to a figure not much greater than what he could cover naturally. With such restrictions no greater reduction in the genetic pool should be evident than that resulting from the continued use of natural covering.

Reduction in the potential income from mare keep fees

A considerable amount of money is generated per year on stud farms from the keep fees charged to mare owners for the keep of their mares on the stallion's stud during their stay for covering. This may be quite a considerable fee in some of the Thoroughbred and Warmblood studs, where mares are habitually brought in a few weeks prior to foaling and foaled down on the stallion's stud, remaining there until they are covered and often until a positive scan has been observed. This stay may be 6–8 weeks long and provides a useful addition to the covering fee. The use of AI does not require the mare to go to the stud. However, many studs could generate significant income if they took advantage of their skills in foaling mares and offered a skilled foaling-down service, which would be popular with many mare owners, especially those with single mares.

Presentation of ethical dilemmas

The ethical question as to whether a stallion should be able to reproduce after his death concerns many individuals and breed societies. Such concerns should not hamper the adopting of AI for all the other advantages outlined above. Due to this concern some breed societies place restrictions on the use of frozen semen and whether it may be used after the stallion's death and, if so, for how long. The potential problem over ownership of the semen in the future has to be addressed.

Presentation of problems over semen ownership

The sale of a stallion or his death may present the problem of semen ownership, but this can be addressed through regulations and agreements. For example, in France the sale of any stallion includes the automatic transfer of all semen stocks to the new owner.

Increasing the opportunity for fraud

Concern is expressed by many that the greater use of AI presents opportunity for fraud as the actual act of a specific stallion covering a certain mare cannot be witnessed. It is feared that the opportunity for swapping semen or the accidental incorrect labelling of samples may lead to problems with the parentage of offspring. The routine use of compulsory blood typing of all offspring practised by many breed societies prior to registration should allow such problems to be highlighted before registration. In addition, careful and regulated semen handling should ensure that such problems are minimized.

Significant variation in the quality of semen available

Significant variation in the quality of semen obtained for inseminating mares is a present problem in the use of equine AI. Many countries (for example,

Germany, France and The Netherlands) do set quality standards that have to be met by all stallions used for AI, as do many individual breed societies. Despite successful evaluation of a semen sample immediately after collection, significant variation exists in the ability of that semen sample to survive storage, especially freezing. This, plus the lack of a simple accurate test that reliably indicates fertility potential in a semen sample, means that the aim of a quality assurance certificate remains an ideal rather than a practical possibility at present.

Increasing the risk of disease transfer

The isolation of the mare and the stallion and the treatment of semen prior to insemination will reduce the risk of venereal and systemic disease transfer. However, evidence suggests that bacteria and viruses may be isolated in semen and may still have the potential to be transferred. This possibility emphasizes the need for strict hygiene and quarantine regulations in order to avoid the very real risk of the spreading of disease worldwide between populations that will have no inherent resistance due to their previous geographical isolation.

Increasing the cost of covering mares

The potential increase in the total costs involved in covering mares by AI compared with natural covering is another area of concern to many breeders. This may well prove a problem to small studs running only a few mares of limited value. In such enterprises, AI may well not be worth the veterinary costs involved for hormonal therapy for the manipulation of oestrus and the induction of ovulation, the monitoring of ovulation and the normal double insemination.

Increases in the rates of returns

When equine AI is compared with bovine AI, it is evident that the number of repeat inseminations required is much higher in the horse. This is largely due to the difficulty in precisely timing oestrus and ovulation in the mare by manipulation of the cycle. As a result, repeat inseminations are not uncommon. This does lead to disappointment and potentially increases the cost if reinsemination is required.

Responsibility for conception lies with mare owner/manager

Traditionally, with natural service, the responsibility for covering the mare was left to relatively few stud workers, who were often highly skilled. However, as mares for insemination are not taken to stud, the responsibility for ensuring that the mare is correctly detected in oestrus, that the semen is ordered to arrive at the appropriate time and that insemination coincides with ovulation now lies with a much larger group of people. These people include mare managers and veterinarians who are likely to have less experience in specifically detecting oestrus and in ascertaining the most appropriate time to cover a mare. As a result, success rates can be disappointing, through no fault of the technique but rather due to the inexperience of personnel.

Requirement for an increased degree of knowledge from both the veterinarian and mare manager

As the responsibility for success of AI lies largely with the mare manager and her veterinarian, education to improve these individuals' knowledge of equine AI is likely to be required. As with all new techniques, initial disappointing results may not be the fault of the technique *per se* but rather of those practising it, a failing that should be rectified with time and experience.

Risks to handlers at semen collection

Some risk does exist to handlers involved in the collecting procedure. Inexperienced stallions and handlers present the greatest risk to themselves and each other. Education and experience of both stallions and handlers should minimize such risks, which in any case should not be significantly greater than those of natural in-hand covering.

1.4. Conclusion

It is evident that, as with all techniques, there are advantages and disadvantages with the use of AI. It is also evident that in the case of equine AI the advantages far outweigh any possible disadvantages, and that many of the latter can be addressed by appropriate regulation and by familiarization with the technique and the associated management required. The industry is becoming increasingly aware of the advantages of AI and it is evident that many of the disadvantages or concerns expressed have been successfully addressed in some countries and within some breed societies. As a result, equine AI is becoming more widespread in its acceptance and will hopefully continue to do so. It is the aim of this book to assist in the widespread acceptance of equine AI largely via the route of education.

2 Historical Development and Present Use of Artificial Insemination

2.1. Introduction

The history of the development and use of artificial insemination in horses is a long and varied one. It mimics closely the rise and fall of the horse from its emergence as a essential tool of warfare, through the horse's heyday as the prime source of power and transport, until its position today mainly in leisure pursuits. The changes in our perception of the horse have been driven largely by its changing role within our society; in turn, this changing role is reflected in society's commitment to research and development into techniques such as AI. In many ways, it is our present perceptions and use of the horse that are the main hindrances to today's equine research.

2.2. Early History of the Technique

The first reference to the use of AI in horses is reputed to be in Arabic texts dating from as early as 1322. The story, apocryphal or not, is that one of two feuding sheikhs stole semen from his enemy's prize stallion for use on one of his mares. No details are given on the methods used but the operation was reported to have been successful (Perry, 1945, 1968). In 1677, Anton van Leeuwenhoek and Johan Hamm first identified spermatozoa using a microscope. They subsequently were described as animalcula – innumerable minute bodies with the power of active forward motion (Perry, 1968). However, it took another 100 years before artificial insemination was documented for the first time. This report was of the work done in 1780 by the Italian physiologist Spallanzani who, after some encouraging successes with amphibians, attempted the insemination of a bitch with freshly collected semen, kept at room temperature prior to immediate insemination directly into the uterus. This procedure was successful and resulted in the birth of three puppies. He later went on to evaluate his technique in horses (Perry, 1945, 1968; Varner, 1986). Work by Rossi and Branchi in 1782 verified

Spallanzani's successes (Perry, 1968). Spallanzani was able, during the course of his research, to establish that it was the sperm (he termed them spermatic vermiculi) that had the power of fertilization, rather than the associated fluid (seminal plasma) which he removed by filtration. Initial experiments into semen cooling and, therefore, prolonging of their life span were also conducted by Spallanzani (1803). He used snow to freeze the 'spermatic vermiculi' and found that though they became motionless when in contact with the snow, they did not die, and could be revived by warming. Before the late 19th century, AI in all species remained largely of scientific concern, until interest in the technique grew from commercial lay personnel. Programmes to apply the use of AI to livestock on a more commercial basis began to be developed in Russia and China during the late 19th century (Foote, 1982).

During the latter years of the 19th century there were reports of the use of AI in several European countries (Boyle, 1992). At the same time, Walter Heape, at the University of Pennsylvania, was building upon his recent success in the insemination of bitches (Heape, 1897) and reported the successful use of AI in mares on a number of farms (Perry, 1945). France was also developing the technique, spearheaded by a vet, Repiquet, who advised on the use of AI in Europe for the first time, to overcome infertility (Perry, 1945). Around this time Professor Hoffman of Stuttgart produced a report, *A Description, Instruments and Techniques for AI as a Supplement to Natural Service*, which, as the title suggests, described the use of AI as an additional safeguard after natural service, in order to try to improve the low conception rates that are characteristic of the horse. In this work, semen was collected from the mare *post coitum,* from the depression in the lower vaginal wall, by means of a speculum or spoon. The semen collected was diluted with cow's milk and inseminated into the mare's uterus using a specially designed syringe. Professor Hoffman only used the technique in addition to natural service or as a back-up to natural service, rather than in isolation. He did not investigate the use of semen in mares that had not already been covered naturally (Perry, 1968). Later on the use of sponges to collect semen deposited in the vagina at natural service was investigated. Though some success was reported, their use proved to be rather inefficient, with low spermatozoa recovery rates and high rates of contamination from vaginal secretions, bacteria and debris (Boyle, 1992). By the very end of the 19th century Lideman (in 1895) and Izmailov and Enisherlov (in 1896) were using AI for the mass production of horses in Russia. Similar techniques were also being used by Chaechowski in 1894 and Kaldrovics in 1902 in the Ukraine and Hungary, respectively, for the mass production of horses on large studs on a commercial basis (Chaechowski, 1894; Tischner, 1992b). In 1902 Sand and Stribolt reported at the Northern Livestock Conference in Copenhagen a 50% success rate (four pregnancies out of eight mares inseminated) for equine AI and identified a use for AI in the increased commercial use of valuable stallions (McKinnon and Voss, 1993).

This view marked the realization of the potential of the technique to improve farm livestock breeding efficiency rather than just as a treatment for infertility. This objective was first seriously considered in Russia at the beginning of the 20th century when the first equine AI programmes were organized by the pioneer I.I. Ivanoff, on Russian government-controlled stud farms, from 1899 onwards (Berliner, 1947; Gordon, 1983). At this time, many Russian studs employed AI as a means of serving their mares, but results were highly variable. It was reported that in inseminations carried out by Ivanoff himself conception rates equivalent to those of natural service were obtained. In one particular piece of work carried out by Ivanoff, 39 mares were inseminated, out of which 31 became pregnant; this compared very favourably with a similarly kept group of mares that were covered by natural service, in which only 10 out of 23 conceived (Ivanoff, 1922; Perry, 1945, 1968). Ivanoff's work was not restricted to horses; he also inseminated birds, cattle and sheep and led to the setting up of the veterinary laboratories under the Ministry of Agriculture in Russia.

Apart from research, these laboratories trained veterinary surgeons and AI practitioners. In the years leading up to the First World War, 300–400 trained AI practitioners went out from these laboratories to work on stud farms, and so significantly increased the number of artificially bred horses in Russia at this time. After the war a central experimental station was set up in Russia under the directorship of Ivanoff. These series of laboratories were directly responsible for the development of the modern artificial vagina in 1930, and for the proliferation of collection techniques including the use of the dummy mare (Olbrycht, 1935; Perry, 1945, 1968; Tischner, 1992a). In 1938, 120,000 mares were reported to have been covered by AI in Russia. This figure significantly increased in the following years to an average of 300,000–400,000 mares covered by AI per year.

Along with this significant interest in Russia, other countries – Japan in particular – started to develop an interest in the use of AI. Between 1913 and 1917, 323 Japanese mares were covered by AI (Perry, 1968). Interest was also developing in China, where 600,000 mares were inseminated in 1959, with a reported pregnancy rate of 61%. In 1960, China's two most popular stallions were used to inseminate 4415 and 3039 mares, respectively, with reported pregnancy rates of 76.9% and 68.1%, respectively (Cheng, 1962). Between the two world wars there was also extensive use of AI in European and Balkan countries (Boyle, 1992). Around this period and immediately after the Second World War, Hanover was particularly prominent in its use and development of AI techniques (Gotze, 1949). The development of AI for horses in the USA was slower than that in Europe, with significant work there being carried out on the use of AI in cattle, and in 1938 the first AI organization was established in the USA for cattle (Foote, 1982). In 1938 only 50 calves were born per sire per year by AI but, by 1981, this figure had increased to 50,000. Unfortunately, there was no synchronous development in the use of AI in goats, swine, sheep or horses. The stated reasons included management problems, prejudice, economic viability and less well developed technology (Graham *et al.*, 1978; Foote, 1980, 1982; Voss *et al.*, 1981).

By the 1930s the techniques for using AI in horses were well established and its use was spreading worldwide. However, from the 1940s onwards the use of AI in horses failed to continue to develop on a par with that of cattle and even of sheep (Table 2.1). As such, its development came to a standstill until a resurgence of interest several decades later.

2.3. Decline in the Horse Population and Equine Research

The most significant reason for the decline in use of AI in horses was the decline in horse numbers associated with the advent of the combustion engine (Boyle, 1992). A decline in horse numbers led to a lack of interest from horse breeders and from those who supported research and development. The traditional large-scale military studs of Russia and the Eastern bloc countries, which provided the most economical way of developing and using AI, were being replaced by smaller private studs. As a direct result of this the sophistication of the techniques for the use of AI in horses lagged behind those for cattle, in which there was a surging interest. Equine AI became an unfashionable subject and suffered as such for many years.

In addition to the problems associated with the decline in general interest, some of the original excuses for the lack of development were proved to be correct – in particular, the mare's notoriously variable oestrous cycle, and along with this the significant variation evident in the absolute and relative timing of oestrus and ovulation. Oestrus detection, without the use of a male of the species, also proved not to be as easy in the mare as with the cow (Voss *et al.*, 1981; Foote, 1982; Boyle, 1992). Finally the restrictive legislation brought in by many breed societies due to their opposition to the technique significantly curtailed the expansion of equine AI. The most notable of these breed societies was the UK Thoroughbred Breeders (Schell, 1948).

Despite the general decline in the use of AI in horses after the Second World War, some sectors of the industry retained their interest and kept the

Table 2.1. Estimate of the extent to which AI was used in domestic animals, worldwide in 1992 (Boyle, 1992).

Species	Total number of females inseminated	Use of frozen semen
Cattle	90,000,000	> 95%
Sheep	50,000,000	Experimental
Goats[a]	> 15,000	Experimental
Swine	6,000,000	5%
Horses	c. 1,000	Experimental

[a] China had an extensive goat population on which AI was used but no numbers are available.

technique alive. In the 1950s there was widespread use of AI in Standardbred breeding in the USA: more than 20,000 Standardbred mares were being inseminated per year (McKinnon and Voss, 1993). In 1953 Russia inseminated 450,000 mares as part of breed improvement schemes (Swire, 1962). In general, however, non-eastern-bloc countries saw a significant decline in the use of AI, with the exception of Japan where, in 1956, 10,000 (7.67% of all horses) were bred by AI. Spain, Brazil and Greece also continued to use the technique (Nishikawa, 1959a). In the late 1950s Denmark was one of the leaders in the commercial use of AI, with the setting up of three AI centres each offering semen from three to six stallions by AI; the majority use was for the covering of riding-school horses. By 1962 the worldwide insemination of horses was estimated to be 750,000, of which 80% were in China. This compares with more than 58 million cows which were being inseminated annually at this time (Boyle, 1992).

The birth of the first foal as a result of insemination with frozen semen, collected by means of an artificial vagina, did not occur until 1968 in the USA, but within the next 6 years 600 foals were born after conception using frozen semen (Ginther, 1979). Though this heralded a new breakthrough in equine AI, the resurgence of interest did not last long, due to failure to find a reliable cryopreservation agent for equine spermatozoa and the variability between the semen of different stallions in surviving the freezing and thawing process. The industry was also slow to realize the advantages of the technique and its widespread use continued to be hindered by opposition from many breed societies. Much of this opposition is gradually being overcome (Boyle, 1992) and in the last 20 years there has been a gradual acceptance of the technique by many national and international authorities throughout Europe. The Thoroughbred General Stud Book in the UK remains a notable exception (Barrelet, 1992).

Countries now using equine AI in significant numbers include China, Germany, Poland, The Netherlands, the Scandinavian countries and Japan (Gordon, 1983). In the 1970s and 1980s the Standardbred breeders remained one of the leaders in the use of the technique, especially in America and Australia, using primarily raw or diluted fresh semen at the place and time of collection in preference to storage and transportation (Gordon, 1983). Similar systems were used in Quarterhorse and Arab breeding. In fact, many large studs in these areas now exclusively breed by AI with the complete exclusion of natural covering (Allen et al., 1976b; Stabenfeldt and Hughes, 1977). The opposition to the technique showed by other societies has now been largely eroded. In the USA in 1973 only six of the 75 associations registering equines accepted stock bred as a result of conception using frozen semen; in 1993 the picture was very different, with only three failing to permit the registration of such progeny, namely the Thoroughbred Breeders Association, the Standard Jack and Jenny register of America and the American Miniature Horse Association (McKinnon and Voss, 1993). However, there is still considerable variation in the allowances and regulations stipulated by each society regarding the use of AI and in particular the storage and transport of semen (McKinnon and Voss, 1993).

2.4. Development of Techniques for the Collection of Semen

2.4.1. Early methods of collection

During the early work on AI, semen was collected largely from the vagina of the mare after she had been covered, using a spatula or spoon to scoop out the semen (Perry, 1968). Alternatively, a syringe fitted with a piece of rubber tubing attached was used to aspirate the semen from the vagina. A further development was the use of a vaginal sponge which would soak up the semen on ejaculation (Boyle, 1992). All these methods were successful in obtaining semen, but the volumes obtained were low and many of the sperm were reported to be damaged as a result of the method of collection. Semen contamination rates were also very high, with the accompanying risk of disease transfer; semen also came in contact with vaginal secretions for varying periods of time, exposing spermatozoa to the adverse effect of the acidic conditions within the vagina.

Due to the high risk of contamination and the loss of the majority of the sample, alternative methods of collection were investigated. One such method was the use of a urethral fistula, which was reported to yield high quality semen, though it never became accepted as a practical method of collection except for the odd stallion which was unable to serve a mare (Zivotkov, 1939). The use of a rubber semen collector was also investigated (Polozoff, 1928; Berliner, 1940; Nishikawa, 1959a; Swire, 1962). This consisted of a rubber bag with a wide neck and a round-bottomed end. For an average 15.1 hh (150 cm high) stallion, the bag would be 40 cm long with a 10 cm neck. It was inserted into the vagina of an oestrous mare or (later on) into a purpose-built lumen within a dummy mare. The stallion was allowed to cover the mare as normal and the ejaculate was collected in the rubber bag. One of the problems encountered with this system initially was the dislodging of the rubber bag as the stallion withdrew from the mare. The addition of a rubber ring, which fitted inside the vaginal commissure of the mare, held the bag in place (Lambert and McKenzie, 1940; Rasbech, 1959). This method was popular until the 1960s in many countries, including Argentina, Germany (Gotze, 1949), Japan (Nishikawa, 1959a,b), the USA (Lambert and McKenzie, 1940), Denmark (Rasbech, 1959) and the USSR (Milovanov, 1934; Parsutin, 1939). However, with the increased use of the artificial vagina and the associated risk of losing the ejaculate from dislodging of the bag, it lost popularity from the 1960s onwards.

Working along the same lines, the condom was introduced. This was a large latex rubber cylinder, placed over the erect penis, which found popularity in Denmark (Rasbech, 1959) and the USSR (Parsutin, 1939). Unfortunately, it had similar disadvantages to the rubber bag collector, with a high risk of becoming dislodged and the sample being lost. Both these methods also yielded samples which tended to be highly contaminated due to the prolonged contact between the sample and the exterior of the penis while in the bag or condom (Skatkin, 1952).

The technique of electro-ejaculation was also developed in the 1940s, primarily for use on bulls (Gunn, 1936; Laplaud and Cassou, 1948; Thibault *et al.*, 1948). It was subsequently refined and modified for use on rams and goats (Rowson and Murdoch, 1954; Edgar, 1957), boars (Dzuik *et al.*, 1954), chickens (Serebrovsky and Sokolovskaya, 1934) and ducks (Watamabe and Sugimori, 1957). However, electro-ejaculation was largely unsuccessful in stallions and its use was never implemented in practice, nor was the technique developed and refined.

2.4.2. Development of the artificial vagina

Throughout this period, work was also being carried out on the collection and insemination of semen in other species, including the dog. It was as part of his work with dogs that Professor Amantea, Professor of Human Physiology at the University of Rome, in 1914 developed the forerunner of today's artificial vagina (AV). The development of the first artificial vagina is therefore attributed to Amantea (Bonadonna and Caretta, 1954).

Transfer of this technique for use in stallions, bulls and rams did not occur until the 1930s. The development of the first stallion AV is attributed to the workers, including Salzmann, in Ivanoff's laboratories in Moscow in 1930. At the same time this laboratory introduced a famous lecture programme on the subject of AI which continued from 1930 to 1939 under the leadership of Professor Olbrycht (Milovanov, 1938; Tischner, 1992b). This laboratory continued to lead the way in semen collection and insemination, creating new techniques including the first reported use of a dummy mare in 1935 (Olbrycht, 1935). Two types or models of AV emerged in the 1940s: the Cambridge or Russian model and the American or Missouri model.

The Cambridge model, described by Walton and Rawochenski (1936), Day (1940) and Anderson (1945) and in the Russian literature by Milovanov (1934) and Parsutin (1939), was a large metal tube surrounded by an inner and outer rubber liner, which together formed a water jacket. There was an additional inner rubber liner to which was attached a collecting vessel into which the semen passed for collection immediately after ejaculation. The initial model had many problems in maintaining adequate pressure within the AV to encourage ejaculation, and the importance of this was first recognized by Milovanov (1934). These models were very large (54 cm in length and 13 cm in diameter), which meant they were exceedingly cumbersome when filled with adequate water to maintain the pressure and temperature required. Some of the early prototypes required two men to hold them as the stallion ejaculated. This significant disadvantage was rectified somewhat by the introduction of an aluminium outer casing with a standard rubber Cambridge liner (Skatkin, 1952). Plastic was also investigated as a lighter alternative to the metal casing, along with a shortening of the AV and the addition of an elastic ring to imitate the vulval sphincter (Berliner, 1940; Day, 1940).

The Missouri model emerged from America at about the same time (Lambert and McKenzie, 1940; Anderson, 1945). This AV was more complex than the Cambridge: it consisted of an outer rubber cylinder, 15 cm in diameter and 45 cm in length, with a thick inner rubber lining tube which was vulcanized to the outer tube at both ends. Again, in an attempt to mimic the vulval sphincter, this AV had a constriction made of a sphincter-like rubber band at the proximal end (nearest the entry). The distal end, where the semen was deposited, tapered to a collecting bottle. Due to the realization of the importance of pressure on the penis for ejaculation, modifications were made to this prototype to try to imitate more closely the conditions of natural service (Nielsen, 1938; Gotze, 1949). Further modifications were made by Berliner (1940, 1947) to imitate vaginal pressure, especially in the area of the glans penis (Love, 1992). This forms the basis of the Missouri type AV used today.

There is now a third popular type of AV: the Nishikawa, which was developed in Japan in the 1950s. It is the smallest of the three types and was introduced in an attempt to overcome the cumbersome nature of the alternative AVs. It is made of an aluminium casing with a removable rubber liner and a doughnut-shaped sponge that can be placed at the distal end of the AV to increase pressure on the glans penis (Nishikawa, 1959a). This model remains largely unchanged but the additional doughnut-shaped ring is not normally used.

The major advantage of all of these AVs was that they resulted in semen samples with significantly reduced contamination rates, because collection was into a separate collecting vessel, attached to the end of the AV, and the sample was never in contact with the external part of the penis or the vaginal secretions of the mare. Their use was also reported not to be associated with a significant loss in libido, which had originally been feared. In addition, ejaculation time – that is, the time between intromission and ejaculation – was only slightly delayed in stallions using an AV (16 s, compared with 13 s). The number of mounts per ejaculate was not significantly affected by the use of an AV (2.2 mounts per ejaculate for an AV, compared with 1.4 mounts per ejaculate for natural covering) (Wierzbowski, 1958). It was also encouraging that the vast majority of stallions were reported to be happy to ejaculate into an AV, and in work by Bielanski (1951) it was reported that, out of a sample of 6000 stallions, only 3.4% refused to ejaculate into an AV. The use of a real teaser mare as opposed to a dummy was associated with fewer refusals, but most stallions accepted the use of a dummy mare quite readily (Parsutin and Rumjanceva, 1953).

These three basic types of AV are the forerunners of the three major types used today. Most of the more recent modifications have been to the material from which they are constructed, in order to reduce their weight and ease their handling, rather than any need to change the basic principles of their design. The development of lighter AVs has allowed semen collection by a single operator. This was further helped by the development, in the 1930s, of the dummy mare (Olbrycht, 1935). The dummy mare was designed to mimic the body of the teaser mare: a stallion could be encouraged to mount the dummy and ejaculate into an AV. The obvious advantage of such a system was the removal of the risk to the teaser mare of accidental conception or injury to what was often a valuable stallion.

The next major development in the AV was the use of plastic disposable liners, which were introduced on a wide scale in the 1980s. They became very popular, as they precluded the need to disassemble and clean the AV after each collection (a very time-consuming job). Once it had been demonstrated that the use of polythene liners had no detrimental affect on spermatozoan quality, specifically motility, their use became widespread (Merilan and Loch, 1987; Silva *et al.*, 1990).

2.5. Development of Techniques for the Evaluation of Semen

Initial evaluation of semen involved visual assessment of the sample's gross appearance, volume, etc. Later the use of light microscopes allowed the anatomical features and dimensions of spermatozoa to be recorded and hence assessed (Bretscheider, 1949; Bielanski, 1951; Nishikawi and Wade, 1951; Nishikawa *et al.*, 1951; Bonadonna and Caretta, 1954; Hancock, 1957). Work on Japanese horses by Nishikawi and Waide (1951) and Nishikawa *et al.* (1951) suggested the following measurements for normal spermatozoa:

Total length	60.55 μm	(57.4–63.0 μm)
Head length	7.0 μm	(6.1–8.0 μm)
Head width	3.9 μm	(3.3–4.6 μm)
Length of connecting piece	9.8 μm	(9.2–11.5 μm)

Results obtained by Bielanski (1951), using Polish horses, largely agreed with these figures for normal spermatozoa:

Total length	55.0 μm
Head length	5.0 μm
Length of connecting piece	8.0 μm

In addition, both sets of work assessed the number of normal and abnormal spermatozoa within their samples. Nishikawa and Waide suggested that the average percentage abnormal spermatozoa within a stallion semen sample was 16% (range 6–27%) whereas Bielanski came up with a higher level of 26%, stating also that the majority of abnormalities were the result of coiled tails and/or the presence of protoplasmic droplets. In the USA Macleod and McGee (1950) suggested, from their work on 23 Thoroughbred stallions, that the average expected volume of semen is 60 ± 20 ml, with a spermatozoan concentration of $234 \times 10^6 \pm 108 \times 10^6$ ml^{-1}. Further work by Wagenaar and Grootenhuis (1953), using 72 ejaculates on 3000 mares, indicated that in order to obtain acceptable fertilization rates a sample with a minimum spermatozoan concentration of 10,000 spermatozoan ml^{-1} was required and that 65% of those sperm should be morphologically normal.

The need for more accurate assessment of spermatozoan concentration in order to work out the volume and dosage of inseminates led to the first reported use of the spectrophotometer in the assessment of spermatozoan concentration. The use of the spectrophotometer has now become widespread in many AI laboratories, though it is not without its drawbacks.

Initial investigations into the possibility of a link between seminal plasma characteristics and spermatozoan quality and concentration were first conducted in the 1950s. Initial work was not encouraging and several workers failed to show any correlation between seminal plasma pH and spermatozoan concentration or spermatozoan motility (Macleod and McGee, 1950; Pecnikov, 1955). In addition, no correlation was evident between pH and spermatozoan survival, number of spermatozoan per ejaculate or the volume of ejaculate (Pecnikov, 1955). However, Mann *et al.* (1956) did find a correlation between the concentration of ergothionine and citric acid and the concentration of spermatozoan, but this was only observed over the course of a year and may have been due to seasonal changes rather than a reflection of differences in spermatozoan quality. This line of investigation has now largely ceased.

The subsequent development of complete semen and spermatozoan analysing systems, via computer analysis, is discussed in Chapter 6.

2.6. Development of Techniques for the Handling and Storage of Semen

The importance of maintaining semen at, or about, body temperature in order for fertilization to be successful was appreciated from the very beginning of work on semen and AI. Significant research on the handling of semen was carried out by Walton (1936), based upon the many trials he conducted. Cooling semen, as mentioned earlier, was first investigated as a means of storage by Spallanzani as early as the beginning of the 19th century (Spallanzani, 1803; Perry, 1945; Varner, 1986). However, the transport of cooled semen was not really successful until Walton and Rawochenski (1936) transported to Poland, by air, semen collected from a Suffolk ram in Cambridge. The semen was cooled to 10°C and stored for transit in a Thermos flask containing crushed ice. On arrival in Poland the semen was inseminated into five ewes, of which two became pregnant (Perry, 1968).

Today storage of semen by cooling is commonplace in the majority of animals from which semen can be collected. Prior to storage, semen extenders are used in order to prolong the life of the spermatozoa and provide them with some protection against the significant drop in temperature at cooling. Many extenders have been investigated for use with stallion semen, starting with mare's milk by the Russians through to the more sophisticated extenders of today, with varying success. Initial extenders for use with horse semen were made up of distilled water, into which some of the following were dissolved: glucose and/or lactose to provide a source of energy and hence prolong the lifetime of the spermatozoa; egg yolk, sodium potassium citrate, potassium chloride or sodium bicarbonate, to act as preservatives (Hejzlar, 1957; Nishikawa, 1959a; Vlachos, 1960). The results obtained from the use of such extenders was quite promising, as far as motility was concerned. In the 1960s the use of milk-based diluents became increasingly popular as a result of their successful use in cattle. Both mare's milk (Kamenev, 1955; Mihailov, 1956; Kuhr, 1957) and cow's milk (Dorotte, 1955; Vlachos, 1960) were used, raw or skimmed, and with or without the addition of an extra source of energy in the form of glucose or lactose. Honey was also used as an extender with some success.

The development and use of these extenders not only allowed spermatozoa to survive at body or room temperature for prolonged periods but also opened up the possibility of cool storage of semen extending the life span of spermatozoa even further. These investigations culminated in the 1980s in the development of the Equitainer, a specially designed insulated container for the transport of semen which permits a controlled initial cooling rate of 0.3°C min^{-1} and subsequent maintenance at 4°C for 36 h (Douglas-Hamilton *et al.*, 1984).

The relatively fragile nature of equine spermatozoa was first appreciated some time ago and attributed to the weak lipid capsule of the cell. The sensitivity of equine spermatozoa had been demonstrated by Milovanov (1934), who had demonstrated their susceptibility to both crude rubber and ebonite when these were used as the main components of semen storage containers.

The long-term storage of semen marked a most significant advancement in AI. The success of such a technique opened the door to numerous further possibilities, including the long-term use of the semen after the death of the donor and worldwide transportation. The first long-term storage of semen was demonstrated by Buiko-Rogalevic (1949), who successfully preserved semen at 0°C in an egg-yolk-based extender (0.6–0.8%) for 8–13 days. Further techniques for freezing semen were developed later in that year when Polge *et al.* (1949) discovered a practical method for the long-term preservation of semen by freezing to a temperature of −79°C using dry ice. This work by Polge and his co-workers was carried out using bull's semen and followed the encouraging results obtained from cooling fowl sperm. In fowl, a glycerol-based diluent had been demonstrated to protect spermatozoa at low temperatures. Polge and Rowson (1952) therefore investigated the efficacy of freezing bull semen, buffered in egg yolk and sodium citrate and equilibrated with glycerol-based diluent. The conception rates achieved using spermatozoa stored in this manner were reported to be satisfactory. Indeed, in 1954, the first breeding organization to operate a 100% frozen semen programme was set up by the Waterloo Cattle Breeding Association, Ontario, from where significant research into the processing, storage and field use of frozen bull semen has emanated. The development of these techniques has led to the commonplace use of frozen bull semen, now frozen to −196°C in liquid nitrogen.

Development of freezing as a means of long-term storage of semen has also been quite successful for other species, such as the ram, initially developed in the 1950s and with a reported fertilization rate of 67% (Smirnov, 1951; White *et al.*, 1954; Kuznetsov, 1956; Colas, 1979; Langford *et al.*, 1979), the goat, developed initially at the same time as the techniques for ram semen storage but with a reported higher success rate of up to 80% (Waide and Niwa, 1961; Lyngset *et al.*, 1965), and the boar, developed in the 1970s (Larsson, 1978; Watson, 1979).

The development of suitable techniques for the freezing of stallion semen has been somewhat delayed, as seen with many areas of equine AI, despite the fact that initial work on freezing equine semen was carried out in the 1950s. Smith and Polge (1950) demonstrated a percentage motility rate of 25% after freezing and storage at $-79°C$. The spermatozoa used were separated from seminal plasma after collection, and re-suspended in glycerol/glucose-based buffer prior to storage at $-79°C$. Similar work by Szumowski (1954) resulted in a 50% sperm survival rate for equine spermatozoa diluted in a glycerol, egg yolk, glucose and streptomycin extender and again frozen to $-79°C$ for 4 months. Hjinskaja (1956) also obtained good survival rates with a glucose/glycerol diluent stored at $-70°C$.

The first pregnancy as a result of insemination with frozen stallion semen was reported by Barker and Gandier (1957). The semen used was collected from the cauda epididymis of a stallion, rather than via an AV, and frozen in heated whole-milk extender plus 10% glycerol. One of the seven mares that were inseminated foaled. Zmurin (1959) successfully used a similar diluent to that used by Szumowski, but he stored the semen at $-20°C$. Nagase *et al.* (1966) froze stallion semen in the form of pellets, but the process was not successful. However, later work by Muller (1982) demonstrated that freezing stallion semen in pellet form in aluminium tubes could be successful, provided that a concentrated sample is used and a strict freezing protocol adhered to. The first foal born as a result of using frozen semen collected after ejaculation into an AV occurred in 1968 (Ginther, 1968). The 1970s and 1980s saw the development of more sophisticated techniques (Tischner *et al.*, 1974; Muller, 1982; Naumenkov and Romankova, 1982) and the first international symposium on the freezing of stallion semen was held at the Horse Breeding Institute in Rybnoje (USSR) in 1978 (Tischner, 1992a).

Despite an upsurge in interest and research, success rates still remain relatively low, in the region of a 40–60% conception rate, which is still not comparable with those achieved using bovine AI. Hence frozen semen has not gained widespread use in the equine industry to date, though further interest is now developing (Amman and Pickett, 1987).

2.7. Development of Techniques for the Insemination of Semen

It is well beyond the scope of this book to detail the development of hormonal manipulation of the oestrous cycle of the mare which has enabled, for example, the development of fixed-time AI, though Chapter 7 will address this subject in part. As far as historical development is concerned, initially AI in horses required the accurate detection of oestrus in the mare and the continued availability of the stallion to provide fresh semen, thus restricting the scope for use of the collected sample. Early attempts at timing oestrus and ovulation involved, among other things, a starvation regime. Mintscheff and Prachoff (1960) demonstrated that a starvation regime, carried out when the ovarian follicles were 3–4 cm in diameter, significantly reduced the length of oestrus. Presumably this also advanced ovulation, and so allowed insemination to be timed more accurately, as it could be carried out during a more defined and shorter oestrous period. This technique has obvious welfare implications and never gained favour. Subsequent development of hormonal manipulation of the oestrous cycle was much more promising and demonstrated that the synchronization of oestrus in the mare using synthetic progestogens and prostaglandins was possible, though it was relatively inefficient when compared with hormonal manipulation carried out in the cow (Berliner, 1947). Results obtained are also very variable between and within mares. It is only more recently that more sophisticated regimes, using additional hormones such as human chorionic gonadotrophin (hCG) and gonadotrophin releasing hormone (GnRH), have been developed to allow accurate synchronization and normal fertility rates (Sullivan *et al.*, 1973; Voss *et al.*, 1975; Hyland and Bristol, 1979; Squires *et al.*, 1981d).

Insemination techniques developed in line with collection techniques, in both horses and cattle. A significant contribution to the development of today's techniques was made by Danish veterinary surgeons in 1937 with their development of the recto-vaginal or cervical fixation method of AI in cattle. The technique involves the manual manipulation of the cervix via the rectum and hence guiding the insemination tube deep into the cervix. This technique is fast and efficient, and at the time was reported to increase conception rates in cattle by 10% (Hendrikse and Van Der Kaay, 1950; Van Denmark, 1952). It was subsequently developed for use in horses and replaced the previously favoured method of insemination in the horse of depositing semen gelatin capsules into the vagina. The use of manual insemination per cervix also considerably increased the fertilization rates obtained in mares (Swire, 1962). Attention in the 1960s was turned to the protection of sperm from cold shock and sunlight during the insemination process and hence the use, in many countries, of insulating coverings for the AV and insemination equipment. The development of clear plastics also had a bearing on insemination, allowing easy confirmation that the insemination process has been carried out correctly (Swire, 1962).

2.8. Present Use of Equine AI

It is very difficult to find evidence to indicate how widespread, on a worldwide scale, is the use of equine AI. Surveys have been conducted in an attempt to arrive at some sorts of figures. Tischner (1992a) reported attempts in 1987 to discover the extent of the acceptance of equine AI at that time and stated that 26 out of 40 countries surveyed did use equine AI; however, the percentage of mares covered by AI within those countries varied significantly from 0.002 to 37%. Based on this survey, he estimated that 730,000 mares worldwide were being covered by AI and that the most widespread use was in China. Fifteen of the 26 countries surveyed which stated that they did use AI were analysed further. This analysis is summarized in Table 2.2. The significance of China's use of equine AI is evident in Table 2.3, which gives the same figures but with those for China omitted (14 countries).

It is evident that the acceptance of equine AI has moved on since 1987, though it is likely that China is still the biggest player. In 1988 it was reported that in the north of China 35% of the 1,531,832 broodmares covered were inseminated, the vast majority with chilled semen but a significant number (31,832) with frozen semen. It is estimated that, currently, more than 550,000 mares per year are inseminated in China (Tischner, 1992a,b). The use of AI in Europe is considered in Table 2.4. Outside Europe the other major users of equine AI are Australia and New Zealand (21% of all mares were inseminated in 1987) (Tischner, 1992a), the USA (36% of all mares were inseminated in 1987) (Tischner, 1992a), and China (25% of all mares were inseminated in 1987). Brazil and South Africa have also significantly increased their use of AI.

The current use of AI depends upon the breed societies accepting progeny conceived by the technique. In the last 15 years there has been a real expansion in the number of breed societies that have come to accept AI. This is particularly evident in countries like the USA, where, in 1969, equine AI was reported to be unpopular, with most breed societies' rules severely restricting or prohibiting its use (Hughes and Loy, 1969b). By 1973 the situation had improved and six out of 75 breed societies surveyed accepted progeny

Table 2.2. Estimated worldwide use of equine AI in 1987, based upon a survey of 15 countries stating that they use equine AI (Tischner, 1992a).

Treatment of semen	Number of mares inseminated	Percentage of all mares inseminated
Fresh	171,000	23%
Chilled	510,200	70%
Frozen	48,700	7%
Total	729,900	100%

Table 2.3. Estimated worldwide use of equine AI in 1987, based upon a survey of 14 countries (those in Table 2.2 minus China) stating that they use equine AI (Tischner, 1992a).

Treatment of semen	Number of mares inseminated	Percentage of all mares inseminated
Fresh	170,300	93%
Chilled	10,200	6%
Frozen	1,700	1%
Total	182,200	100%

Table 2.4. The major users of equine AI in Europe (Barrelet, 1992; Tischner, 1992a).

Country	Percentage of mares inseminated	Date	Reference
Belgium	9.2%	1987	Tischner, 1992a
Czechoslovakia	9.1%	1987	Tischner, 1992a
Finland	37%	1987–1988	Finland, Central Association, AI Society, 1988; Tischner, 1992a
France	6.3%	1987–1988	Tischner, 1992a
Germany (Hannover region)	80%	1991	Barrelet, 1992
The Netherlands	3.5%	1987	Tischner, 1992a
Switzerland	33%	1991	Barrelet, 1992
Poland	<1%	1987	Tischner, 1992a
Spain	<1%	1987	Tischner, 1992a
UK	<1%	1987	Tischner, 1992a

conceived by AI using frozen semen, with other breed societies accepting AI but only using fresh semen, which precluded many of the advantages of storage and transportation (Bartlett, 1973). A considerable improvement is seen by 1993, with a 90% acceptance rate for use of AI reported by Bailey *et al.* (1995). This had improved again by 1995 when, according to a survey carried out by Dippert (1997), 93.8% (76 of the 81 breed societies surveyed) accepted AI and many of the previous restrictions regarding storage and transportation had been removed. The first breed societies that accepted AI in the USA were those of the Standardbreds, Quarterhorses and Arabians (Stabenfelt and Hughes, 1977). The Warmbloods of USA soon followed and these, plus the American Saddlebred Horse Association, the North American Shagya Society, the United States Lipizzan Registry and the United States Trotting Association, are the major organizations which now routinely use AI (Varner, 1992; Loomis, 1993; Ghei *et al.*, 1994; Dippert, 1997).

A similar but more rapid expansion in the use of equine AI has been seen in Australasia. AI was initially accepted by the Australian Stock Horse Society in 1982, closely followed by the Standardbred, Warmbloods, Arabians, Quarterhorses and Clydesdales (Woodward, 1987; Dowsett *et al.*, 1995), all of which, along with many others, now accept AI with fresh, chilled or frozen semen. Many studs, especially Standardbred studs, today use AI exclusively for covering all their mares.

On a more specific level, the change in attitudes to AI is demonstrated well in Germany and in particular with the Hannoverian Stud Book, initially reluctant but now a great advocate for AI. In 1978 all matings for registration in the stud book were natural. Demand by members resulted in acceptance of AI in 1986, and 7% of registered progeny were conceived by AI in that year. In 1991 the figure rose to 52.5% and in 1993 it was expected to be 90% (Klug, 1992; Uphaus and Kalm, 1994).

Countries like China and Russia have a long history of using AI and do not seem to have had the problems of recognition by registering authorities that have dogged the development of AI elsewhere. Despite all the advances and advantages in using AI, some countries are still reluctant to accept the technique. A prime example is the UK. According to Tischner (1992a), his survey in 1987 demonstrated that fewer than 1% of UK mares were covered by AI. The most significant factor affecting the willingness of the UK breed societies to accept progeny conceived by AI is the refusal of the Thoroughbred breed society to do so. Historically the equine industry of the UK and also of Ireland has been dominated and led by the Thoroughbred industry, whose standards and attitudes are reflected by the numerous other breed societies. Due to the largely unique nature of the equine industry in the UK and Ireland, there has not been the economic pressure put upon them to adapt to AI as was seen in the cattle industry, which, was also initially reluctant to take AI on board (Boyle, 1992). Mainly due to the introduction of European Directives and associated Decisions 95/294/EC, 95/307/EC, 96/539/EC and 96/540/EC, which relate to free trade and govern the transportation of semen both within the Community and outside, the UK industry has been forced to accept the use of AI in order to maintain a level playing-field with regard to the breeding of horses across the Community. Hence, in 1997, 76% of breed societies did accept AI, with a variety of different restrictions. The Thoroughbred remains among those that do not accept the technique. All the societies that accepted AI in 1997 allowed the exportation of semen and its use within the UK, but some reticence is shown in allowing the use of imported semen, with only 74% of the breed societies that accept AI registering such stock (Wetheridge, 1996).

2.8.1. Regulations governing the use of equine AI

Regulation of equine AI is very variable and differs widely with both the countries and the breed societies involved. There are three main methods of regulation or control.

State control

In some countries control is entirely by the State, which directly runs the registers for all horses and controls breeding policy. This is evident in France, where all AI and breeding is regulated by the State through the French Ministry for Agriculture via the *Service des Haras des Courses et Equitation* (National Equitation Centres) and the *Haras Nationaux* (National Studs). By this means the Ministry controls and sets standards for the veterinarians, technicians and premises involved. Semen collection and insemination may only be carried out on one of two types of licensed centres, both of these being under the control of, and often situated at, either a national equitation centre or a national stud. The main centre, the *Centre de Production de*

Semence, is licensed to collect, process and store semen and to inseminate mares. Satellite centres (the *Centres de mise en place de semence*) associated with each main centre are licensed only to inseminate mares. All these centres and stations are regularly inspected by the Ministry, which sets minimum standards for the stabling of stallions, laboratories, storage containers, rooms and equipment maintenance. The Ministry also sets out the standard service charge for both natural service and AI and sets a limit to the number of mares a stallion can serve each year. This limit depends upon the performance index of the stallion, again a system run by the Ministry. The best stallions – those with a performance index of greater than 150 – have a limit of 100 mares per stallion per year; the remaining approved stallions have a limit of 60 mares per year (Magistrini and Vidament, 1992; Tainturier, 1993; Mesnil du Buisson, 1994). This system is highly controlled, giving very little scope for private enterprise, though some is now becoming evident. The big advantage of such a system is that the State financially supports all the centres involved. As an offshoot of this, research into many areas (including AI) is widely supported by the Ministry. Initial interest in equine AI was shown by INRA in Tours, and with the support of the Ministry the technique became commercially available, in the 1980s, for use in French Thoroughbreds and draught horses. Interest in its use with draught horses stemmed from the excessive weight of such animals and the consequent difficulties in transporting them, as well as dwindling numbers and hence the distances of travel required (Domerg, 1984; Tischner, 1992a).

As a result of this control and support, France is one of the leaders in the field of equine AI research and commercial availability. It is now widely available for use on all breeds except the English Thoroughbred. In 1993 more than 15,000 mares were inseminated, which included 34% of all French Saddlebreds and 42% of all Trotter mares (Mesnil du Buisson, 1994).

Collaboration between State and designated breed societies

Many countries conform to the collaborative form of regulation, which is a mix of State control and privately run enterprises under licence. Such a system is evident in Germany, where AI is controlled by the federal law concerning animal husbandry, which defines the responsibilities of regional administrators and breed authorities. AI stations are then licensed by regional authorities in collaboration with breed societies. The breed societies are specifically responsible for the zootechnical aspects of regulation. The German FN (Equine Federation) then represents the external interests of these various breed societies. The Federation controls the licensing of AI centres and the training of non-veterinarians, and all AI centres must nominate a veterinarian in charge. Semen can only be exported from, imported to, or inseminated at, a regional licensed centre (Barrelet, 1992). Some satellite insemination centres along the lines of those in France have also been set up, inseminating mainly with fresh semen (Klug and Tekin, 1991). However, the Federation does not control which stallion can be accepted for AI. This is determined by the breed societies, but they must comply with the general directives of the FN. The breed societies are also responsible for setting the limits for the number of mares per stallion per year (Barrelet, 1992). Again this controlled system has led to significant advances in the use of equine AI, with 80% of mares bred from within the Hannover region in 1991 being serviced by AI.

The Netherlands runs a similar system of collaborative control between the Ministry of Agriculture and the breed societies, with the breed societies controlling the zootechnical aspects and the Ministry of Agriculture (Landbouwschap) controlling the technical aspects. The two collaborate in the registering of stallions for use with AI. The Ministry produced a series of minimal technical standards (Reglement KI bij Paarden, 1977; Verordening KI bij Dieren, 1985; Reglement KI bij Paarden, 1985; Nadere Regelen Paarden, 1991) which govern the health requirements of stallions, the regulation of the collecting and insemination centres, and the processing and transportation of semen as well as the administration. Minimal standards for semen evaluation *in vitro* have also been defined. Stallions put forward for breeding need to be approved and meet minimal semen quality parameters. Ejaculations of stallions using AI are examined periodically by the Ministry and records of each ejaculate, semen quality before and after export and the mares inseminated must be kept and reported to the Ministry (Barrelet, 1992).

Sweden has very similar regulations to those of Germany and The Netherlands but with a greater emphasis on health, with strict regulations aimed at minimizing the transfer of systemic and venereal disease via infected semen. Strict records on semen movement, *in vitro* quality and mares inseminated must be kept (Barrelet, 1992).

Switzerland is only a small horse-breeding nation and has largely adopted the regulations of its neighbours, running a system very similar to that of Germany and The Netherlands. Policy is regulated and administered by the Federal (National) Stud Farm, which also licenses stallions, regularly screens stallions and controls on an individual basis the importation of semen. The Federal Veterinary Bureau issues health regulations and general requirements for AI stallions both used within Switzerland and for import/export. The stallions used for AI may be either State or privately owned (Barrelet, 1992).

Private or independent control

In some countries there is no central control or regulation: it is entirely up to the stud owner, veterinarian in charge and the breed society to lay down any criteria required. Such a system exists in the UK, where government support for the industry is more or less non-existent. All premises should work within the general guidelines produced by the Ministry of Agriculture, Fisheries and Food. These regulate the health aspects of imported and exported semen and largely reflect the EC Directives (90/426/EC; 90/427/EC; 90/428/EC) which govern equine zootechnical, competition and transport matters, and which regulate all countries within the EC. For movement within the UK, there is no central administrative body, no control or regulations and no standards set out, except those laid down by the individual breed societies – and these vary widely.

Regulation of practice

In addition to regulation of the use of AI, many countries set restrictions governing who can carry out semen collection, processing and insemination. The majority of countries insist that a veterinary surgeon or trained individual must carry out or be involved in all procedures. In France, courses are run by the Ministry to train collectors and inseminators. These trained personnel then work under the control of the centre chief, who is normally a veterinarian. In Germany, appropriately trained non-veterinarians may practise under the control of a centre manager, who is a veterinarian. The Netherlands runs a similar system, with scope for trained lay personnel to practise collection and insemination under the guidance of a veterinarian. The same applies in Hungary. In Switzerland, insemination is restricted to veterinary surgeons, but collection and processing may be carried out by trained personnel (Barrelet, 1992; J. Vilmos, Hungary, 1996, personal communication). Similar restricted semi-veterinarian practice is evident in Australia, the USA, South Africa and many other parts of the world. Again, the UK stands out as different, with no restrictions as yet in place, though all guidelines suggest that all processes should be restricted to veterinarians only.

2.9. Current Reasons for Using Equine AI

There are many and varied reasons why equine AI is practised today, and some of these depend upon, and are limited by, the regulations of the countries involved and the breed societies responsible for registering stock. In general, under appropriate management, conception rates to AI are equivalent to those achieved by natural service. In fact, work by Pickett and Shiner (1994) suggested that the use of AI achieves 10.3% better conception rates than natural service when comparing normal mares. This apparent greater success with AI is even more evident in problem mares, with conception rate differences of 30.1% in favour of AI being reported (Pickett and Shiner, 1994).

Some of the reasons for current use of AI are discussed in the following sections. These are not exhaustive and the relevance of some will depend upon the aims of the establishment and the country and breed involved.

2.9.1. Removal of geographical restrictions

The most obvious and arguably the most important reason for using AI is that it is far easier to transport semen over large distances than it is to transport animals, be it the stallion to the mare or vice versa. It is now quite possible to collect from a stallion in one part of a country and transport the semen, cooled, several hundred miles in order to inseminate a single or a number of mares. This system has been well developed and is regularly practised in Australia, China, USA, Germany and Scandinavia, among other countries, where semen is often transported in purpose-made thermoregulatory flasks (Equitainer, see Chapter 7) using regular parcel or postal services. Transportation of chilled semen is restricted by the journey time, as survival of chilled semen is 72 h. This limits its transportation to within a country or between relatively close neighbours. The development and increasingly widespread use of frozen semen adds an extra dimension, allowing semen to be transported worldwide and enabling mares and stallions to mate in a way that would normally be beyond the realms of possibility.

2.9.2. Reduction in transfer of infection

There is no doubt that the risk of transfer of systemic and venereal disease between mare and stallion is considerably reduced by the use of AI. There are many systemic diseases that may be passed by contact at covering that are not transferred by the venereal route, examples of which are equine influenza, strangles (causative agent *Streptococcus equi*), glanders (causative agent *Pseudomonas mallei*), *Actinomyces bovis* and *Brucella abortus* (Jaskowski, 1951). The use of AI, with the enforced remoteness of the mare and stallion, means that the transfer of such diseases is not possible.

Venereal diseases are transferred via direct sexual contact and again the use of AI potentially avoids this situation. The more prevalent venereal disease bacterial infections include those caused by *Taylorella equigenitalis* (formally known as *Haemophilus equigenitalis*, and the causative agent for contagious equine metritis) (Crowhurst *et al.*, 1979; Rossdale *et al.*, 1979; Tainturier, 1981); *Klebsiella aerogenes* (Crouch *et al.*, 1972; Rossdale and Ricketts, 1980); *Pseudomonas aeruginosa* (Hughes *et al.*, 1966; Hughes and Loy, 1975; Danek *et al.*, 1993, 1994b); *Streptococcus zooepidemicus* (Rossdale and Ricketts, 1980; Bowen, 1986; Le Blanc *et al.*, 1991; Couto and Hughes, 1993; Danek *et al.*, 1993, 1994b); *Staphylococcus aureus* (Rossdale and Ricketts, 1980); *Escherichia coli* (Bowen, 1986; Couto and Hughes, 1993) and *Proteus* (Couto and Hughes, 1993). The use of AI can potentially avoid the transfer of many of these infections. There is evidence to show that *Klebsiella pneumoniae* and *Pseudomonas aeruginosa*, which are the agents of the two true venereal diseases (i.e. transferred primarily via the venereal route), may be isolated in semen. Further evidence suggests that *Taylorella equigenitalis* may also be present in semen (Vaissaire *et al.*, 1987; Philpott, 1993; Madsen and Christensen, 1995). There is, therefore, a possibility that these bacteria may still be transferred by AI.

Other contact diseases include viral infections such as equine viral arteritis (Timoney and McCollum, 1985; Timoney *et al.*, 1988a), equine coital exanthema (genital horse pox), transferred primarily at coitus but also by instrument, and equine herpes viruses (EHV3 and possibly EHV1) (Pascoe *et al.*, 1968, 1969; Krogsrud and Onstad, 1971; Gibbs *et al.*, 1972; Pascoe and Bagust, 1975; Jacob *et al.*, 1988). Also, the transmission of dourine caused by the sexually transmitted protozoan *Trypanosma equiperdum* (Couto and Hughes, 1993), summer sores (habronemiasis lesions) caused by *Habronema* larvae (Jaskowski, 1951; Stick, 1981; Philpott, 1993) and fungal infections such as mycotic endometritis caused by the *Candida* fungus (Zafracas, 1975) may be significantly reduced by using AI.

This potential of AI to avoid disease transfer was one of the main reasons for its early development in countries such as Poland, where its use in the late 1940s and in the 1950s and 1960s was primarily by State studs, which at that time had a problem with contagious venereal diseases (Tischner, 1992b). However, there is considerable debate on the transfer of some venereal diseases via the semen used in AI. It is known that up to 23% of semen samples from reproductively active stallions contain facultatively pathogenic organisms, with considerable annual fluctuations being evident (Autorino *et al.*, 1994; Danek *et al.*, 1994b; Golnik and Cierpisz, 1994; Scherbarth *et al.*, 1994). What is not clear is whether there is a significant risk that these pathogens may be transferred via AI and cause disease. As indicated above, it is suspected that *Klebsiella pneumoniae*, *Pseudomonas aeruginosa* and *Taylorella equigenitalis* may be passed via AI along with equine viral arteritis (Roberts, 1986a,b; Timoney *et al.*, 1987; Golnik *et al.*, 1991; Chirnside, 1992) and possibly equine infectious anaemia (Roberts and Lucas, 1987). It is, therefore, possible that worldwide transport of semen could potentially increase the worldwide spread of disease. In order to reduce such risks to a minimum many countries have strict regulations, restrictions and certification that have to be satisfied in order that semen can be imported or exported. For example, the movement of semen within the European community and its importation from outside are governed by European Directives and associated Decisions 90/426/EEC, 90/427/EEC, 90/428/EEC, 95/294/EC, 95/307/EC, 96/539/EC and 96/540/EC that regulate health restrictions and certifications that have to be satisfied prior to the transport of semen. In general, the restrictions are laid down by the country into which the semen is to be imported.

2.9.3. Improvement of native stock

Transportation of semen has had a significant effect on the genetic improvement of equine stock in many countries, the genetic pool having previously been significantly restricted, due to the decline in horse numbers during the middle of this century. Some of the smaller east European countries (for example, Hungary and former Yugoslavia) used imported semen extensively to improve their native stock, in particular for sports horses. Hungary, in 1989, started to import semen from west European stallions such as the Trakehner to improve the native Kisbér Halfbred horses with the aim of producing top quality sports and driving animals. Today, AI is widely used with stallions standing at one of three State studs for semen collection and insemination. Strict rules exist regarding the registration and subsequent standing of all stallions for AI, both home-bred and imported. They must all pass a series of tests related to health, conformation, venereal disease, behaviour, fertility and all aspects of a four-day performance test. In addition, stallion semen exchange systems have been set up with the UK, the USA and Slovakia. As a result the native horse has been significantly improved and 75% of all their sports horses are now conceived by AI (J. Vilmos, Jr, Hungary, 1996, personal communication). In 1982, former Yugoslavia used semen imported from the former German Federal Republic to improve their native stock with the aim of increasing the available gene pool and producing quality sports horses (Sukalic *et al.*, 1982). In the UK importation of semen from Warmblood stallions has played its part in improving the sports and performance horse of today. India has imported stallion and donkey semen from France in an attempt to improve the size and strength of its native mules, a prime source of locomotion and agricultural power (Ghei *et al.*, 1994).

2.9.4. Development of gene banks

AI has allowed the development of the concept of a gene bank, to store valuable semen for posterity. This idea has been developed in particular in Poland (Tischner, 1992b), Greece (Zafracas, 1994) and, more recently, Hungary (J. Vilmos, Hungary, 1996, personal communication). All three have taken active steps to organize a centrally controlled storage facility to preserve semen collected from valuable stallions. This is not supported by all breed societies, some of which do not allow the use of such semen or place restrictions on the use of a deceased stallion's semen.

2.9.5. Breeding from difficult mares

A mare may be classified as a difficult breeder for either physical or psychological reasons. In considering whether or not AI will allow that mare to be bred, it is very important to ascertain why these problems have occurred.

In the case of a mare with inherited problems the question should really be asked as to whether she should be covered at all and so risk perpetuating the problem in subsequent generations. If, however, the problem can be identified as non-genetic in origin, AI may present an acceptable alternative method for breeding from that mare.

Physical problems

There are numerous reasons why a mare may have physical problems that affect her ability to conceive by natural service. Some of these will be briefly discussed in the following sections.

INJURY

One of the reasons for using AI is that a mare has been injured in some way that makes it very difficult or uncomfortable for her to be covered naturally. Hindleg injuries and chronic lameness are the most common causes. Arthritis, navicular disease and laminitis are also problems that may suggest the use of AI in order to avoid the pain of holding the stallion's weight at covering. Nevertheless, there are suggestions that such conditions are inherited and, therefore, careful consideration must be made before breeding from such mares. Mares with back problems can also be covered painlessly using AI. In considering the use of AI in mares with such injuries it must be remembered that although AI takes away the strain of holding the stallion's weight at covering, the mare must still be fit enough to carry a pregnancy to full term, the weight of which can be quite considerable.

INFECTION

AI has been successfully used to cover mares that are susceptible to post-coital endometritis. Some mares are unable to mobilize the normal defence mechanisms by which the natural uterine microflora can be reinstated after the bacterial challenge of coitus. In these mares, as a result of covering, there is a reduction in the normal neutrophil and myometrial activity response to the invariable occurrence of post-coital acute endometritis. If this acute endometritis is not dealt with efficiently and so persists, as is the case with these mares, there is a significant reduction in the chance of the conceptus implanting (Evans et al., 1986, 1987; Watson, 1988; Kotilainen et al., 1994). In an attempt to overcome the problems of these immunologically compromised mares, AI can be performed using semen diluted in extenders containing antibiotics, or alternatively the uterus can be flushed with 100–300 ml antibiotic (for example, 50 mg l^{-1} penicillin or gentamycin) semen extender immediately prior to insemination. Such methods are reported to reduce significantly the occurrence of post-coital endometritis (Kenney et al., 1975; Clement et al., 1993).

PERINEAL CONFORMATION

The perineal conformation of an increasing number of mares is so poor that the chance of natural conception is low. This is primarily due to upper reproductive tract infections resulting from an unacceptable bacterial challenge to the outer reproductive tract from faeces passing on to to vulva lips and being aspirated into the vagina, through an incompetent vulva seal, during movement. The significance of perineal conformation in the reproductive potential of a mare is discussed in more detail in Chapter 8 (section 8.2.5). The severity of the condition depends upon the angle of slope of the perinium and the height of the pelvic floor in relation to the vulva opening (Caslick, 1937; Pascoe, 1979). Using these measurements mares can be allocated a Caslick index figure, which will allow them to be classified into one of three categories (Pascoe, 1979). All mares in Type 3 and most in Type 2 show some degree of problem, and the routine treatment for these mares is a Caslick operation – the suturing of a proportion of the vulval lips (Caslick, 1937). Most mares that have a Caslick operation are unable to breed naturally unless an episiotomy is performed and the Caslick operation re-performed immediately *post coitum*. This is not only an expensive and inconvenient process to go through every time a mare is covered; it also has welfare considerations as there is a limit to the number of times the operation may be performed due to the development of scar tissue. However, the use of AI allows such mares to be covered without the need for an episiotomy. Again care must be practised in selecting such mares for breeding as the incidence of poor perineal conformation within the equine population and the use of Caslick operations has significantly increased in recent years, enabling mares with poor conformation to breed. Human selection of mares for athletic performance rather than their breeding ability is a prime contributor to this problem. Today, the majority of Thoroughbred mares, especially in the USA, require Caslick operations.

Psychological problems

It is not unknown for a mare to have an aversion to a particular type of stallion. This aversion is often associated with colour and may be traced back to an unpleasant covering experience. However, it may not be that specific as the feeling of fear or being threatened by horses of a specific colour may be evident in situations other than covering. The use of AI obviously eliminates this. Temperament may also be a problem in some mares. A very nervous or flighty animal may not be able to cope with the additional stress in the environment presented by a strange stallion, especially when she may be already in a highly agitated state due to the dominance of oestrogens within her system at oestrus.

Some mares never seem able to show adequate signs of oestrus to allow the time for covering to be determined accurately by man or by a teaser, especially under today's highly intensified management systems. Such mares show no inclination to stand for the stallion, even though they may technically be in oestrus and near to ovulation. Natural covering of such mares is dangerous and can be impossible. However, follicle maturation and the timing of ovulation can now be ascertained with the aid of an ultrasonic scanner or rectal palpation and so these mares can be inseminated without the need for a stallion's presence at all.

Breeding mares with foal at foot can also present problems of behaviour, especially mares that are very foal proud. The use of AI in such circumstances removes the need for direct contact between the mare and the stallion and so avoids the trigger for the behavioural problems.

Again, when considering the use of AI to overcome physical or psychological problems it is imperative that the cause is ascertained. If such problems can be attributed to a genetic propensity, then breeding such mares should be avoided to reduce perpetuation of the problem in subsequent generations.

2.9.6. Breeding from difficult stallions

As with the mare, stallions may be difficult to breed from naturally due to physical or psychological problems. Difficult stallions are less evident in a population, because of the higher selection placed on stallions and also the reluctance of stallion owners to divulge any such information which might prove detrimental to their nomination rates. However, in several countries, including Hungary (J. Vilmos Jr, Hungary, 1996, personal communication), Finland (Finland Central Association of AI Society, 1988), France (Barrelet, 1992; Tainturier, 1993), Germany (Klug and Tekin, 1991), The Netherlands (Barrelet, 1992) and Switzerland (Barrelet, 1992), strict regulations are laid down regarding the criteria (physical and psychological as well as performance) that a stallion must satisfy before he can be licensed. These restrictions are often even more rigorous when it comes to licensing stallions for use with AI. It is essential that stallions with problems that can be identified as genetic in origin should not be bred from, as the influence of a single stallion on the population is far greater than that of a single mare.

Physical problems

The procedure for the collection of semen for AI by means of an artificial vagina puts very similar strains and stresses on the animal to those encountered during the natural act of covering a mare. It is likely, therefore, that a stallion unable to cover naturally would have some problems in using the more conventional semen-collecting methods. However, the use of a dummy mare and AV provides a known safer environment for the stallion, as opposed to the rather unpredictable nature of even the best mares. Semen collection without mounting is also possible (Schumacher and Riddell, 1986; Love, 1992).

INJURY
Aged stallions suffering from arthritic conditions, especially in the hindquarters or back region, may find it easier to use an AV than natural service. AI also allows more mares to be covered per ejaculate, minimizing the number of mounts required. Accidental damage to the same areas of the hindquarters, the back and the pelvic region may also affect a stallion's ability to cover a mare naturally. In severe cases of lameness manual stimulation may be successful, producing semen of a normal quality and quantity, without the need for the stallion to mount (Schumacher and Riddell, 1986; Crump and Crump, 1989; McDonnell and Love, 1990; McDonnell *et al.*, 1991). Several other alternative methods for semen collection are discussed in Chapter 5 (section 5.7). The collection and insemination of semen from a stallion with erectile dysfunction has been reported (Love *et al.*, 1992).

INFECTION
AI has been successfully used in helping to prevent the transfer of both systemic and venereal diseases. Examples of the infections that may be

controlled by the judicious use of AI and the risks involved have already been discussed (section 2.9.5).

INADEQUATE SEMINAL CHARACTERISTICS

Several semen parameters need to be met for a sample to be classified as normal (Chapter 6). Many stallions fail to reach one, or several, of these ideals. The problem may not be immediately apparent, especially with stallions with a low workload, and may only manifest itself if the stallion's nominations increase in number. Therefore, the collection of semen for AI may be the first time a problem has become apparent. The problem may originate from numerous causes, many of which are discussed in Chapter 4 (section 4.3.2) and Chapter 6. Regardless of their cause, modern technology allows a considerable amount of semen manipulation to take place. The most common technique used is the concentration, by centrifugation, of semen samples with low spermatozoa counts, allowing the ideal insemination of $500,000 \times 10^6$ spermatozoa per covering. Semen may also be treated with antibiotics to minimize the transmission of venereal diseases (Hughes *et al.*, 1967). At the other extreme, techniques such as *in vitro* fertilization may also be considered.

Psychological problems

Psychological problems in stallions normally manifest themselves as low libido – an unwillingness to cover a mare for whatever reason. Physical problems, as discussed above, may also manifest themselves as a reduction in libido. Psychological problems can be due to inherent temperament abnormalities but they often stem from incorrect management and training, especially in the stallion's formative years. It is not within the remit of this book to discuss stallion management in detail but suffice to say that consistent discipline and a stimulating environment from a young age are essential. Introduction to covering may also be a source of behavioural problems: the experience needs to be pleasurable and not associated with stress. Covering rough, difficult mares, especially in the early years, discourages many stallions and can manifest itself as a general reduction in libido or a specific aversion to a particular type or colour of mare.

Many stallions today find that they must change their careers in mid life from that of a performance horse to a breeding stallion. This can have long-term effects on libido as the stallion, during his early life, will have been punished for showing any sexual interest in mares, but is subsequently expected to cover mares and exhibit behaviour that has previously been suppressed. The effect on libido is variable and many of these stallions, given enough encouragement, will cover mares naturally. However, it may take several unsuccessful attempts after prolonged teasing. This is both labour intensive and frustrating for stallion and handler. The use of AI has the potential to bypass some of these problems. Semen can easily be collected from stallions while they are still continuing their performance careers. For example, as a regular routine, semen can be collected from performance stallions at a set time (usually early in the morning) on set days of the week during the season. The semen can then be prepared for immediate insemination or storage and transportation. Thus semen collection and the insemination of mares can become part of the regular routine of the yard and so minimize interference with the stallion's performance career (Boyle, 1992). It may also prove easier and more reliable to collect semen for AI from ex performance stallions than it is to use them for natural service.

At the other end of the scale, over-zealous or vicious stallions present obvious problems with natural mating. Some stallions are notorious for savaging their mares, necessitating the use of neck guards, poll guards and biting rolls in an attempt to reduce the physical damage to the mare. Care should be taken with such stallions as a vicious temperament is more likely to be due to inherited behavioural problems rather than management and so there is the risk of the trait being passed to his offspring.

2.9.7. Reduction in labour costs

The potential for reducing stud management labour costs and time is a good reason for using AI. It allows a single ejaculate to be split and used to cover several mares, either immediately, or after storage. If semen is collected regularly at a set time each day, then the sample can be used on mares as and when they arrive on the yard as visiting mares or as they are seen in oestrus. The stallion will only be required for a short period each day, thus avoiding prolonged teasing and covering of numerous mares at all times of the day, and the associated disruption to yard routine (Boyle, 1992). The use of fixed-time AI is another potentially significant labour-saving technique. Unfortunately, as discussed in Chapter 8 (section 8.3.1), synchronization and the exact timing of oestrus and ovulation in the mare are not as successful as in cattle. However, such hormonal regimes are still of particular value in mares that tend to have very variable oestrous cycle characteristics; they reduce the need for the time-consuming continual monitoring of the mare's oestrus.

The potential savings in management and labour costs are most significant in large studs running out herds of mares – for example, in Australia, South Africa and the USA. In such yards the use of AI can allow operations to be streamlined. Mares can be run through in large groups and inseminated in batches on set days following manipulation of their oestrous cycles.

2.10. Conclusion

It may be concluded that, despite the horse being at the forefront of the initial development of AI techniques, the use of AI within the horse industry today and the sophistication of the technology available lag far behind those for other farm livestock. Several reasons are put forward for this, including the significant decline in the horse population evident at the beginning of this century – it was inevitable that investment in equine research in general would be less forthcoming. In spite of the subsequent increase in horse numbers and interest in all aspects of equine science and welfare, many governments have been slow to reinvest in equine research, and some are still failing to do so, despite the economic potential and real need. Regardless of this, the use of equine AI is becoming more acceptable. However, the industry in many ways remains its own worst enemy. It is slow and reluctant to take on board new ideas and, in the case of equine AI, reinvest time and interest in developing old ones. The availability of funding is largely driven by the industry, and by consumers, and until they show the commitment, equine science, and equine AI in particular, will remain underfunded.

3 Stallion Reproductive Anatomy and Control

3.1. Introduction

A good understanding of the reproductive anatomy and control mechanisms of stallion reproduction is an essential prerequisite for studying the applied aspects of semen collection, evaluation and storage. Figure 3.1 shows the reproductive system of the stallion after slaughter and dissection. All the major structures are evident, with the exclusion of the accessory glands and associated structures (these are too well embedded within the body cavity to allow easy removal and examination). Figure 3.2 provides an illustration of the reproductive system given in Fig. 3.1, and Fig. 3.3 illustrates the entire reproductive system as it would appear *in situ*.

3.2. Stallion Anatomy

For ease of description the stallion's reproductive tract will be described in structure order, starting with the penis. Further accounts of stallion reproductive anatomy are given by Sisson and Grossman (1975), Rossdale and Ricketts (1980), Amann (1981a, 1993a,b), Setchell (1991), Thompson (1992), Varner *et al.* (1991b) and Samper (1997), among others.

3.2.1. The penis

The penis consists of three identifiable parts: the roots, the body or the shaft, and the glans penis. In the resting position the penis lies retracted and held within its protective sheath, with the prepuce surrounding the glans penis, which is out of sight (Fig. 3.4).

The penis is held in this retracted position by the retractor penis muscle and the bulbospongiosus muscle, which run ventral to the penis along its length. Innervation by the sympathetic nervous system of these two muscles

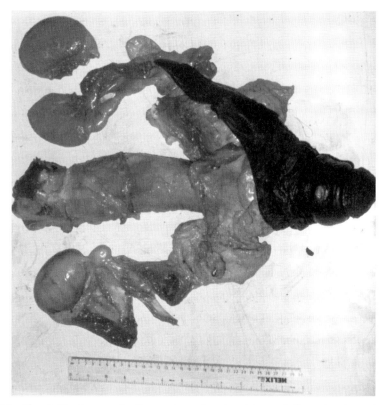

Fig. 3.1. The stallion's reproductive system after slaughter and dissection (see also Fig. 3.2).

results in their contraction, drawing the penis up within the sheath. The stallion's penis is classified as haemodynamic (reacting to an increase in blood pressure) and is of a musculocarvenosus type (muscle tissue surrounding air cavities), in common with humans. During sexual excitation the penis undergoes considerable enlargement in both width and length as a result of engorgement of the areas of erectile tissue with blood. This is unlike the bull, ram and boar, whose penises have a sigmoid flexure ('S'-shaped bend) which straightens at sexual excitation, giving rise to an increase in penis length.

Glans penis

The glans penis, sometimes referred to as the rose penis, is the outer portion of the penis, normally retracted out of sight and held within its protective sheath or prepuce. The whole of the area is richly innervated and hence highly sensitive. The area becomes highly engorged and erect during sexual excitation, increasing in size three- to fourfold. Protruding from the end of the glans penis by approximately 5 mm is the exit to the urethra. Surrounding this protruding exit are the folds of the urethral fossa. This area is of particular significance as it is often filled with smegma (red-brown secretions

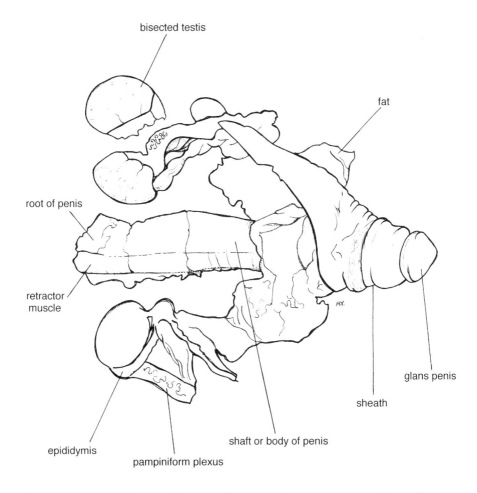

Fig. 3.2. A labelled diagram of Fig. 3.1 illustrating the major features of the stallion's reproductive system after dissection from the body post slaughter.

of the glans penis's epithelial glans plus sloughed cellular debris). This provides an environment ideal for the harbouring of bacteria, especially those responsible for venereal disease.

In cross-section, as illustrated in Figs 3.5 and 3.6, the glans penis can be seen to consist primarily of two areas of erectile tissue: the corona glandis, found dorsal to the urethra and outside the fibrous tunica albuginea; and the corpus carvenosus penis, also found dorsal to the urethra but within the tunica albuginea. The major area of erectile tissue is the corona glandis, which, as it is outside the tunica albuginea, has no fibrous capsule restricting the extent to which it can expand. Closer towards the end of the penis this area becomes increasingly more significant, at the expense of the corpus carvenosus penis. The significant expansion of the glans penis apparent at ejaculation is achieved

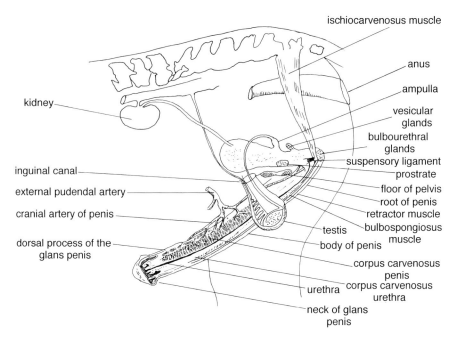

Fig. 3.3. The entire reproductive system of the stallion as evident *in situ* (with the penis in its erect form).

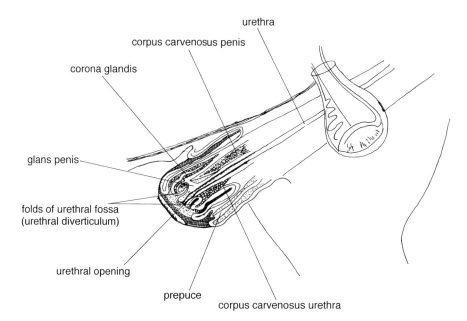

Fig. 3.4. Structure and protective folds surrounding the glans penis (the area has been enlarged to demonstrate the protective folds).

by engorgement of the unrestricted corona glandis tissue rather than the corpus carvenosus penis tissue, which is primarily responsible for the increase in size seen with the body of the penis.

Body of the penis

The body of the penis is the major section running from the glans penis, externally, to the roots, internally. The major erectile tissue of the penis is situated in this body section, the main area of which is the corpus carvenosus penis, which runs along the dorsal aspect of the penis above the urethra and is enclosed by the tunica albuginea. This tough, fibrous sheath confers some rigidity to the penis, giving it support at full erection but, due to its elastic properties, also allowing for expansion of the penis, by up to twofold (Figs 3.7 and 3.8).

The corpus carvenosus penis is made up of a dense network of trabeculae with associated muscle tissue forming cavities, which become engorged at erection. A further, much smaller area of erectile tissue runs ventrally to the corpus carvenosus penis and is termed the corpus carvenosus urethra; it lies ventral to and around the urethra. This area, though not enclosed within, nor restricted by, the tunica albuginea, plays only a minor role in erection of the body of the penis. However, it is this area that extends to form the corona glandis of the glans penis. A third major area can be identified: the bulbospongiosus muscle, running along the ventral side of the penis, enclosing the corpus carvenosus urethra. This muscle and the retractor muscle (which can be identified as a relatively small, self-contained muscle running externally along the ventral side of the penis) are responsible for retraction of the penis and its retention within its sheath.

Roots of the penis

The base of the penis consists of two root areas, emerging laterally. These two root areas are attached to the pelvic tuber ischii via the paired ischiocarvenosus muscles running along either side of the penis. The root of the penis is stabilized by paired suspensory ligaments, again running along either side of the root from the base of the ischiocarvenosus muscle to the base of the pelvis.

3.2.2. Deposition of spermatozoa

There are three stages involved in the deposition of spermatozoa into the genital tract of the mare: erection, emission and ejaculation. Each will be considered in turn.

Erection

Erection is the first reaction observed after sexual stimulation. It is the result of both physical stimulation, via nerve-end stimulation in the highly innervated area of the glans penis, and psychological stimulation from the

(a)

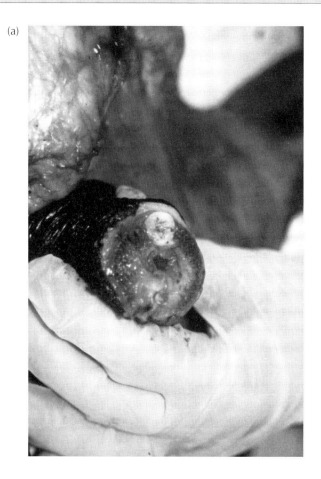

(b)

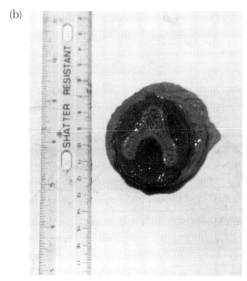

Fig. 3.5. Transverse cross-sections through the glans penis of the stallion (see also Fig. 3.6).

(a)

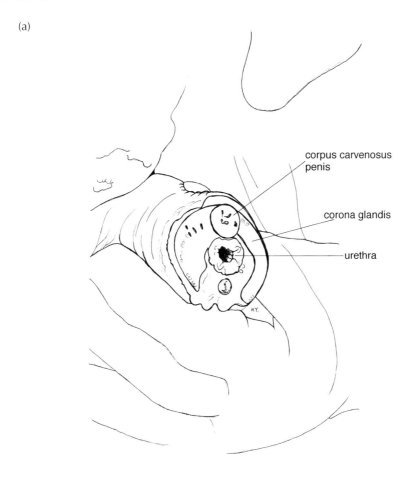

(b)

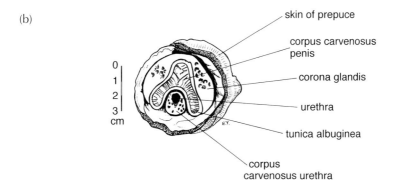

Fig. 3.6. (a) Transverse cross-section through the glans penis of the stallion, very close to the opening of the urethra, as shown in Fig. 3.5(a). (b) Transverse cross-section taken slightly higher up the penis through the glans penis, as shown in Fig. 3.5(b).

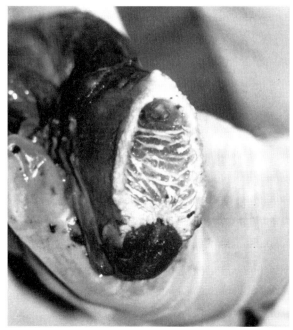

Fig. 3.7. Transverse cross-section through the body of the stallion's penis (see also Fig. 3.8).

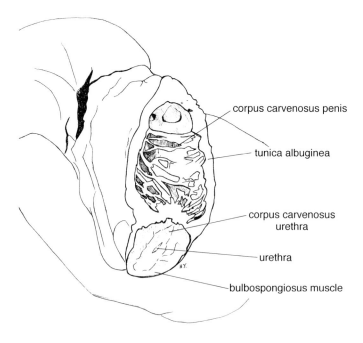

Fig. 3.8. Transverse cross-section through the body of the stallion's penis, as shown in Fig. 3.7.

sight or smell of an oestrous mare or an area associated with sexual contact. The initial stage of erection is the relaxation of the penile muscles normally responsible for holding the penis retracted within its sheath. This is accomplished via parasympathetic stimulation of the splanchnic nerves to the penis, overriding the normal sympathetic stimulation and partly constricting the arterioles of the penis. This allows engorgement of the penile arterioles, increasing the volume of blood within the erectile tissue. The initial effect is evident in the corpus carvenosus penis, followed by the corpus carvenosus urethra. At the same time the sympathetic stimulation of the retractor muscle is overridden, allowing relaxation of this muscle and extrusion of the penis. The subsequent penile relaxation and engorgement of the erectile tissue result in the lengthening and gradual stiffening of the penis. Erection leads initially to a state of turgid pressure within the air spaces of the penis. However, it is not adequate for intromission to occur successfully; for this, intromission pressure must be achieved. As a development of, and partly concurrent with, the above changes, continual central nervous system (CNS) stimulation activates the muscles in the root of the penis, namely the ischiocarvenosus and bulbospongiosus, and the urethralis muscles surrounding the urethra. As a result of stimulation, these muscles contract, drawing the base of the body of penis up against the ischial arch of the pelvis, which largely prevents the venous return of blood from the penis via the deep dorsal veins. This retention of blood within the penis increases the blood pressure further, achieving intromission pressure. At the same time the stallion's heart rate, and hence the cardiac output, increases, further enhancing the effect (Beckett *et al.*, 1973, 1975).

The penis returns to its normal flaccid state once sexual stimulation has been removed. The penile muscles relax and the arteries return to their normal constricted state as dominance of the sympathetic stimulation recurs (Pearson and Weaver, 1978).

Emission

Emission is the mixing of spermatozoa with seminal plasma and the passage of both to the urethra penis. This is achieved by a series of pulses. These pulses are achieved by nervous system stimulation, which causes strong peristaltic contractions of the walls of the epididymis, vas deferens and accessory glands, followed by contraction of the bulbospongiosus and the urethralis muscle. As a result, semen is ejaculated in a series of jets.

Ejaculation

Ejaculation occurs in association with and as a result of emission. Both emission and ejaculation result from muscle contraction, primarily of the walls of the vas deferens and the urethra. During ejaculation there is additional help from contraction of the penile muscle fibres. As already noted, ejaculation occurs in a series of jets and is accompanied by a series of pelvic thrusts. There are normally six to nine jets in total per ejaculate, lasting on average

6.15 ± 2.98 s and with a decrease in volume with each successive jet. These jets can be grouped into those that make up the pre-sperm fraction, the sperm-rich fraction and the post-sperm fraction. The pre-sperm fraction is a minor secretion often evident before intromission. The major fraction is the sperm-rich fraction, normally the first three jets, in which 70% of the biochemical components of the sample are contained (Tischner *et al.*, 1974; Kosiniak, 1975; Weber and Woods, 1993). These initial three jets are produced at high pressure, when engorgement of the penis is at its maximum and with the stallion thrusting hard. At this time the stallion's urethral process is pressed in close to the mare's cervix, which, as a result, is forced open by the fully engorged glans penis, which also acts as a seal within the vagina. These three jets are, therefore, deposited directly into the uterus. The remaining jets may be deposited in the vagina as detumescence begins (Boyle, 1992).

3.2.3. The accessory glands

The accessory glands are a series of four glands (sometimes reduced to three with the omission of the ampulla) responsible for the production and secretion of the majority of the fluid component of semen, termed seminal plasma. Different glands are responsible for the secretion of different fractions or jets. The epididymis of the testis is the other (minor) contributor to seminal plasma. The accessory glands are situated between the end of the vas deferens and the root of the penis (Fig. 3.9).

Access to (and, therefore, evaluation of the function and products of) the accessory glands is particularly difficult, due to their position within the pelvic cavity of the stallion. Hence little information is available on these glands, their relative importance and the characteristics of their secretions. Some information can be gleaned from ultrasonic examination – for example, the order of gland secretions and gland activity during ejaculation (Weber and Woods, 1993) – and from ^{1}H nuclear magnetic resonance (Magistrini *et al.*, 1995). Such examination confirms the involvement of the accessory glands in seminal plasma production and indicates that sexual preparation and ejaculation are associated with an increase, and subsequent decrease, in the size of the bulbourethral glands, prostate gland and vesicular glands (Weber *et al.*, 1990). Table 3.1 illustrates the major secretions of the various accessory glands in some farm livestock; however, it is evident that considerable variation exists between and within stallions in the amounts of the components produced and their exact site of secretion (Magistrini *et al.*, 1995). With recent developments, information on these glands is increasingly available (Weber and Woods, 1992, 1993). The following section will discuss each gland in turn. A more detailed account of seminal plasma is given in Chapter 4.

Bulbourethral glands

The bulbourethral glands, also known as the cowpers glands or glandula bulbourethralis, are situated nearest the root of the penis. They are paired;

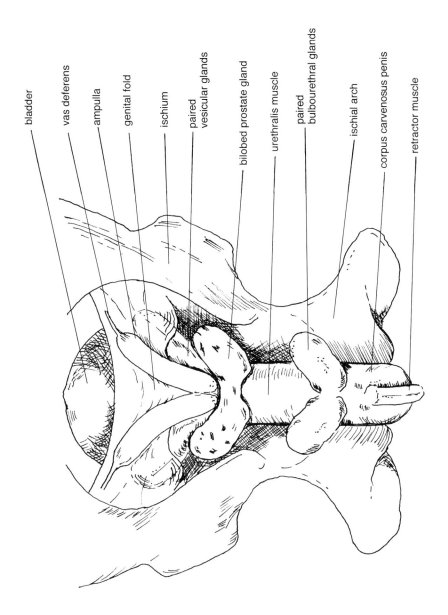

Fig. 3.9. Ventral view of the stallion's reproductive system, illustrating the relative positions of accessory glands.

Table 3.1. Major secretions of accessory glands in selected farm livestock (Mann, 1964; Mann and Lutwak-Mann, 1981; Setchell, 1991).

Gland	Stallion	Bull	Ram	Boar
Seminal vesicles				
Major secretions	Citric acid	Fructose	Fructose	Citric acid
	Potassium	Citric acid	Citric acid, Prostaglandins	Ergothionine Inositol
Minor secretions	Protein	Inositol	–	Fructose
Virtually absent	–	Ergothionine	–	–
Prostate				
Major secretions	Protein	–	–	–
	Citric acid			
	Zinc			
Minor secretions	–	–	–	Citric acid
Virtually absent	–	–	–	–
Bulbourethral				
Major secretions	Sodium chloride	–	–	Sialoproteins
Minor secretions	–	–	–	–
Virtually absent	–	–	–	Citric acid Inositol
Ampulla				No gland
Major secretions	Ergothionine	–	–	
Minor secretions	Inositol	Fructose	–	
		Citric acid	–	
Virtually absent	Fructose	–	–	
	Citric acid			

they measure, on average, 19 mm by 32 mm and weigh in the order of 11 g each (Gebauer *et al.*, 1974b; Little and Woods, 1987). They lie near the ischial arch on either side of the urethra into which they exit. They are normally oval in shape, but become increasingly flattened and ovoid during ejaculation (Weber and Woods, 1993). The secretions produced by the bulbourethral glands are clear, watery and thin in appearance, and high in sodium chloride (Mann, 1975). These secretions are mainly produced during the main sperm-rich part of the ejaculate (Weber and Woods, 1993) but possibly also contribute to the pre-spermatozoan fraction (Mann, 1975). The volume of the bulbourethral gland after sexual preparation, assessed by ultrasonography, is reported to correlate with the number of false mounts attempted by the stallion during sexual preparation (Weber *et al.*, 1990).

Prostate gland

The prostate gland is a bilobed, single gland, each lobe being approximately 70 mm by 40 mm by 20 mm, weighing in its entirety 40–50 g (Gebauer *et al.*, 1974b). Both lobes of the gland are interconnected via an isthmus 30 mm in length with a diameter of 6 mm, and exit via a single opening into the urethra near the bulbourethral glands (Little and Woods, 1987). The secretions of the prostate gland are thin and watery, with an alkaline pH, and are high in protein, citric acid and zinc. The significance of these levels is unclear. The secretions of the prostate gland make a significant contribution to the pre-spermatozoan fraction, whose likely importance is the flushing out of the system prior to the passage of the spermatozoa. The lobular and isthmus thickness of the prostate gland increase significantly after sexual preparation and significantly decline post ejaculation (Weber *et al.*, 1990).

Vesicular glands

The vesicular glands, or seminal vesicles, are paired, multi-lobed glands lying either side of the bladder and exiting again near the exit of the bulbourethral glands. They are, on average, 160–200 mm in length, 25 mm in width and 13 mm in height, and weigh 25–30 g (Gebauer *et al.*, 1974b; Little and Woods, 1987). The main function of these glands seems not to be (as first suggested) spermatozoan storage, but rather the production of the gel-like last fraction of semen ejaculated. Work by Weber *et al.* (1990) indicated that the decrease in size of the vesicular glands post ejaculation is positively correlated with the volume of gel evident in that ejaculate. The vesicular glands' function, however, is apparently not exclusively gel production. Weber and Woods (1993) demonstrated that there was fluid in the vesicular gland's excurrent ducts in only six out of 17 ejaculates that resulted in gel. It is likely that the vesicular glands are also responsible for secretions within the main sperm-rich fraction of the ejaculate. Vesiculectomy results in a significant decline in seminal volume not entirely accounted for by a lack of gel fraction secreted (Klug *et al.*, 1979; Webb *et al.*, 1990). The secretions of the vesicular glands are reported to be high in potassium and citric acid (Rossdale and Ricketts, 1980; Mann and Lutwak-Mann, 1981). The vesicular glands are the most testosterone dependent of all the accessory glands, their secretions being significantly reduced as a result of declining testosterone levels – for example, during the non-breeding season (Thompson *et al.*, 1980).

Ampullae

The ampullae are not always classified as accessory glands as they do not strictly conform to the definition 'discrete structures separated from their site of secretion, that is the vas deferens, by duct(s)'. Rather they are paired out-foldings of the vas deferens, at its junction with the urethra. They form a pouch approximately 12 mm in diameter, compared with the normal vas deferens diameter in this area of 4–5 mm. Their average length is 16 mm (Little and Woods, 1987), making them relatively large compared with those of other farm livestock studied (Leone, 1954; Mann *et al.*, 1956; Mann, 1975). These two pouches consist of a series of crypts and folds, along with the general increase in the diameter of the vas deferens lumen. Little is known of the contribution, in terms of volume, made by the ampullae towards seminal plasma (Amann, 1993a,b), but it is evident that any contribution they make occurs primarily before penile contraction or ejaculation (Weber and Woods, 1993). The secretions of the ampullae are thought to include ergothionine with some inositol; concentrations of fructose and citric acid are negligible (Rossdale and Ricketts, 1980; Mann and Lutwak-Mann, 1981). It has also been suggested that their function may be one of sperm storage, but work done by Gebauer *et al.* (1974a) does not support this theory, at least in regularly used stallions. The secretions of the ampullae vary with season (as do those of some of the other glands): in general they tend to be greater at the beginning and end of the season rather than at the peak. This is indicated by similar seasonal changes in ergothioine concentrations (Mann, 1975).

3.2.4. The vas deferens

The vas deferens, or ductus deferens, runs from the cauda epididymis to join with the urethra in the pelvic region of the stallion. It is 25–30 cm in length and 4–5 mm in diameter, except, as already discussed, where it widens out to form the ampullae near the urethral junction (Gebauer *et al.*, 1974b). The wall of the vas deferens is thick, in comparison with the diameter of the lumen, and there are three layers of smooth muscle fibres: the inner oblique, the middle circular and the outer longitudinal (Fig. 3.10). The thick muscular

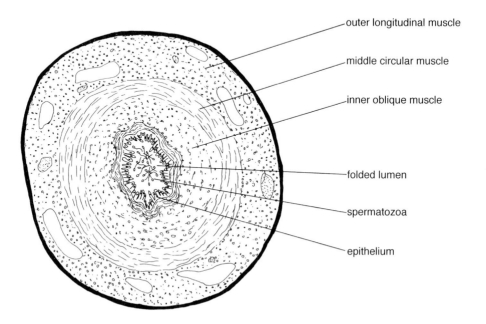

Fig. 3.10. Transverse cross-sectional view through the vas deferens, illustrating the relatively large muscular walls.

wall is responsible for the peristaltic contractions that actively propel spermatozoa and its associated fluids from the testis to join the urethra. The lumen of the vas deferens is lined by folds, and this is especially evident where it nears the caudal epididymis. These folds increase the surface area for storage of spermatozoa.

The vas deferens enters the abdominal cavity of the stallion from the testis, via the inguinal canal. It passes through this canal, in the body wall, along with the testicular blood and nerve supply and in close association with the cremaster muscle (Fig. 3.11). The cremaster muscle, as it passes into the body cavity, is divided into two: the internal cremaster muscle running the length of the canal and lying between the blood and nerve supply and the vas deferens; and the external cremaster muscle, the larger portion, which lies lateral to the canal. Both these cremaster muscle blocks are responsible for supporting the testis and for the retraction of the testis up towards the abdomen in response to cold, fear and shock.

3.2.5. The epididymis

Each testis has an epididymis lying over its dorsal aspect. The tail of the epididymis (joined to the vas deferens) is situated at the caudal end of the testis and the head (joined to the rete testis) is situated at the cranial end. The epididymis consists of long, highly convoluted tubules measuring up to 45 m

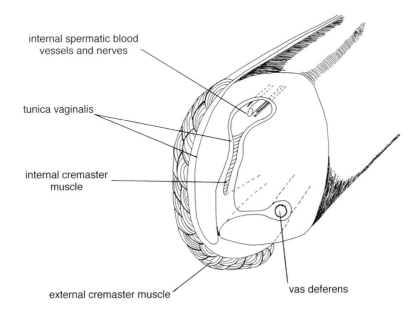

Fig. 3.11. Transverse cross-section through the inguinal canal area of the stallion.

in length and weighing in the order of 25 g each in a mature stallion (Thompson, 1992). They originate from the rete testis as a series of tubules which, over the length of the epididymis, converge to form one single tube continuous with the vas deferens.

The epididymis can be divided into three sections. This may be done on the basis of anatomy; that is, on epithelial cell characteristics – namely, the caput or head, the corpus or body, and the caudal or tail end of the epididymis; or alternatively, on the basis of function – namely, the initial (part of the efferent ducts plus the initial part of the caput), the middle (remainder of the caput plus the corpus) and the terminal (cauda plus the proximal vas deferens). The caudal end of the epididymis lies caudal to the testis, forming the characteristic projection seen on the exterior of the scrotum when viewed in the live animal. It is this caudal end that joins with the vas deferens and, as the terminal section, it is the major site of spermatozoan storage, providing storage for between 53% and 62% of the extra-gonadal spermatozoa reserves in sexually active and rested stallions, respectively (Gebauer et al., 1974a; Amann et al., 1979a). The middle epididymis is responsible primarily for spermatozoan maturation and is the likely site for the secretion of androgen-dependent protein maturation factors enabling sperm to undergo capacitation within the female tract in readiness for fertilization (Merkies and Buhr, 1998); it is also likely to be the site of initiation of spermatozoa progressive motility (Johnson et al., 1980; Setchell, 1991). The initial epididymis is primarily the site for fluid and solute reabsorption and resultant concentration of spermatozoa in readiness for emission.

Throughout the length of the epididymis, the epithelial cell lining of the lumen is highly folded with additional microvilli, thus significantly increasing the surface area available for storage of spermatozoa and for fluid reabsorption (Fig. 3.12). The tubules are surrounded by a well innervated smooth muscle layer which is involved in epididymal contraction and which is increasingly obvious as the cauda end is approached (Setchell and Brooks, 1988).

In order for maturation to occur, spermatozoa need to remain within the epididymis for up to 7 days. Those that are subsequently not ejaculated are reabsorbed, ensuring a continual supply of fresh spermatozoa (Gebauer et al., 1974a,b,c; Thompson et al., 1979a). In general, mammalian epididymal fluid contains significant concentrations of glycerylphosphorylcholine, glycosidases, amino acids, androgen-dependent proteins, potassium, inositol, carnitine, dihydrotestosterone, and alkaline and acidic phosphatases. The specific concentration of these elements has not yet been ascertained in horses, but it is likely that equine epididymal fluid contains similar components (Rossdale and Ricketts, 1980; Setchell, 1991). Spermatozoa spend 4–5 days moving through the epididymis by means of muscle contraction of the epididymal walls and hydrostatic pressure. Spermatozoa can be stored in the caudal epididymis for between 2 days and several weeks. The storage capacity of the cauda is 10×10^9; storage capacities of the caput and corpus epididymis are much smaller, at 12 and 17×10^6, respectively (Samper, 1995a).

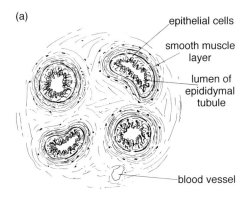

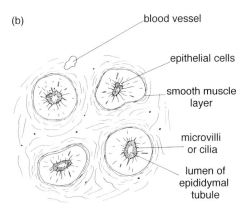

Fig. 3.12. Cross-section through the (a) cauda and (b) caput epididymis. Note the differences in the epithelial cell lining of the lumen, which is highly ciliated in the caput region, and the increased thickness in muscle wall in the cauda epididymis.

3.2.6. The testis

The testes are the site of male gamete or spermatozoa production, and the primary site for androgenic hormone production in the stallion. As such, they play a central role in the reproductive function of the stallion (Figs 3.13 and 3.14).

Scrotum

The testes are held within the scrotal sacks, outside the abdominal cavity, and suspended from the abdomen by the spermatic cord area and the cremaster muscles. The scrotum in the horse is relatively short and non-pendulous in nature, when compared with that of other farm livestock such as the bull or the ram. As in other mammals, it is evident that the positioning of the testes outside the main body cavity is necessary to maintain a testes temperature of 35–36°C

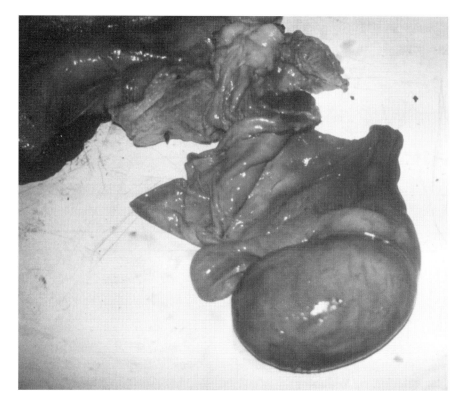

Fig. 3.13. The testes of the stallion after slaughter and dissection (see also Fig. 3.14).

(approximately 3°C below the main body temperature), which is a requirement for optimal production of spermatozoa (Cox *et al.*, 1979). This temperature difference is maintained by means of the cremaster muscle and the pampiniform plexus. The cremaster muscle, as indicated earlier, allows the testes to be drawn up towards or let down away from the abdomen, in response to a reduction or increase, respectively, in environmental temperature. The pampiniform plexus is the term given to the intimate association between the fine network of capillaries branching from the central testicular vein, which returns blood from the testes to the body, and the similarly highly coiled arterial supply entering the testes. This arrangement acts as an arteriovenous counter-current heat exchange mechanism, allowing heat in the warmer arterial supply to be lost to the cooler venous return system. As a result the blood entering the testes is cooled, helping to maintain a temperature difference between the testes and the body core. In addition, the arrangement of the blood vessels in this area of the spermatic cord also acts to eliminate the pulse in the arterial blood supply entering the testes, without significantly reducing the blood pressure. The significance of this is unclear, but it may be important in the passage of diffusible elements between the incoming and outgoing blood supplies (Setchell, 1991).

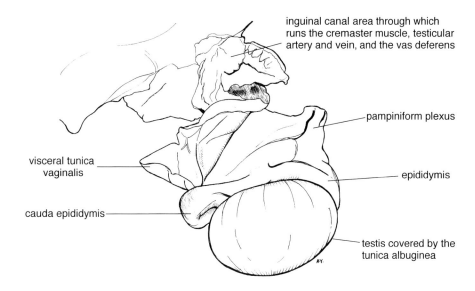

Fig. 3.14. A labelled diagram of Fig. 3.13 illustrating the major features of the stallion's testes after dissection from the body, post slaughter.

The testes of the stallion descend through the body cavity and into the scrotum at birth (or soon after), passing down from a position ventral to the kidneys, through the inguinal ring and into the scrotum. Problems may be encountered with incomplete descent of the testes and inadequate closure of the inguinal ring. Such conditions are not uncommon and often result in cryptorchidism; animals with this condition are sometimes referred to as rigs (section 4.3.2). There is reputed to be a higher incidence of failure of the testis to descend on the left side than the right (Bergin *et al.*, 1970).

The testes of the horse lie along the longitudinal axis; that is, horizontal to the body of the stallion – unlike the bull and ram, where the testes lie in a vertical axis. The positioning of the testes is such that the caput epididymis is directed cranially and the corpus epididymis lies horizontally over the dorsal aspect of the testes, with the cauda epididymis lying caudally. On retraction of the cremaster muscles the testes are drawn up towards the body, turning slightly away from the horizontal plane.

The size of a stallion's testes varies considerably with age during the first 5 years of life. At birth the testes weigh on average 5–10 g in a lightweight riding horse and do not change significantly during the first 10–12 months of life, increasing in size allometrically with general body growth (Thompson, 1992). With the advent of spermatogenesis at 18 months of age, testicular growth is accelerated, resulting in near final testes size at sexual maturation, around 5 years of age (Amann, 1993a,b). The final weight is 300–350 g per testis, with a longitudinal axis length of 6–12 cm, a depth of 4–7 cm and a width of 5 cm. In many stallions there is a tendency for the left testis to be slightly smaller than the right. The significance of such a size difference is unclear (Thompson *et al.*, 1979).

Internal structure of the testis

The testes lie protected within the scrotum, the outer layer of which is the skin. Scrotal skin is of a similar structure to the skin covering the remainder of the horse's body, but with a higher concentration of sweat glands and nerve endings. Within the outer skin layer lies the tunica dartos, a muscle fibre and connective tissue layer, contraction of which again aids in drawing the testes up towards the body. Within the tunica dartos lies the connective tissue layer of the scrotal fascia, and within this is the tunica vaginalis (Figs 3.15 and 3.16).

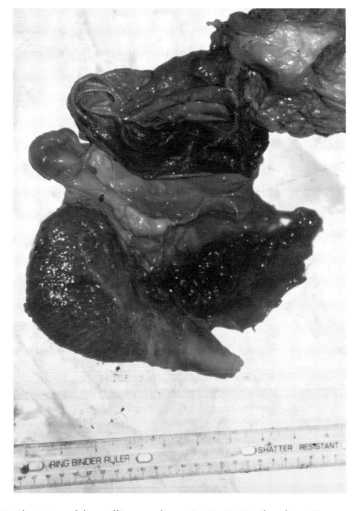

Fig. 3.15. The testes of the stallion, as shown in Fig. 3.13, after dissection.

The tunica vaginalis is divided into two parts: the outermost is the parietal vaginalis and the inner layer is the visceral vaginalis (Fig. 3.17). Adhesions between these two layers may occur, largely as a result of trauma, age and general senescence. Such adhesions restrict the movement of the testis within the scrotum and hence affect the efficiency of the testicular temperature-control mechanisms. Within the tunica vaginalis lies the tunica albuginea, which forms a relatively thick fibrous capsule. Within the tunica albuginea lies the functional area of the testes: the parenchyma. Extending

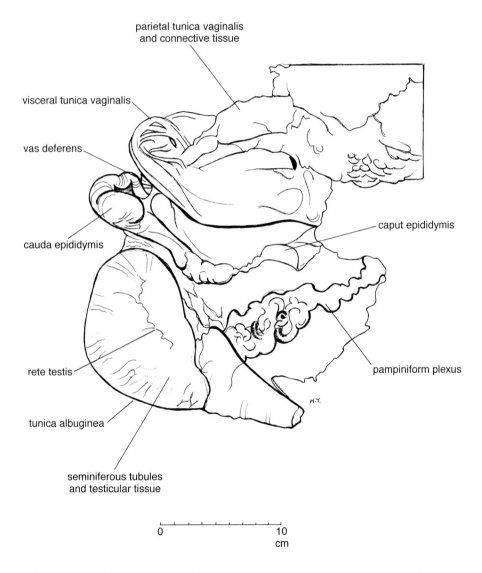

Fig. 3.16. A labelled diagram of Fig. 3.15 illustrating the internal structures of the stallion's testis.

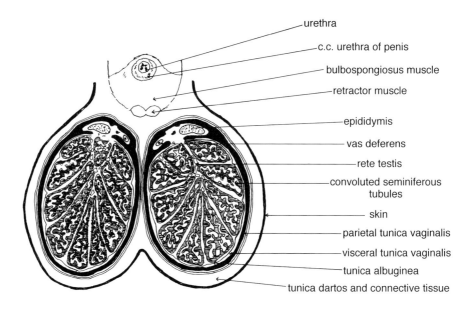

Fig. 3.17. Vertical cross-section through the testes, epididymis and part of the penis of the stallion.

into the parenchyma are strands of connective tissue originating from the tunica albuginea, forming lobes within the testes. The colour of testicular parenchyma is indicative of stallion age. At puberty, as the testes become functional, they become lighter in colour. Dark-coloured testes in stallions under 5 years of age is indicative of small, inactive seminiferous tubules composed mainly of gonocytes and Sertoli cells, with no lumen. Active seminiferous tubules, with all types of germinal cells and Sertoli cells and with a lumen, appear lighter in colour (Clemmons *et al.*, 1995). Post puberty, the testes parenchyma tends to darken gradually with age, due to greater pigmentation of the Leidig cells (Johnson and Neaves, 1981).

The testicular parenchyma may be divided into two main functional compartments: the seminiferous tubules and the interstitial tissue. The relative composition of the two is reported to be 58–72% seminferous tubules and 28–42% interstitial tissue (Swierstra *et al.*, 1974; Christensen, 1975; Johnson and Neaves, 1981; Berndtson *et al.*, 1983).

Seminiferous tubules

Seminiferous tubules are defined by the lamina propria, which consists of fibroblasts, myoid cells (specialized smooth muscle cells that give rhythmic contraction of the tubules) and laminin. The tubules are not penetrated by the blood, lymph or the nervous system. Within the seminiferous tubules is an epithelial cell lining, consisting of germinal cells at different stages of development (Kroning, 1986), plus somatic cells (Sertoli, support or nurse cells). The seminiferous tubules are arched in shape, with a straighter tapered portion at either end of a central convoluted area. It is this central convoluted portion that is primarily involved in spermatogenesis and contains up to 15–20% Sertoli cells, the remainder being germinal cells at various stages of development. The Sertoli cells may be found radiating out into the lumen of the tubules in association with their attached germinal cells. On release into the lumen, these germinal cells, now spermatids, progress, by rhythmic contractions of the tubules, towards either tapered end of the seminiferous tubule. These tapered ends connect to the rete testis situated in the cranial area of the testis. The rete testis penetrates through the tunica albuginea to form the extracellular rete testis. Each of the tubules within the rete testis fuses with the 13–15 efferent ducts that lead on to form the initial part of the caput epididymis (Amann *et al.*, 1977).

Sertoli cells surround all germinal cells, except for spermatogonia, and as such closely control the environment of the developing spermatids. They fulfil three main functions. Firstly they provide a barrier between the spermatids and the stallion's system; this is particularly important in considering the protection of spermatids from rejection by the stallion's immune system. Secondly, they are a source of nutrients for the germinal cells, acting either as transport systems, or by producing the nutrients required themselves. Thirdly, they provide a means of movement for the developing spermatids which, as they develop, move progressively towards the lumen of the tubules in readiness for release at spermiation. In association with this, Sertoli cells also have a phagocytic role, removing degenerating germinal cells plus residual spermatozoan cytoplasmic bodies.

The major function of Sertoli cells – that of immunological protector – is largely conferred by their action as a blood–testis barrier consisting of junctional complexes between neighbouring Sertoli cells. This protection is required to prevent the rejection of germinal cells, which in essence (due to their haploid nature) are 'foreign bodies' within the stallion's system. The blood–testis barrier also acts to control the passage of molecules from the blood plasma to the seminiferous tubules and seems to actively exclude large water-soluble molecules, though allowing the passage of lipid-soluble molecules and water (Setchell, 1991). As a component of this protective blood–testis barrier, the Sertoli cells must also have a communication role allowing communication between themselves, the germinal cells, the lamina propria and the Leidig cells (Amann, 1993a). Damage to this blood–testis barrier, though rare, can occur and results in a significant reduction in reproductive ability, which may be irreversible.

Sertoli cells also produce and secrete lactate, a source of energy to the spermatids, along with inhibin and activin (regulatory hormones) and several proteins. Some of these proteins act as carriers providing a means of transport for iron, copper and vitamin A to the spermatids; some may act to regulate spermatozoa production (Amann, 1993a) and others (transferrin and ceruloplasmin) regulate germinal cell movement (Amann, 1993b). This intimate association and the reliance of germinal cells upon Sertoli cells does have the major disadvantage that any effect on Sertoli cell number or function will have a direct effect on germinal cell development. Indeed it is likely that many of the effectors of spermatozoan production do act via the Sertoli cells, rather than directly upon the developing spermatids (Hochereau-de Reviers *et al.*, 1990; Amann, 1993a,b). The proportion of Sertoli cells to germinal cells is reported to be highly heritable (Hochereau-de Reviers *et al.*, 1987). This, plus the normally observed reduction in Sertoli cell numbers with age, may account for the apparent reduction in spermatogenesis over time and the considerable variation in spermatozoan production capabilities between stallions (Johnson and Thompson, 1983; Blanchard and Johnson, 1997).

Interstitial tissue

The interstitial tissue of the testes is made up largely of Leidig cells (21–57%, depending upon the stallion's age: Table 3.2), blood vessels, lymph vessels, connective tissue and nerves (Fig. 3.18). It is evident that the Leidig cells may make up the majority of the interstitial tissue, and in the horse they appear to be relatively large in size, compared with those of the rat, hamster, ram and bull (Johnson and Neaves, 1981).

Leidig cells are primarily involved in steroid hormone production, in particular the production of testosterone from cholesterol. Two types of Leidig cell have been identified and are associated with stallion age and cellular function. Post-puberty Leidig cells are gradually replaced, with age, by adult Leidig

Table 3.2. Changes in testicular composition evident with age in the stallion (Johnson and Neaves, 1981; Pickett *et al.*, 1989).

	Age (years)		
	2–3	4–5	13–20
Testicular weight (g)			
Parenchyma	105[a]	146[b]	184[c]
Tunica albuginea	12[a]	15[b]	29[c]
Parenchyma composition (%)			
Leydig cells (%)	6[a]	12[b]	18[c]
($\mu l\ g^{-1}$ testis)	50	115	175
Other interstitial tissue	22[a]	16[b]	10[c]
Seminiferous tubules	72	72	72
Seminiferous tubules			
Diameter (m)	212[a]	230[b]	242[b]
Length/testis (m)	2040[a]	2390[ab]	2790[b]
Sertoli cell no. ($\times 10^6\ g^{-1}$ testis)	39	29	21
Daily spermatozoal production			
per testes ($\times 10^9$)	1.3[a]	2.7[b]	3.2[b]

[a,b,c] Means in the same row with different superscripts differ ($P < 0.05$)

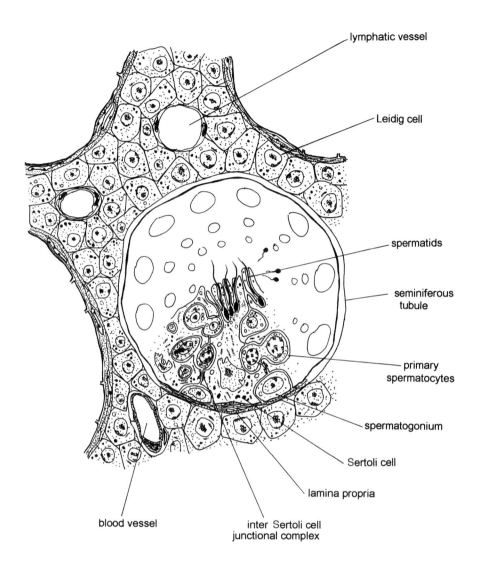

Fig. 3.18. Cross-section through the testicular tissue, illustrating interstitial tissue and an enlarged area of seminiferous tubule with developing spermatids.

cells, which contain a higher concentration of pigmented lipids and have a greater number of interdigitations. Their production of testosterone is also greater (Almahbobi *et al.*, 1988). In addition to changes in the function of Leidig cells, there is also a difference in the ratio of Leidig cells to seminiferous tubules. The older the stallion, the relatively greater is the number of Leidig cells present, with up to a threefold increase being reported to occur between 2–3-year-old stallions and ones of 13–20 years of age (Table 3.2).

3.3. Control of Stallion Reproduction

The reproductive organs of the stallion include not only the anatomical structures previously described, but also structures within the brain. The organs of copulation, spermatozoan production and passage to the exterior may be referred to as the intrinsic reproductive organs and those within the brain as the extrinsic organs. It is the extrinsic organs that largely control the intrinsic organs and ensure that the reproductive system does not work in isolation, but in harmony with other bodily functions and control mechanisms. These extrinsic organs form the basis of the following discussion on the control of stallion reproduction.

3.3.1. Commencement of reproductive activity

Puberty heralds the start of reproductive activity in the stallion. The exact timing of puberty is unclear but has been defined as the time at which a stallion can produce 50×10^6 spermatozoa per ejaculate with greater than 10% progressive motility. Using these parameters, puberty is seen to commence at 21 months (Clay and Clay, 1992). The exact timing of puberty is known to be affected by several factors, including breed, management and the rate of stallion growth and development. Classically puberty is considered to commence between 17 and 22 months of age (Nishikawa, 1959b). Histological examination of testicular tissue indicates that spermatogenesis commences *in utero* with the migration of primordial cells into the fetal gonads (Everett, 1945). These primordial cells undergo division to produce gonocytes, which are found between the Sertoli cells. Puberty then heralds the further division of these to produce spermatogonia. Histological work on the changes within the testes, especially changes associated with the Leidig cells, suggests that puberty may occur between 12 and 26 months of age (Cornwall, 1972; Naden *et al.*, 1990; Clay and Clay, 1992). Circulating hormone levels have also been used in an attempt to identify the timing of puberty. During early life, significant increases in circulating luteinizing hormone (LH) and follicle-stimulating hormone (FSH) levels have been demonstrated, starting as early as 8–10 months. This is followed at 12 months by an increase in testes size; the appearance of spermatogonia and primary spermatocytes is evident at 12–14 months of age, followed by the appearance of round spermatids at 16 months and mature spermatids at 36 months. This heralds the start of spermatozoan production (Cornwall *et al.*, 1973; Curtis and Amann, 1981; Johnson, 1991a). Changes in circulating testosterone levels are also evident, with a significant rise in concentrations being reported to occur between 16 and 21 months (Naden *et al.*, 1990; Clay and Clay, 1992).

Regardless of the exact timing of puberty, it is evident that a stallion does not reach full adult reproductive ability and so is not capable of a full covering season until he is 5 years old. This is demonstrated in work by Johnson and Neaves (1981), Thompson and Honey (1984) and Berndtson and Jones

(1989) involving the assessment of several characteristics, including testicular weight, spermatozoan production, testosterone concentrations, and Leidig cell and Sertoli cell number and volume. Follow-on work by Johnson *et al.* (1991b) demonstrated that stallions of 3 years of age or younger are increasingly active spermatogenically and, therefore, capable of fertilizing mares. However, they only possess a limited sperm-producing capacity. Significant improvement in scrotal width, testicular tone, spermatozoan motility, the percentage of normal spermatozoa and the total number of spermatozoa produced is evident even between 2–3-year-old stallions (Sigler and Kiracofe, 1988). At 4 years of age stallions are reported to be able to produce adequate numbers of spermatozoa to cover the number of mares expected of a fully mature stallion, but they may not perform consistently well under heavy work loads. By 5 years of age they should be capable of consistently carrying out a full covering season (Johnson *et al.*, 1991b). These apparent changes in reproductive activity in early life were also reported by Naden *et al.* (1990), who demonstrated changes in sexual behaviour and semen characteristics between 2 years of age and sexual maturity. However, their work indicated no change in time to erection, time to mount and the percentage of motile spermatozoa. It has been suggested that this improvement in reproductive ability is due not only to an increase in testis size and, therefore, function but also to an increase in the efficiency of spermatozoan production per gram of testicular tissue (Table 3.3).

Post maturity, reproductive function should continue, largely unchanged, until old age. However, reproductive function in terms of semen quality has been reported to decline after 20 years of age (Johnson and Thompson, 1983; Amann, 1993a,b). It has been suggested that this may be due to a reduction in output of spermatozoa (by up to 66%) evident after 15 years of age, accompanied by an increase in the incidence of morphological abnormalities (Van der Holst, 1975). This is disputed by other work which suggested that no such decline in performance is necessarily apparent (Johnson *et al.*, 1991a).

3.3.2. Seasonality

In common with mares and many other mammals, the reproductive activity of the stallion is governed by season. He is, therefore, classified as a seasonal

Table 3.3. Change in daily spermatozoan production (DSP) g^{-1} of testicular tissue with age (Johnson and Thompson, 1983).

Age (years)	DSP g^{-1} testes parenchyma
2–3	6×10^6
4–5	18×10^6
6–20	20×10^6

breeder. The length of the breeding season depends upon many factors, including breed, environmental conditions and management. In the northern hemisphere the breeding season runs on average from March to November, and in the southern hemisphere from September to June. Coldblood and similar types of stallion tend to show shorter, better-defined seasons compared with hotblood and warmblood stallions, but in general the stallion's season is less distinct than the mare's season and, unlike her, given enough encouragement the majority of stallions are capable of breeding all year around. Spermatozoan production (unlike ova production) is a continual process, not governed by cyclical hormonal changes. However, season does affect various semen parameters and behavioural characteristics. Seminal volume, spermatozoan concentration, total number of spermatozoa per ejaculate, the number of mounts per ejaculate and the reaction time to an oestrous mare all suffer in the non-breeding season, making the stallion in general less efficient reproductively (Pickett et al., 1970, 1975b; Pickett and Voss, 1972; Johnson, 1991b; Clay and Clay, 1992). Figures 3.19–3.23 illustrate the effects of season on stallion reproductive performance.

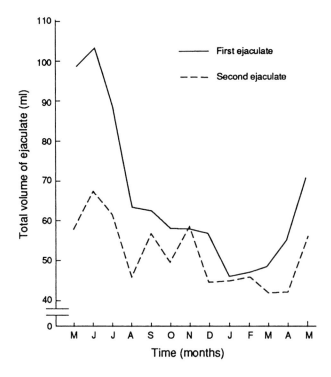

Fig. 3.19. Effect of season on the total volume of ejaculate produced (Pickett and Voss, 1972).

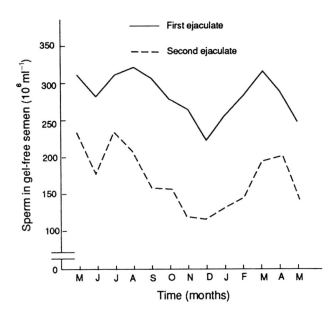

Fig. 3.20. Effect of season on the number of spermatozoa produced in the gel-free portion of a semen sample (Pickett and Voss, 1972).

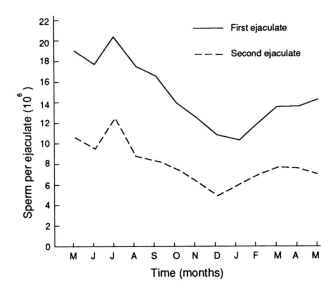

Fig. 3.21. Effect of season on the total number of spermatozoa produced per ejaculate (Pickett and Voss, 1972).

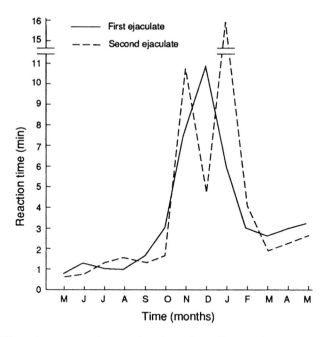

Fig. 3.22. Effect of season on the reaction time of a stallion to the presence of an oestrous mare (Pickett and Voss, 1972).

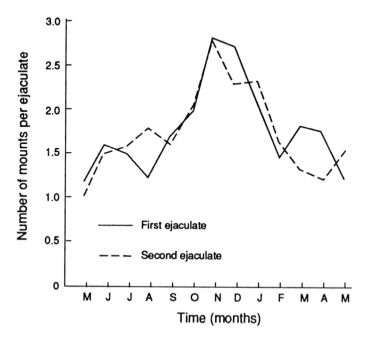

Fig. 3.23. Effect of season on the number of mounts required prior to successful ejaculation (Pickett and Voss, 1972).

In addition to the above effects, breeding outside the natural season is reported to increase the percentage of spermatozoa with morphological abnormalities (Van der Holst, 1975). The reduction in a stallion's libido during the non-breeding season may be accounted for by a decrease in the number of Leidig cells and so by implication a reduction in circulating testosterone levels (Johnson and Tatum, 1989).

The seasonality of stallion reproduction is governed primarily by a daylength effect on the reproductive control axis: the hypothalamic–pituitary–testis axis (Fig. 3.24). This effect is mediated by the pineal gland. Other factors, including pheromones, nutrition and environmental temperature, also play a significant role. The importance of these other factors in the seasonality of reproduction is more evident in stallions kept nearer the equator, where daylength changes are more subtle and the seasonal variations in hormone concentrations less distinct (Lang *et al.*, 1995).

3.3.3. The hypothalamic–pituitary–gonadal axis

The hypothalamic–pituitary–testis axis is responsible for the control of reproduction in the stallion. The system works in a cascade fashion, with appropriate negative feedback loops. The axis is also controlled by secretions of the pineal gland.

Pineal gland

The pineal gland lies above and behind the hypothalamus, held within a fibrous capsule. Septa extend into the gland from separate masses of round epithelial cells. These cells produce the hormone melatonin from tryptophan; melatonin has an antigonadotrophic effect on the hypothalamus (Sharp and Clever, 1993). The function of these cells is affected by daylength: increasing daylength inhibits pineal function, i.e. the production of melatonin (Wesson *et al.*, 1979). The means by which daylength takes its effect is unclear in the horse but it is believed that impulses from the retina in the eye travel via the rostral accessory optic tract to the thoracic spinal cord and thence to the cranial cervical ganglion. Nerve fibres from the cervical ganglion then follow the path of the arteries to the pineal gland epithelial cells, where they take effect (Venzke, 1975; Moore, 1978; Reiter, 1981).

Hypothalamus

The hypothalamus forms part of the diencephalon of the brain. It is served by the CNS and connects, via this, to several areas of the brain and spinal cord. Various sections of the hypothalamus govern hormones which are released by the pituitary. It consists largely of neural tissue containing the nuclei of nerve cell bodies. These are termed the supraoptic, paraventricular, preoptic and rostral nuclei. As such they are involved not only in reproductive activity via sexual behaviour and the release of trophic hormones, but also in the

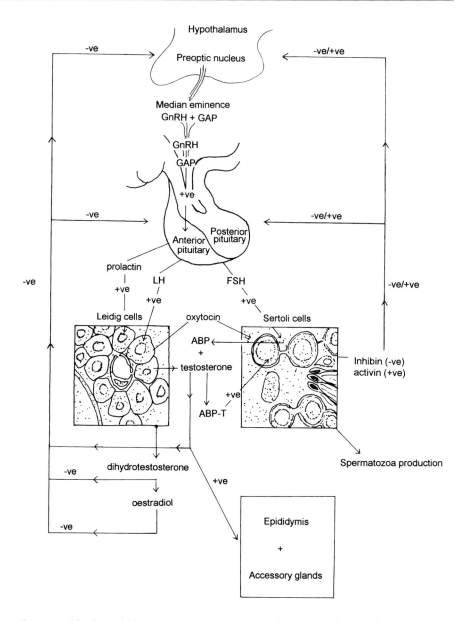

Fig. 3.24. The hypothalamic–pituitary–testis axis, which controls reproductive function in the stallion.

regulation of appetite, thirst, body temperature, vasomotor activity, emotion, use of the body's nutrient reserves and the states of sleep and wakefulness (Amann, 1993b). The nucleus of particular concern in reproduction is the preoptic nucleus, containing the cell bodies of the parvicellular neurosecretory cells, the source of gonadotrophin-releasing hormone (GnRH) (Kainer, 1993).

Pituitary

The pituitary or hypophysis is an oval mass contained within a fibrous capsule, weighing in the region of 1.8 g (Thompson, 1992). It lies below, and is connected to, the hypothalamus. The pituitary is divided into the anterior pituitary (adenohypophysis) and the posterior pituitary (neurohypophysis). The posterior pituitary, containing neurological cells, is largely involved in the storage of hormones produced as a result of neural stimulation. It is the anterior pituitary, made up of follicles or cords of cells, which is involved in reproductive control (Kainer, 1993). The anterior pituitary secretes the following hormones: somatotrophin, prolactin, adrenocorticotrophin hormone, FSH, LH, thyrotrophin and lipoproteins. The connection of the anterior pituitary to the hypothalamus is via parvicellular neurosecretory cells and hypophyseal portal vessels. The parvicellular neurosecretory cells terminate in the medial eminence, a neurohaemal organ that acts as a site of transfer or release of inhibitory or releasing hormones into the blood of the hypophyseal portal vessels which pass to the secondary plexus in the anterior pituitary. This in turn passes the hormones on to the target cells in the anterior pituitary. In response the pituitary produces (among other hormones) FSH and LH, which are of prime importance in the control of reproduction in the stallion (Cupps, 1991; Kainer, 1993).

3.3.4. Hypothalamic and pituitary hormones

Numerous hormones have an effect upon or are produced by the hypothalamus or the pituitary. These will be discussed in turn, but it must be remembered that the hormones do not work in isolation; rather, they form part of an integrated system of whole body control.

Melatonin

As indicated previously, one of the major means of controlling the hypothalamus is via the hormone melatonin, which is produced by the pineal gland. Melatonin is secreted during the hours of darkness and, due to its antigonadotrophic nature, it dominates the system and suppresses hypothalamic activity during the periods of its secretion (Burns *et al.*, 1982; Argo *et al.*, 1991). This response to daylength results in melatonin secretion being divided into two phases: photophase during the daytime and scotophase during the hours of darkness (Grubaugh, 1982). Within this pattern, secretion is episodic in fashion, with episodes of greater frequency and amplitude being evident during the hours of darkness. Additionally, melatonin is secreted in a circadian fashion, maintaining a 24-h pattern of release despite exposure to continuous daylight or darkness. Photorefractoriness to continual long days does occur; seasonal cycles in scrotal width and output of spermatozoa continue despite continual treatment of 16 h light, 8 h dark (Clay *et al.*, 1987). Melatonin secretion, therefore, demonstrates an inherent rhythm which is modified by external factors, such as daylength.

The exact site and mode of action of melatonin is unclear but its effect is upon GnRH production. It is possible that it acts at the level of the median eminence, where the split between GnRH-associated protein (GAP) and GnRH occurs, allowing the release of GnRH (Strauss *et al.*,1979).

Gonadotrophin-releasing hormone

Gonadotrophin-releasing hormone (GnRH) is a decapeptide initially joined to a 56 amino acid peptide called GAP as part of a larger molecule. As a dual structure it is stored, as granules, within the median eminence. When the environment is appropriate it splits into GnRH and GAP and is discharged as such down the hypophyseal portal vessels to the anterior pituitary, where it binds to receptors on the gonadotrophe of the pituitary (Strauss *et al.*, 1979). Eighty per cent of the GnRH so released passes along this route; the remaining 20% passes back through the median eminence and on to the higher centres of the brain driving sexual behaviour (Pozor *et al.*, 1991). GnRH is secreted in both a pulsatile and continuous (tonic) fashion. Pulse frequency during periods of sexual rest are 1–2 per day, both within and outside the breeding season. However, within the breeding season the tonic levels of GnRH are significantly elevated in the absence of melatonin inhibition. Sexual activity increases pulse frequency to two per hour or more (Irvine and Alexander, 1993).

The action of GnRH upon the pituitary results directly in the release of two main gonadotrophic hormones in the male: LH and FSH (Blue *et al.*, 1991; Seamens *et al.*, 1991). The pituitary may also secrete prolactin, which, as will be evident later, may be involved in the control of stallion reproduction. Exogenous treatment with GnRH results in elevated levels of LH and FSH; conversely, active immunization against GnRH has been shown to suppress LH and FSH production (Schanbacher and Pratt, 1985).

Luteinizing hormone and follicle-stimulating hormone

LH and FSH secretion occurs as a direct result of GnRH stimulation of the anterior pituitary. They are both glycoproteins, consisting of two subunits, α and β. The α subunit is species specific and is the same for LH, FSH, thyroid-stimulating hormone (TSH) and equine chorionic gonadotrophin (eCG). The β subunit differs for each of the hormones and determines the biological function of the hormone (Alexander and Irvine, 1993).

LH, previously known as interstitial cell-stimulating hormone (ICSH), is secreted in a pulsatile fashion (Thompson *et al.*, 1985, 1986; Clay *et al.*, 1988). During the breeding season the level of tonic LH secretion is reported to be 4–7 ng ml^{-1} with a normal pulse frequency of 0.8–0.9 pulses per hour, these pulses having an amplitude of 2–4.5 ng ml^{-1} and a duration of 43–48 min (Blanchard *et al.*, 1990). However, there are considerable differences between stallions, in addition to which a circadian rhythm has been reported, with higher concentrations in general being observed during the hours of daylight. This circadian rhythm in LH secretion is mimicked by similar changes in testosterone secretion (Lang *et al.*, 1995). The relatively high tonic levels of

LH in stallions during the breeding season mean that the pulses of LH secretion are not always detected in blood collected from the jugular vein (Thompson et al., 1985). The site of action for LH is the testes, and LH levels are positively correlated with daily spermatozoan production (DSP) per gram of testes (Blanchard et al., 1990).

FSH, in common with LH, is produced in a pulsatile fashion, in response to the pulsatile release of GnRH. FSH tonic level of secretion is reported to be 1–14 ng ml^{-1} during the breeding season, accompanied by a pulse frequency of 0.7–1.0 h^{-1}, a pulse amplitude of 5–7 ng ml^{-1} and a duration of 31–55 min (Blanchard et al., 1990). However, FSH shows a greater variability in secretion patterns than both LH and GnRH (Thompson et al., 1985; Clay, 1988). Each pulse of FSH has been reported to be associated with an episode of LH, though other workers have failed to demonstrate such a close association (Blanchard et al., 1990). Seasonal variation in FSH levels is less dramatic than is the case with LH (Harris et al., 1983; Johnson and Thompson, 1983; Thompson et al., 1986). It is evident, therefore, that the observed variation in testicular function with season is more likely to be due to a variation in secretion of LH rather than FSH (Johnson and Thompson, 1983).

A feedback mechanism is known to exist that controls LH and FSH production in the stallion and which originates from the testes. In the absence of testes (for example, in geldings), LH levels are naturally high; however, the administration of dihydrotestosterone to such animals results in a depression in LH secretion and eventually in LH production (Thompson et al., 1979b). This can be further demonstrated by the treatment of stallions with exogenous testosterone. Such treatment results in a reduction in circulating FSH levels (Ashley et al., 1986). Two other hormones, activin and inhibin, which originate in the testes are also known to be involved; they have, respectively, an activating and inhibitory role on FSH production (Amann, 1993b).

Prolactin

Prolactin is a single-chain protein, made up of 199 amino acids (Nett, 1993). It is produced by the anterior pituitary under the control of dopamine, originating from the hypothalamus (Johnson and Becker, 1987). In stallions, the variation in prolactin concentration is governed mainly by: season, the highest levels being seen during periods of long daylength (Thompson et al., 1986; Thompson and Johnson, 1987; Evans et al., 1991); age, with levels increasing until maturity at 5 years of age; and sexual activity (Thompson and Johnson, 1987; Rabb et al., 1989).

The function that prolactin plays in the stallion is unclear. In other animals, prolactin is associated with the function of the accessory glands and Leidig cells, enhancing the effect of LH on spermatogenesis. Such an association has not been found as yet in the stallion, though treatment with bromocriptine is reported to decrease the volume of seminal plasma, and a link has been established between sexual activity, and elevated concentrations of prolactin and cortisol (Nett, 1993; Thomson et al., 1996).

3.3.5. Testicular hormones

The final level in the hypothalamic–pituitary–gonadal axis in the stallion is the testes. Within the testes there are two major cell populations or target organs: the Leidig cells and the Sertoli cells. These two cell populations have two major roles: steroid hormone production and the production of spermatozoa. A close feedback association must exist between these various functional structures within the testes. The exact nature of their control is unclear in the horse, much evidence being extrapolated from other mammals. However, the primary endocrine function of the testes is the production of testosterone by the Leidig cells. Other hormones are also produced, as secondary endocrine products, their importance becoming increasingly evident with current research (Roser, 1997). These secondary hormones include activin and inhibin, produced by the Sertoli cells, as well as oxytocin, produced by the Leidig cells, and oestrogens, the production site of which is unclear but is likely also to be the Leidig cells (Roser, 1997).

Testosterone

Testosterone is the major androgenic steroid produced by the Leidig cells of the testes (Savard and Goldziecher, 1960; Oh and Tamaoki, 1970). Its effect is both local, within the testis, and systemic. Its local effect is on the neighbouring Sertoli cells, where it controls the final stages of spermatogenesis, the process being initially driven by FSH. Its effect upon spermatozoan production is apparently one of decreasing the rate of germinal cell degeneration and increasing the production of spermatogonia, rather than by decreasing the length of the spermatogenic cycle (Setchell, 1982; Johnson, 1991a). Testosterone is essential to drive spermatogenesis in the adult stallion (Amann, 1981b). Its passage to the Sertoli cells is active, rather than passive, via attachment to androgen-binding protein (ABP). This active transfer is evident from the elevated concentrations of testosterone bathing the seminiferous tubules (up to 70 mg g^{-1} of testicular parenchyma) (Johnson and Thompson, 1987), which is significantly higher than that in blood serum (300 pg ml^{-1}) (Johnson and Thompson, 1983). Its major systemic functions include: the development and maintenance of the male genitalia; the exhibition of characteristic stallion behaviour, including libido; and differential muscle growth. The average concentration of circulating testosterone varies considerably with season (Pickett and Voss, 1972) (Fig. 3.25).

Figure 3.25 shows that testosterone concentrations are at their highest during the spring and summer months and lowest during the autumn and winter (Berndtson *et al.*, 1974; Johnson and Thompson, 1983). If Fig. 3.25 is compared with Figs 3.19–3.23, it is evident that testosterone concentrations are correlated with sexual behaviour and reproductive efficiency (Byers *et al.*, 1983). This variation in plasma testosterone concentrations is evident at the testes level as well as the whole body plasma level (Berndtson *et al.*, 1983). Testosterone concentrations not only vary with season but also show a diurnal variation (Sharman, 1976; Pickett *et al.*, 1989): testosterone plasma

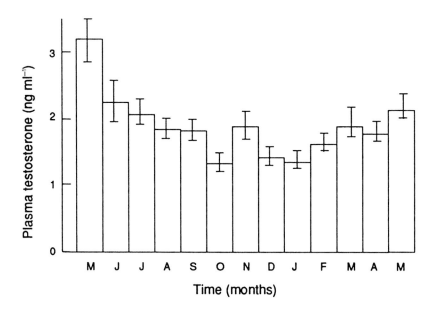

Fig. 3.25. Effect of season on testosterone concentrations in the stallion (Pickett and Voss, 1972).

concentrations were significantly higher at 6.00 hours and 18.00 hours, the average increase being 100% (Pickett *et al.*, 1989). Sexual activity shows a similar diurnal variation and it may be postulated that, in the wild, mating activity is increased at dusk and dawn, as at these times the horse is at least risk from predators. Testosterone levels are also reported to vary with testis size and to increase with age. It has been postulated that this may account for the variation in spermatozoan production evident between stallions within a single breeding season (Berndtson and Jones, 1989).

Superimposed upon these major variations in testosterone concentration are changes associated with the episodic nature of testosterone release. In general, testosterone pulses are linked with LH pulses; therefore, testosterone characteristically shows a continuous level of tonic secretion, with the occasional superimposed episodic increase as a result of an LH pulse. This continuous tonic level of testosterone secretion between episodes in the absence of LH pulse secretion and the considerable variation in the response seen to LH pulses indicate that pulsatile LH release may not be the sole requirement for testosterone release (Squires *et al.*, 1977; Thompson *et al.*, 1985; Thompson, 1992; Roser, 1997), though elevation of endogenous LH concentration by administration of GnRH or the use of human chorionic gonadotrophin (hCG) does result in elevated testosterone concentrations (Roser, 1995).

Inhibin and activin

Inhibin and activin are closely related glycoproteins, sharing common subunits termed α, β_A or β_B. Activin is a combination of the two β subunits and inhibin is a combination of the α plus one of the β subunits. The exact configuration of the form of activin and inhibin produced by the stallion's testis is yet to be ascertained (Bardin, 1989). Both inhibin and activin are produced by the Sertoli cells and act upon the anterior pituitary, inhibiting (inhibin) and activating (activin) its function. Both hormones act primarily upon FSH production, modulating the response of the pituitary to GnRH stimulation (Amann, 1993b; Roser, 1997).

Oxytocin

Oxytocin is a neuropeptide, synthesized as part of a larger molecule, neurophysin (OT-NP) (Nett, 1993). Oxytocin is known to be produced by the Leidig cells of the testis of the ram and the rat and to pass into the interstitial fluid (Knickerbocker *et al.*, 1988; Pickering *et al.*, 1989). From here it passes to the Sertoli cells and on to the lumen of the seminiferous tubules; it seems that it may also be secreted into the lumen of the seminiferous tubules by the rete testis. Its function is unclear, but it may well be involved in the contraction of the seminiferous tubules and hence the movement of spermatozoa towards the epididymis and beyond. Although these conclusions have been drawn from work in animals other than the horse, it is likely that there is a similar function for oxytocin in the horse's testis. Burns *et al.* (1981) demonstrated a link between the level of sexual activity and oxytocin concentrations in stallion's plasma.

Dihydrotestosterone

Dihydrotestosterone is a derivative of testosterone, produced in, and released from, the Leidig cells. Its function is likely to be as an additional negative feedback on pituitary function. It also has an effect on the development of male characteristics, but to a lesser extent than testosterone.

Oestrogen

The stallion's testis contains significantly higher concentrations of oestrogens than would be expected from work carried out on other mammals (Ganjam and Kenney, 1975). The major oestrogens are oestradiol and oestrone plus the two unique equine oestrogens: equilin and equilenin. It is not clear where their site of production is, but Amann (1993b) suggests that the Leidig cells are responsible for some of the oestrogen production. The function of these oestrogens is also unclear. However, the fact that concentrations of oestrogen in venous return from the testis are elevated, coupled with the discovery that the administration of oestrogen in the presence of testosterone results in an enhanced decline in LH and FSH production, indicates that oestrogen may act as a negative inhibitor of hypothalamic–pituitary function acting in union with testosterone (Thompson et al., 1979b; Seamens et al., 1991). Seminal plasma concentrations of total oestrogens (4447 pg ml^{-1}) are even higher than those observed in blood serum (2497 pg ml^{-1}). If specifically oestrone sulphate concentrations are measured, it becomes apparent that the majority of oestrogens in seminal plasma are oestrone sulphate (4116 pg ml^{-1}), with total free oestrogens a minor component (330.5 pg ml^{-1}). The concentration of oestradiol 17β has been reported by some workers to be higher in seminal plasma (73.4 ± 87.4 pg ml^{-1}) than in blood plasma (40.0 ± 27.6 pg ml^{-1}) (Claus et al., 1992). This difference is disputed by others (Landeck, 1997). It is interesting to note that the work by Landeck (1997) suggested that oestradiol 17β is higher in the spermatozoan fraction of semen than in seminal plasma.

3.3.6. Behavioural control

Testosterone is the major hormone associated with the behavioural patterns exhibited by stallions. However, other hormones, such as dihydrotestosterone, do have a contributory effect. Both these hormones drive the normal libido of the stallion, possibly via their conversion to oestradiol and, through this, exert an effect on the higher centres of the brain (Johnson, 1991a). In addition to these two hormones, GnRH apparently directly affects behaviour, especially at the time of mating. At the sight of a mare, or at sexual stimulation, the frequency and amplitude of GnRH pulse release increase. The result is twofold: an increase in LH and FSH pulse frequency and amplitude, leading to a rise in testosterone secretion and a corresponding enhancement of libido (Irvine and Alexander, 1991); and a direct effect on the higher centres of the brain, stimulating the CNS and so acting as an additional drive to mating behaviour (Pozor et al., 1991) (Fig. 3.26).

3.3.7. Hormonal regulation and control of spermatogenesis

In summary, the hormone regulation of stallion reproduction and spermatogenesis involves a three-tier system: the hypothalamic–pituitary–gonadal axis. Acting as an overriding influence upon this axis are environmental factors such as daylength. The effect of daylength is mediated via the pineal gland and melatonin secretion, which governs the equine breeding season and, to some extent, also sexual maturity and development (Marusi and Ferroni, 1993). Under appropriate conditions, melatonin secretion declines and its inhibitory effect upon the hypothalamus is removed. In response, the hypothalamus secretes GnRH, which acts on the higher centres of the brain to affect libido and passes to the anterior pituitary via the hypophyseal portal vessels. In response, the anterior pituitary produces LH and FSH. These two hormones in turn act, respectively, upon the Leidig cells, to induce testosterone production, and on the Sertoli cells, to initiate spermatogenesis. A positive correlation between LH and testosterone pulse frequency has been demonstrated.

Fig. 3.26. A stallion demonstrating Flehman. This typical behavioural response to sexual stimulation is driven largely by testosterone but also partly by dihydrotestosterone and GnRH.

The functions of the hypothalamus and the pituitary are also affected by testicular hormones. Testosterone has a negative feedback effect on both, reducing the concentration of GnRH released and the sensitivity of the anterior pituitary to GnRH stimulation (Thompson *et al.*, 1979b; Irvine *et al.*, 1986). In response to this long negative feedback loop, the pituitary produces less LH. This negative feedback may also be attributed to the effect of dihydrotestosterone and possibly oestrogens and maybe progesterone (Amann, 1993b). As a result of this decline in LH levels, the stimulation of the Leidig cells is depressed, so reducing the production of testosterone. The decline in circulating testosterone levels reduces the negative feedback loop and allows activation of the hypothalamic–pituitary axis and further pulsatile release of LH and a corresponding increase in testosterone. Testosterone also exerts its effect on spermatogenesis, allowing completion of the process and spermatozoan production and release. Testosterone, therefore, acts as the major control over stallion reproduction, resulting in an equilibrium within a finely balanced system.

FSH is also produced by the anterior pituitary and its production controlled by GnRH. However, it is increasingly apparent that the control of LH and FSH is partly separate, as episodic release of LH does not necessarily correspond with a similar release of FSH (Thompson *et al.*, 1985; Clay *et al.*, 1988, 1989). In other species it has been demonstrated that FSH acts exclusively upon the Sertoli cells during the initial stages of spermatogenesis (Flink, 1988). Sertoli cells also secrete inhibin and activin, which again act upon the hypothalamic–pituitary axis and so also control the release of GnRH and the pituitary's response as far as FSH secretion is concerned. It is evident, therefore, that there are two main feedback loops that control stallion reproductive function: one primarily controlling LH and hence testosterone secretion, and the other controlling FSH and hence spermatogenesis. However, this is a considerable oversimplification of a highly complex and as yet poorly understood area of control.

In addition to the above there is evidence for the involvement of oxytocin, prolactin and oestrogen. It is also evident that there must be a continual cycle of messages between the Leidig cells, the germinal cells and the Sertoli cells and then between the testis and the body as a whole.

3.4. Conclusion

The reproductive anatomy of the stallion has been well documented for some time, though fewer details are known about the more inaccessible parts of the tract, such as the accessory glands and associated structures within the pelvic region. However, the use of increasingly sophisticated ultrasound equipment has allowed these areas to be studied in more detail. The area of reproductive control is less well understood. Historically, considerable effort has been geared towards understanding reproductive control in the mare, as she is statistically more likely to be the limiting factor in any breeding programme. It is evident that much is yet to be learnt about the control mechanisms governing stallion reproduction. Much of the information available to date is in part extrapolated from other species.

Production of Spermatozoa

4.1. Introduction

Semen has two main components, seminal plasma and spermatozoa, both of which will be discussed in some detail in the following sections. A summary of the normal parameters of stallion semen is given in Table 4.1, which shows that there is considerable variation in the acceptable values for the various parameters demonstrated by semen from different stallions.

By use of an open-ended artificial vagina, four main fractions can be identified within a sample of stallion semen: the pre-sperm, sperm-rich, post-sperm and post-coital fractions. It is assumed, therefore, that these are evident in natural service, though the rare use of electro-ejaculation in the stallion results in a variation in the relative components of these fractions depending upon the placement of the electrodes (Mann and Lutwak-Mann, 1981).

Identification of the origin of the secretions which make up the various fractions is particularly difficult due to their inaccessibility. Conflicting reports

Table 4.1. Range of normal values for various parameters of stallion semen (Davies Morel, 1993).

Parameter	Expected normal value
Volume	30–300 ml
Concentration	30–600 ml^{-1}
Morphology (minimum physiologically normal)	65%
Live:dead ratio	6.5:3.5
Motility (minimum progressively motile)	40%
Longevity (at room temperature)	45% alive after 3 h
	10% alive after 8 h
pH	6.9–7.8
White blood cells	< 1500 ml^{-1}
Red blood cells	< 500 ml^{-1}

occur and a clear pattern is not, as yet, evident, though more recent research using ultrasonography has aided investigations (Little, 1998). The pre-sperm fraction acts as a lubricant and is often seen before the stallion enters the mare. There are virtually no spermatozoa present within this fraction, and those that are present may be old and non-viable. This fraction is thought to originate largely from the prostate and bulbourethral glands, and possibly the ampulla (Weber and Woods, 1992). It is watery in consistency and low in ergothionine, glycerylphosphorycholine (GPC) and citric acid, which are major components of stallion semen. It is high in sodium chloride, known to be secreted by the bulbourethral glands (Mann, 1975). The pre-sperm fraction is normally 10–20 ml in volume, and its function is likely to be the flushing out and lubrication of the urethra, so removing stale urine and bacteria in readiness for the major deposition of spermatozoa. When collecting a semen sample for AI, it is best to try to avoid collecting this fraction, due to its high bacterial count and the resultant risk to collected spermatozoa.

The sperm-rich fraction is the next to be emitted and is the major fraction, containing 80–90% of the spermatozoa and 80–90% of the biochemical components of semen (for example, ergothionine and GPC); it is relatively low in citric acid and sodium chloride (Mann, 1975; Setchell, 1991). It is a milky deposit, containing several jets of ejaculate (up to nine in total), and is naturally deposited into the upper cervix of uterus immediately after the glans penis has become engorged to block the exit via the vagina. This fraction is in the order of 40–80 ml in volume in an average stallion, the fluid component of which largely originates from ampulla and bulbourethral glands, with possible contributions from the prostate and vesicular glands. GPC from epididymis secretions, along with the spermatozoa, is also a major component of the sperm-rich fraction (Amann and Graham, 1993).

The post-sperm fraction, sometimes termed the gel fraction, is deposited after the major deposition of spermatozoa. The concentration of spermatozoa in this fraction is very low, as is the concentration of ergothionine, but the concentration of citric acid is relatively high (Setchell, 1991). This fraction is thought to originate largely from the vesicular glands and its volume varies enormously from zero to in excess of 80 ml. It is highly dependent upon season and the individual stallion, the two factors generally being closely correlated (Pickett and Voss, 1972). Season and age also affect the volume of the gel fraction, which is lower in the non-breeding season and in younger stallions (Table 4.2). Previous use of the stallion may also affect the volume of the gel fraction. A second ejaculate, collected closely following a first (within 1–2 h), will have a smaller volume of gel fraction; a difference of up to 50% has been observed between the volume of gel produced in the first ejaculate when compared with a second collected 1 h later (Pickett and Voss, 1972). The gel fraction appears to be high in potassium, the significance of which is unclear, but it has been postulated that the gel fraction is (or is the remnants of) a physical or biochemical barrier that prevents the spermatozoa from a second covering competing with those from a first, i.e. acting as a paternity assurance mechanism (Evans and Davies Morel, unpublished data; Evans,

1996). It is known that potassium has an inhibitory effect on spermatozoa movement in other animals (Burkman *et al.*, 1984). Such protection may be afforded via a physical or biochemical barrier but the gel fraction does not appear to act as a chemical inhibitor of spermatozoa: the gel fraction from one stallion apparently has no adverse affect on the spermatozoa from another (Evans and Davies Morel, unpublished data; Evans, 1996). The alternative possibility is that it acts as a physical barrier in a similar way to that reported for mice, but the rather fluid consistency of the gel fraction in stallion semen would make this seem unlikely.

The fourth fraction, the post-coital fraction, is only a few millilitres in volume. It is watery in consistency and contains virtually no spermatozoa, with a very high concentration of ergothionine (Tischner *et al.*, 1974). This fraction was thought, therefore, to originate from the ampulla, but more recent evidence from Weber and Woods (1993), using ultrasonography, indicates that the ampulla glands empty primarily before ejaculation and not after.

As previously discussed, the major deposition of stallion semen occurs in a series of six to nine jets. Successful ejaculation is marked by the flagging of the stallion's tail. The first three jets contain the majority of the spermatozoa and the biochemical components of seminal plasma; successive jets decline in both (Fig. 4.1).

The total volume of semen produced varies with age (Table 4.2) (Squires *et al.*, 1979a), season (see Fig. 3.19) (Pickett and Voss, 1972), breed and frequency of use (Tables 4.3 and 4.4) (Dowsett and Pattie, 1987; Pickett *et al.*, 1988a,b; Oh *et al.*, 1994; Gunnarsson, 1997). Table 4.5 gives the comparative analysis of semen from several different domestic animals (Mann, 1964b; Mann and Lutwak-Mann, 1981).

Further details on the many factors that specifically affect spermatozoa production are given later in this chapter. When collecting samples for analysis very varying results may be obtained, especially with regard to spermatozoa concentrations and the total number of spermatozoa ejaculated, which is, amongst other things, highly dependent upon the previous use of

Table 4.2. Effect of age on seminal characteristics in the stallion (Squires *et al.*, 1979a).

	Age (years)			
Characteristics	2–3 ($n = 7$)	4–6 ($n = 16$)	9–12 ($n = 21$)	P
Total seminal volume (ml)	16.2^a	31.4^{ab}	43.2	< 0.01
Gel (ml)	2.1	5.1	13.3	NS
Gel-free (ml)	14.2^a	26.2^b	29.8^b	< 0.05
Spermatozoa concentration ($\times 10^6$)	120.4	160.9	161.3	NS
Total number ($\times 10^9$)	1.8^a	3.6^{ab}	4.5^b	< 0.05
pH	7.47	7.41	7.39	NS

^{a,b} Means in the same row with different superscripts differ.
NS = no significance.

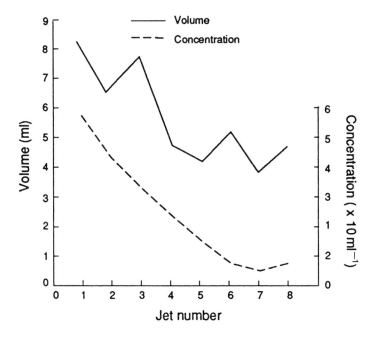

Fig. 4.1. The volume and spermatozoa concentration evident in successive jets of a stallion's ejaculate. (Kosiniak, 1975.)

Table 4.3. Effect of stallion breed on total volume of semen produced.

Breed	No.	Volume (ml)			Spermatozoa concentration ($\times 10^6$)	Total sperm no. ($\times 10^6$)
		Total	Gel	Gel-free		
[a]Thoroughbred	141	31.0	2.7	28.3	114.29	5,027.47
[a]Standardbred	111	33.3	3.1	30.2	97.24	4,737.85
[a]Arab	73	37.2	1.0	36.2	286.82	12,661.22
[a]Quarterhorse	30	27.8	4.0	23.8	171.66	5,371.53
[a]Palomino	44	24.9	1.1	23.8	138.48	4,016.26
[a]Shetland	8	57.5	13.1	44.4	101.25	1,720.70
[a]Appaloosa	18	25.3	2.0	23.3	90.42	3,331.89
[b]Paint horse	–	42.0	8.0	34.0	304.00	8,500.00
[c]Cheju	5	38.0	–	–	38.0	826.00
[d]Icelandic	38	–	–	36.4	232.8	7,789.70

[a]Dowsett and Pattie, 1987
[b]Pickett et al., 1988a, b
[c]Oh et al., 1994
[d]Gunnarsson, 1997.

Table 4.4. Effect of frequency of use of a stallion on total volume of semen produced (Dowsett and Pattie, 1987).

Service frequency	No.	Volume (ml) Total	Gel	Gel-free	Spermatozoa concentration ($\times 10^6$)	Total sperm no. ($\times 10^6$)
0–1 h	38	27.7	1.5	22.2	69.84	1903.98
1–6 h	47	28.8	2.6	22.2	89.92	2318.24
6–24 h	100	33.1	3.2	30.9	92.09	3399.80
1–7 days	140	37.2	1.7	35.5	163.35	6579.79
> 7 days	178	37.5	2.4	35.1	193.34	7783.87
1st service	24	43.4	1.0	42.4	246.69	9646.98

Table 4.5. Comparative analysis of semen from several different domestic animals (Mann, 1964b; Mann and Lutwak-Mann, 1981).

Animal	Total volume (ml)	Spermatozoa concentration ($\times 10^6$)
Stallion	50–100	100–150
Boar	100–150	100–200
Bull	3–5	800–1200
Ram	0.3–1.0	1200–2000
Cock	0.10–0.3	5000–9000
Dog	2–25	60–540
Guinea pig	0.4–0.8	5–20
Rabbit	0.4–0.6	50–350
Goat	0.5–2.5	1000–5000

the stallion. Ideally, collections for analysis should be made daily for 2 weeks, all the samples being analysed. For up to 2 weeks the spermatozoa ejaculated may originate from, or be affected by, previous use, but thereafter a stable output of spermatozoa per ejaculate should be reached, giving a good indication of daily spermatozoa production for that individual stallion. This system is highly labour intensive and, though ideal, is far from practical in many situations. Variations on this theme are, therefore, regularly used (Gebauer *et al.*, 1974a; Amann, 1993b).

4.2. Seminal Plasma

Seminal plasma is the fluid fraction of semen. It may be isolated for analysis by centrifugation or filtration of the semen sample. This allows the fluid fraction to be isolated from the spermatozoa. There have not been many comprehensive studies carried out on the biochemical components of stallion seminal plasma but it is likely that it may not, in general, differ significantly from that of other domestic animals, though some differences do occur (Table 4.6).

In general, seminal plasma contains a number of substances that may not be evident in the rest of the body or may be present in much higher concentrations. The absolute concentration of many of the components varies with individual stallions and with circulating levels of testosterone, as well as the amount of use of a stallion, the time of year and the semen fraction collected. Age is reported to affect specific components of seminal plasma, namely GPC and total protein concentrations (Abou-Ahmed et al., 1993). Osmolarity is also reported to vary with season (142 ± 61.6 mOsm in June compared with 329 ± 23.6 mOsm in December; $P < 0.05$) (Oba et al., 1993). The breed of stallion is reported not to have an effect upon fructose, citric acid and phosphorus concentrations nor upon semen pH (Oba et al., 1993).

The functions of seminal plasma are many. It provides a substrate for the conveyance of spermatozoa to the mare. It allows for, and initiates, maturation of spermatozoa as indicated by motility (spermatozoa being immotile prior to

Table 4.6. Composition of semen of some domestic animals.

Component	Stallion	Bull	Boar	Dog	Rabbit	Ram	Cock
Protein[a]	1.2–12	30–80	37	21–37	–	50	18–28
Fructose[a]	0.02–0.08	1.2–6.0	0.09–0.4	< 0.01	0.4–1.5	1.5–6.0	0.04
Glucose[a]	0.82						
Sorbitol[a]	0.2–0.6	0.1–1.4	0.06–0.18	< 0.01	0.8	0.26–1.7	0–0.1
Citric acid[a]	0.08–0.53	3.57–10	0.36–3.25	0.04	–	1.1–2.6	0.0
Inositol[a]	0.19–0.47	0.25–0.46	3.8–6.3	–	–	0.07–0.15	0.16–0.2
Ascorbic acid[a]	–	0.09	–	–	–	0.05	–
Ergothioneine[a]	0.03–1.1	< 0.01	0.06–0.3	–	–	0.0	0–0.02
Glycerylphos-phorylcholine[a]	0.4–3.8	1.1–5.0	1.1–2.4	1.8	2.5–3.7	11–21	0–0.4
Glutamic acid[a]	–	0.35–0.41	–	–	–	0.76	–
Sodium[a]	2.57	2.25	5.87	3.32	–	1.78	3.52
Potassium[a]	1.03	1.55	1.97	0.31	–	0.89	0.61
Phosphorus[a]	0.02–0.07	–	–	–	–	–	–
Calcium[a]	0.26	0.4	0.06	0.05		0.06	0.1
Magnesium[a]	0.09	0.08	0.05–0.14	0.04	–	0.06	0.14
Chloride[a]	4.48	1.74–3.2	2.6–4.3	4.4	–	0.86	1.47
Bicarbonate[b]	–	7.1	–	2.9	–	7.1	–
α-Mannosidase[c]	–	400	–	–	–	50	–
β-N-Acetylglucos-aminidase[c]	625	15,000	–	–	–	16,000	–
pH	6.2–7.8	6.48–7.8	6.85–7.9	6.1–7.0	6.59–7.5	5.9–7.3	7.2–7.6
Osmolarity[d]	142–334						

[a] All values in mg ml^{-1}
[b] All values in mmol l^{-1}
[c] All values in units ml^{-1}
[d] All values in mOsm kg^{-1}.

Sources: Mann, 1964a,b; Lake, 1971; Foote, 1980; Gilbert, 1980; White, 1980; Mann and Lutwak-Mann, 1981; Polakoski and Kopta, 1982; Serban et al., 1982; Amann et al., 1987; Abou-Ahmed et al., 1993; Augusto et al., 1992; Amann and Graham, 1993; Oh et al., 1994; Lawson, 1996.

mixing with seminal plasma, though the stimulus for the induction of motility is unclear). It enhances survival and motility of spermatozoa by the provision of energy, mainly in the form of glucose. Seminal plasma also provides protection from fluctuations in osmotic pressure between the spermatozoa and the surrounding fluid in the form of citrate or citric acid. It is essential that the osmotic pressure within the spermatozoa is the same as that of the surrounding fluid, in order to prevent leakage of water across the semi-permeable plasma membrane, which would result in bursting or shrinkage of the spermatozoa. Seminal plasma also acts as a preservative agent, preventing the oxidation of other biochemical components by means of ergothioine (Mann, 1964a,b). Finally it acts as a clotting agent by means of the gel component in the post-spermatozoa fraction.

Seminal plasma is secreted by the accessory glands at the time of ejaculation. In the case of the stallion the epididymis also makes a contribution in the form of GPC. The main components of stallion seminal plasma are discussed in the following sections.

4.2.1. Protein

The total concentration of protein in a stallion semen sample is reported to be as high as 12 mg ml^{-1} (Amann *et al.*, 1987). The protein component largely originates from the prostate gland and the vesicular glands. Twenty-seven proteins have been identified in total, with molecular weights varying from 13,000 to 122,000 Da, of which 13 vary in their relative proportions between stallions (Amann *et al.*, 1987). Work on identifying the amino acid sequence of one of the major proteins, HSP-1, has recently been carried out (Calvete *et al.*, 1994, 1995). The function of these proteins is unclear, but it has been suggested that they may be involved in capacitation (Nessau, 1994; Samper, 1995b). Indeed, among the proteins identified, there are low molecular weight heparin-binding proteins. Similar (but not identical) heparin-binding proteins are found in other mammalian semen, leading to the hypothesis that these proteins may well constitute species-specific capacitation factors (Calvete *et al.*, 1994). Proteins are apparently responsible for providing a protective coating for spermatozoa and hence increase their survival time within the female tract, and it might be that this coating may also be a prerequisite for capacitation (Myles and Primakoff, 1991; Samper and Gartley, 1991). Part, if not all, of this coating is assumed to occur within the epididymis, as a protein of similar amino acid sequence to that found in rat, mouse and human epididymal spermatozoa has been isolated. The proteins present also differ with different areas of the spermatozoon head (Calvete *et al.*, 1994). It is interesting to note that a significant proportion of the protein within stallion semen must be associated with the spermatozoa, as the protein concentration of spermatozoa alone is similar to that of semen (Samper, 1995b). Despite this possible protective function, proteins within the seminal plasma of boars and bulls are reported to increase susceptibility to cold shock (Matousek, 1968; Moore *et al.*, 1976). However, stallion spermatozoa appear to be comparatively resistant to cold shock and so differences in the effects of such proteins may occur (Watson and Plummer, 1985).

4.2.2. Glucose

Table 4.6 shows that glucose is the major glycolysable sugar found in stallion semen (glucose concentration 0.82 mg ml^{-1}, fructose concentration 0.02–0.08 mg ml^{-1}), unlike the bull and the ram in which fructose is the major energy source (glucose concentration negligible, fructose concentration 1.2–6.0 mg ml^{-1} and 1.5–6.0 mg ml^{-1}, respectively). The reason for the difference is unclear. The presence of significant concentrations of sorbitol (an intermediary product of the conversion of glucose to fructose) and lactic acid (normally arising from the glycolytic process) indicates that at least some of the enzymes necessary for the conversion of glucose to fructose are present within the stallion's reproductive tract. However, stallion spermatozoa show limited ability to use fructose as an energy source, at least under anaerobic conditions, illustrated by the slow rate of conversion of fructose to lactic acid evident in stallion semen (Mann *et al.*, 1963).

4.2.3. Sorbitol

Sorbitol, a sugar alcohol, is also present in significant quantities in equine semen. Its function is to ensure an equilibrium in osmotic pressure between seminal plasma and the spermatozoa (Mann, 1975). It may also provide an additional source of energy. It is a normal intermediary product of the conversion of glucose to fructose. However, in the stallion the major energy source is glucose and, therefore, it is unlikely that sorbitol is present due to this conversion (King and Mann, 1959).

4.2.4. Lactic acid

Significant quantities of lactic acid have also been reported in stallion semen, possibly as a by-product of anaerobic respiration and the glycolysis of glucose, to produce energy in the form of adenosine triphosphate (ATP) (Amann and Graham, 1993). ATP is required for spermatozoan metabolism, survival and motility, and may itself be used as an additional source of energy. Lactic acid concentrations do increase with time but the rate of build-up is much slower in stallion semen than in bull and ram semen, where fructose is the major source of energy (Setchell, 1991).

4.2.5. Citric acid

Significant concentrations of citric acid can also be identified in stallion semen. It is an organic acid, derived mainly from the vesicular glands. The concentration of citric acid in the secretions of the vesicular glands has been reported to be as high as 4 mg ml^{-1} but it is generally nearer to 1 mg ml^{-1}. The concentration of citric acid in the whole semen sample varies with the contribution made by the vesicular glands, which in turn is affected by (among other things) season. The function of citric acid within semen is not clear. It has been suggested that, in common with sorbitol, it may be responsible for maintaining an equilibrium in the osmotic pressure between the seminal plasma and the intracellular fluid of the spermatozoa. It may also form complexes with calcium, or act as an antiperoxidant, a known activity of citrates (Mann, 1975).

4.2.6. Inositol

Inositol is an organic compound, present in stallion semen in concentrations of between 0.19 and 0.47 mg ml^{-1}. It is secreted by the epididymis along with GPC. Its function in seminal plasma is unclear (Mann, 1975).

4.2 7. Ergothionine

The concentration of ergothionine in stallion semen is relatively high at 0.03–1.0 mg ml^{-1} (Leone, 1954; Mann *et al.*, 1956; Mann, 1975). Ergothionine is largely secreted by the ampulla glands and is apparent in the sperm-rich portion of an ejaculate. Its function is as an antioxidant, preventing the oxidation of, and protecting the integrity of, other biochemical components within seminal plasma. Ergothionine acts in a reducing fashion similar to the action of cysteine or ascorbic acid (components not found in stallion seminal plasma but present in other mammals), protecting spermatozoa from the immobilizing action of oxidizing and peroxidizing agents (Mann, 1975). It has been reported that the secretions of the ampulla may contain concentrations of up to 1 mg ml^{-1}, though concentrations of 0.2–0.6 mg ml^{-1} are more common. The concentration of ergothionine in a semen sample may, therefore, be used as an indicator of the percentage contribution of ampullary secretions. Season is known to affect ergiothionine concentrations, with the highest concentrations being reported during early spring and late autumn (Mann, 1975).

4.2.8. Glycerylphosphorylcholine

GPC is a base found in concentrations of 0.4–3.8 mg ml^{-1}. It is produced largely by the epididymis, and hence it is secreted as part of the sperm-rich fraction (Dawson *et al.*, 1957; Chow *et al.*, 1986). A significant amount of the total acid-insoluble organically bound phosphate of seminal plasma is found in GPC. GPC concentrations in epididymal secretions are reported to be in the order of 10 mg ml^{-1} or greater. The function of GPC is unclear: the fact that it is produced in the epididymis leads to the belief that it might be involved in spermatozoan maturation. Alternatively it may be involved in the biosynthesis or metabolism of lipids in epididymal semen (Mann, 1975).

4.2.9. Minerals

Little work has been done specifically on the minerals found in stallion semen. However, it is evident that sodium (Na), potassium (K) and phosphorus (P) may be isolated in stallion semen, along with lower concentrations of calcium (Ca), magnesium (Mg) and zinc (Zn) (Oba *et al.*, 1993). The work by Oba *et al.* (1993) illustrated a breed difference in the concentrations of Na and P and an age effect on the concentration of K. No suggestion was made as to why these differences may be evident. Work by Evans (1996) suggests that potassium concentrations are significant and may be up to four times the levels found in blood plasma.

4.2.10. Glycosidases

Glycosidases are a group of enzymes, of which at least four are identifiable in stallion epididymal secretions and occur in higher concentrations in the stallion than are evident in other males. These are α-mannosidase, β-mannosidase, β-galactosidase and β-*N*-acetylglucosaminidase (Mann, 1975). The significance of their relatively high concentrations is unclear.

4.2.11. Additional minor seminal plasma components

Several other components have been isolated more recently in stallion seminal plasma, including acetylhydrolase, which is reported to increase with time and with protein concentrations. A positive correlation with Ca concentrations has also been suggested (Hough and Parks, 1994). Heparin-binding components and zona pellucida-binding proteins have also been reported to be present (Calvete *et al.*, 1994).

4.2.12. Seminal plasma abnormalities

Many factors have been identified as having the potential to affect the quality and quantity of seminal plasma; these include season, age and nutrition, which have been discussed previously. In addition, anatomical abnormalities may occur, and the use of drugs can also be detrimental. The identification of the components that make up the various fractions of stallion semen, their site of production, and the specific order in which they are deposited, provides a useful tool for identifying the existence of a problem and, if so, which gland is likely to be responsible through a lack of its secretion or an aberration in the normal strict sequence of events.

4.3. Spermatozoa

The spermatozoon is a unique cell within the stallion's body, being haploid in nature and with the unique function of fertilization. The processes involved in the production of a spermatozoon are collectively termed spermatogenesis and will be considered in detail later in this chapter. It is important to remember that the process does not end until the fusion of the nuclear material of the spermtaozoon and the oocyte takes place, rather than when the spermtozoa are released from the epididymis. Change and development are continually evident in spermatozoa from their origins as somatic cells, the result being specialized cells with three functional areas: condensed nuclear material, an energy-providing area and a propulsive area. Spermatozoa then mature further while they are within the epididymis, followed by further development induced firstly by contact with seminal plasma and then by the secretions of the female tract. The final stages of spermatozoon development are induced by the immediate environment of the oocyte and its enclosing investments. These changes result in a highly specialized unique cell, designed for one set function – fertilization. This specialization is achieved at a cost, reducing the spermatozoon's ability to repair itself. This loss of biochemical mechanisms for cell repair results in a greater susceptibility to environmental change (Bedford and Hoskins, 1990; Johnson, 1991a). Any interruption in this natural course of events, either by human interference or by an abnormal event, can have a detrimental effect on the development of spermatozoa and hence on conception rates.

4.3.1. Anatomy

A spermatozoon may be divided into five main sections: the head, the neck, the mid-piece, the principal piece and the end-piece (Fig. 4.2) (Dott, 1975; Johnson *et al.*, 1980; Johnson, 1991a; Amman and Graham, 1993; Christensen *et al.*, 1995).

Head

The head of the stallion spermatozoon is illustrated in Figs 4.3–4.6. Several authors have reported various dimensions for the head. Its length has been reported to vary from 4.9 to 10.6 µm, with a maximum width of 2.5–3.91 µm.

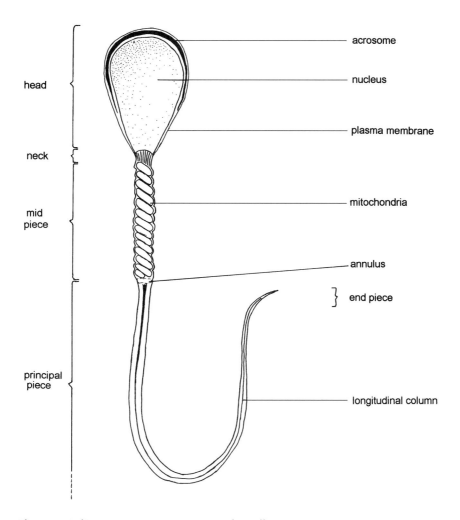

Fig. 4.2. A diagrammatic representation of a stallion spermatozoon.

The length of the post-acrosomal region (post-acrosinal lamina or cap) has been reported to be 1.65 µm, and the width of the base of the head to be 1.45–1.8 µm, with a total surface area for the spermatozoon head of 10.3–16.5 µm^2 (Nishikawa *et al.*, 1951; Dott, 1975; Johnson *et al.*, 1978b, 1980; Bielanski and Kaczmarski, 1979; Massanyi, 1988; Gravance *et al.*, 1996, 1997; Casey *et al.*, 1997). The work by Gravance *et al.* (1996) and Casey *et al.* (1997) reported an association between large heads and subfertility.

The head of the spermatozoon may be divided into three sections: the nucleus, the acrosome and the post-acrosomal region.

The nucleus, the contents of which are electron dense with a few nuclear vacuoles, is contained within a double-layered nuclear membrane, an envelope

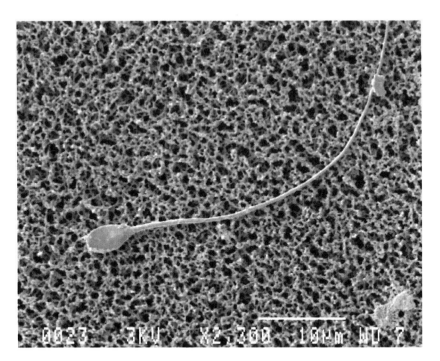

Fig. 4.3. Scanning electron micrograph of a fresh stallion spermatozoon. (Bar = 10 μm.) (Photograph by Jennifer Jones.)

with a few pores (Bielanski and Kaczmarski, 1979). The nucleus contains highly condensed chromatin, which is DNA complexed with protamine, the genetic material for fusion with the corresponding oocyte nuclear material. The shape of the head is determined by the shape of the nucleus and tends to be broad, flat and asymmetrical (Figs 4.3, 4.4 and 4.5). When viewed from the side (narrowest aspect) it has a total length of up to 10.6 μm and is thicker at the caudal end, tapering to 0.6 μm at the cranial end (Johnson *et al.*, 1978b, 1980; Massanyi, 1988). When viewed from the front (broadest aspect) the head is similarly tapered at the cranial end, with the widest (2.36 μm) section being in the centre aspect (Johnson *et al.*, 1978b, 1980). This shape is at variance with other mammalian spermatozoa, which tend to be widest at the base (Fawcett, 1970, 1975).

The acrosome region of the spermatozoon head (Fig. 4.6) lies over the caudal two-thirds of the nucleus and can be subdivided into the apical ridge, the principal segment and the equatorial segment. The apical ridge tends to be thicker (that is, a greater distance separating the inner and outer acrosomal membranes) than the principal segment, which in turn is thicker than the equatorial segment (Bielanski and Kaczmarski, 1979). The acrosome also appears to be smaller and more uniform in shape in the stallion than thickened to one side, a characteristic of bull, boar and ram spermatozoa (Massanyi, 1988; Amann and Graham, 1993). Occasionally a perforatum is seen

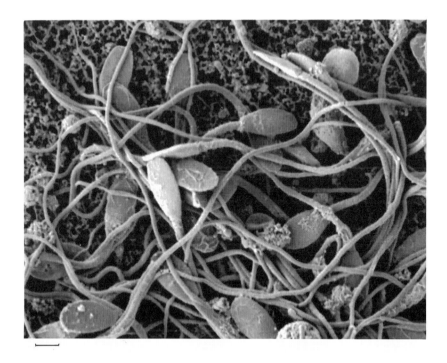

Fig. 4.4. Scanning electron micrograph of fresh stallion spermatozoa illustrating the broad flat shape of the head with both perpendicular (central) and abaxial (non-central) attachment of the head to the neck region. (Bar = 2 µm.) (Photograph by Jennifer Jones.)

at the cranial end of the acrosome, lying in the region of the apical ridge. This is postulated to be a stabilizing structure (Amann and Graham, 1993). The membrane of the acrosome comprises an inner and an outer membrane. The inner membrane is in juxtaposition to the nuclear membrane. The outer membrane lies under the plasma membrane, with which it fuses during the acrosome reaction to release the acrosomal contents. The acrosome region contains glycolipids and hydrolytic enzymes, including hyaluronidase, proacrosin/acrosin and lipases, all of which are released during the acrosome reaction and are required for subsequent passage of the spermatozoon through the surrounding investments of the oocyte (Goodpasture *et al.*, 1981; Eddy, 1988; Bedford and Hoskins, 1990; Johnson, 1991a). The membrane of the acrosome region is particularly high in cholesterol, when compared with other parts of the spermatozoon plasma membrane (Lopez and Souza, 1991). The equatorial segment of the acrosome region does not contain enzymes and so is not involved in the acrosome reaction, but its membranes and the associated plasma membrane do fuse with the oocyte membrane at fertilization.

The post-acrosomal region is the region of the head not covered by the acrosome; it lies caudally to the acrosome region and runs down to the

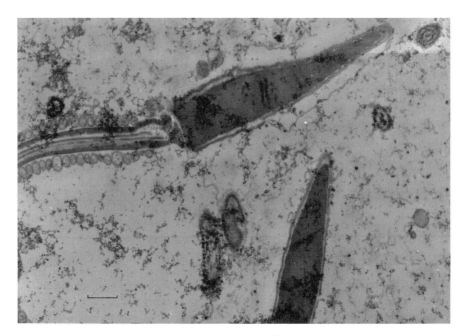

Fig. 4.5. Transmission electron micrograph of a stallion spermatozoon showing the narrowest aspect of the head, neck and beginning of the mid-piece. The tear apparent in the plasma membrane on the right-hand side of the spermatozoon is a common artefact of transmission electron microscopy of stallion spermatozoa. (Bar = 0.36 µm.)

connection with the neck. It is composed of characteristically tight lamellae of high electron density (Bielanski and Kaczmarski, 1979) but its function is unclear (Fig. 4.5). The plasma membrane in the post-acrosomal region may also be involved in fusion with the oocyte plasma membrane. Small protrusions from the base of the nuclear membrane may be evident at the junction between the post-acrosomal region and the neck. These are remainders of nuclear membrane left after the contraction and condensation of nuclear material during spermatogenesis (Bielanski and Kaczanaski, 1979).

Neck

The spermatozoon neck is the connection between the mid-piece and the head. Its connection is at the implantation (or articular) fossa, at which there is a thickened area of the double nuclear membrane, forming a basal plate for articulated attachment (Bielanski and Kaczmarski, 1979; Katz, 1991) (Figs 4.5, 4.7 and 4.8).

Attachment in the stallion spermatozoon differs from other spermatozoa in two ways. Firstly, the attachment seems to be particularly fragile, accounting for the high numbers of detached heads and tails, or inappropriate attachments of the two, seen in stallion spermatozoa. Secondly, in about 50% of

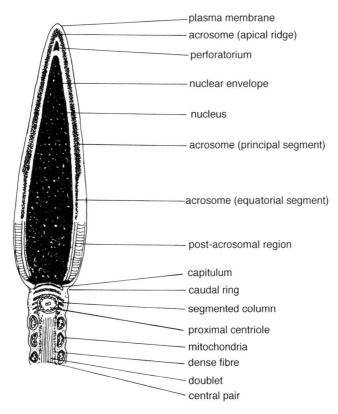

Fig. 4.6. Longitudinal cross-section through neck and head region of a stallion spermatozoon, taken through the narrowest aspect, as seen in Fig. 4.5.

stallions, the attachment of the head and the mid-piece in the neck region is not perpendicular but abaxial; that is, the neck is attached off-centre (Fig. 4.9) (Bielanski, 1951; Dott, 1975; Pickett *et al.*, 1987).

The neck itself is only a short area, 0.8 µm in diameter, containing the connecting piece, the proximal centriole and several small mitochondria. The connecting piece is the area where the connection between the head and the neck is seen. The attachment area lying on the neck side of the basal plate of the head is called the capitulum (Figs 4.7 and 4.8). The capitulum is formed primarily from the two major, segmented columns within the neck. In total, the neck region contains nine segmented columns formed from fibrous protein and arranged like a pile of 15 plates. Four of these segmented columns fuse into two pairs, forming an area termed the two major segmented columns. These two segmented columns fuse and largely form the capitulum. There is a difference between these two major segmented columns. The primary one gives rise to the major portion of the capitulum, its two segmented columns being fused along the entire length. The secondary major segmented column

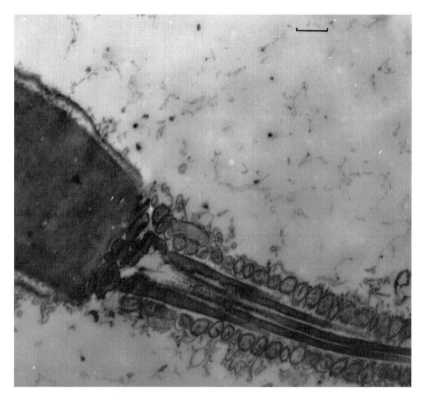

Fig. 4.7. Transmission electron micrograph of a stallion spermatozoon showing a longitudinal cross-section through the broadest aspect of the neck region, to illustrate the attachment of the head and mid-piece. (Bar = 0.25 μm.) A diagrammatic representation of this area is given in Fig. 4.8. (Photograph by Jennifer Jones.)

also gives rise to, and attaches to, the capitulum, but its two component segmented columns do not fuse until they are cranial to the proximal centriole. The remaining four unpaired segmented columns also merge with the capitulum at right angles to the attachment of the two major segmented columns (Fig. 4.8) (Amann and Graham, 1993; James, 1998).

The proximal centriole is seen between the two major segmented columns, towards the cranial region of the connecting piece. It is situated oblique to the long axis of the spermatozoon at an angle of 45–60° to the mid-piece axis, with one end near the plasma membrane. It is typically larger in stallion spermatozoa than observed in other mammals. The distal centriole is not evident, having been lost during the development of the connecting piece.

Small mitochondria along with numerous microtubules may also be seen in the area of the connecting piece and proximal centriole (Bielanski and Kaczmarski, 1979). The presence of numerous microtubules is of interest, as their presence has not been reported in other mammals and their function is

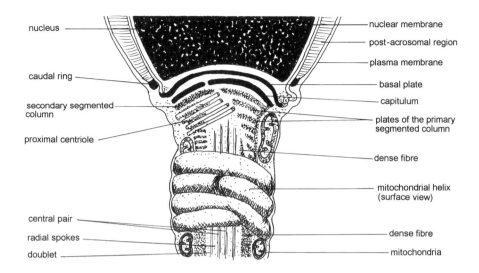

Fig. 4.8. Longitudinal section of neck and initial part of the mid-piece region of a stallion spermatozoon, taken through the broadest aspect, as seen in Fig. 4.7. A surface view (lower portion) is given to illustrate the helical arrangement of mitochondria around the central inner dense fibres and axenome of the mid-piece of the spermatozoa.

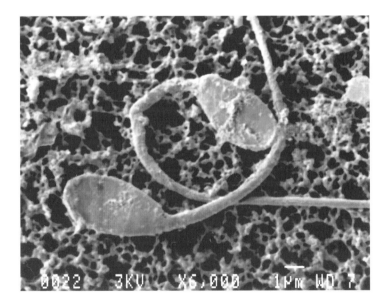

Fig. 4.9. Scanning electron micrograph of stallion spermatozoa, illustrating the abaxial attachment of the mid-piece to the head of the spermatozoon in the neck region. (Bar = 1 µm.) (Photograph by Jennifer Jones.)

unclear. On leaving the testis the spermatozoon neck is surrounded by cytoplasmic droplets, remnants of old spermatozoon cytoplasm that remains from the original gamete cell prior to concentration. This cytoplasmic droplet, which migrates down the mid-piece and is normally lost by the time the spermatozoa are ejaculated, is rich in hydrolytic and glycolytic enzymes (Harrison and White, 1972; Dott, 1975).

The attachment of the neck to the mid-piece is via dense fibres. The caudal end of each segmented column within the neck is fused, but not continuous, with one of nine dense fibres originating in the mid-piece.

Mid-piece

The mid-piece of the spermatozoon connects the neck and the principal piece and is characterized by a high concentration of mitochondria. Its length, including the neck region, is reported to be 10.5 ± 1.27 µm (excluding the neck, 9.83 ± 3.3 µm), with a diameter of 0.6 ± 0.11 µm (Nishikawa *et al.*, 1951; Bielanski and Kaczmarski, 1979). The mitochondria are arranged circumventrally around the outer limits of the mid-piece (Figs 4.8, 4.10, 4.11 and 4.12), in a double layered spiral, each mitochondrion being half the

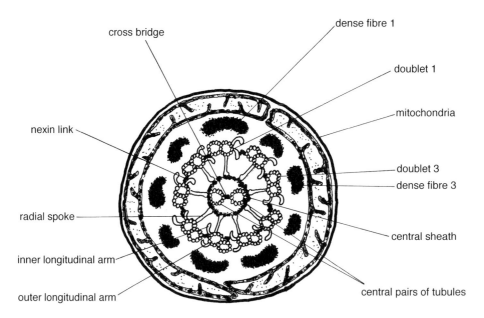

Fig. 4.10. Transverse cross-section through the mid-piece of the stallion spermatozoon. Both the doublets and dense fibres within the mid-piece are numbered 1–9. Number 1 dense fibre and doublet is the one bisected by a line running at right angles to the cross bridge connecting the two central tubules. The remaining dense fibres and doublets are numbered in order in the direction in which the arms of the doublet point. In this diagram, dense fibre and doublet 1 are at 12 o'clock; the remainder are numbered in a clockwise fashion.

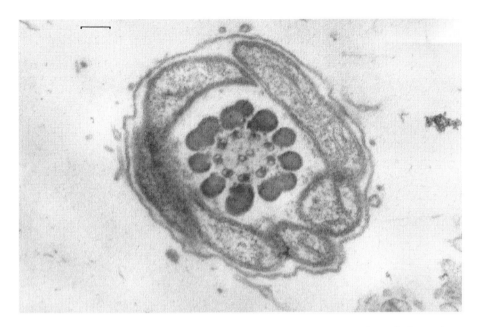

Fig. 4.11. Tranmission electron micrograph of a transverse cross-section through the mid-piece of the spermatozoon, illustrating the outer helical arrangement of the mitochondria and the inner dense fibres and axenome. Using the convention given in Figure 4.10, doublet and dense fibre 1 are situated at 10 o'clock; the remaining doublets and dense fibres are evident in an anticlockwise fashion. The large dense fibres 5 and 6 are evident at 6 and 4 o'clock, respectively, the intermediate-size dense fibre number 9 is evident at 12 o'clock. (Bar = 0.09 µm.) (Photograph by Jennifer Jones.)

circumference (that is, half a turn) in length. The junction between two mitochondria in one layer is positioned directly above and below the centres of the adjacent mitochondria (Fig. 4.8).

In the stallion's spermatozoon the mitochondria are typically arranged in 50–60 helical turns (Bielanski and Kaczmarski, 1979; Amann and Graham, 1993). These mitochondria are responsible for the production of ATP and as such contain the enzymes and cofactors necessary for this process (Gibbons and Gibbons, 1972).

In the centre of the tube of mitochondria are nine dense fibres involved in fusion with the segmented columns of the neck (Figs 4.8, 4.10 and 4.11). These provide support, by way of their tough keratin-like fibrous structure, and probably provide stability and so reduce the wobble factor that the whipping of the spermatozoon tail would normally generate, but still allowing some flexibility to be maintained. These dense fibres are not equal in size, numbers one, five and six being the largest, and number nine intermediate in size, compared with the others.

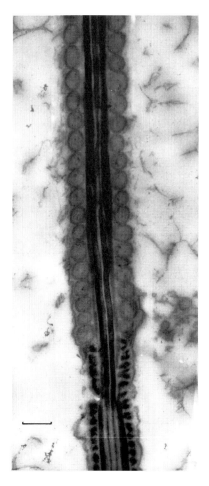

Fig. 4.12. Transmission electron micrograph showing a longitudinal cross-section through the mid (upper) and principal (lower) piece of the stallion spermatozoon, illustrating the helical arrangement of the mitochondria around the central core of the mid-piece. No such arrangement of mitochondria is observed in the principal piece region shown in the lower part of the photograph. Also shown is the annulus region at the junction of the mid-piece and the principal piece. (Bar = 0.25 µm.) (Photograph by Jennifer Jones.)

Running down the centre of these dense fibres is the axoneme (Figs 4.8, 4.10 and 4.11). The axoneme consists of a central pair of microtubules surrounded by a ring of nine doublets. Each of the doublets is made up of two microtubules, A and B. The A tubule is the major one, with the 'C'-shaped B microtubule attached to it. Each doublet has four arms or attachment points from the A microtubule. Two of these are longitudinal arms pointing towards the next doublet; a third arm, called the radial spoke, attaches each doublet to the central pair (nine radial arms in total); and finally a nexin link attaches each doublet to its neighbour around the circumference (Fig. 4.13).

The longitudinal arms contain dynein, rich in ATPase, which transduces chemical energy into mechanical action. The nexin links are likely to regulate the relative displacements of the doublets during their sliding and contraction and so maintain symmetry. The radial spokes also contain dynein, and are also likely to provide structural support and stability. The microtubules themselves are made up largely of tubulin molecules arranged in 13 and nine or ten protofilaments (Gibbons and Gibbons, 1972; Kimball, 1983; Beford and Hoskins, 1990; Amann and Graham, 1993).

The end of the mid-piece is marked by the annulus or Jensen's ring, an electron-dense area marking the junction between the mitochondria of the mid-piece and the fibrous sheath of the principal piece. At this point the spermatozoon plasma membrane is firmly attached to the underlying annulus (Fig. 4.12) (Bielanski and Kaczmarski, 1979; Amann and Graham, 1993).

Principal piece

The principal piece is in essence a continuation of the mid-piece, resembling it internally with the continuation of the dense fibres and the axeneme, but having a smaller diameter (0.45 µm) (Johnson *et al.*, 1978b, 1980; Amann and

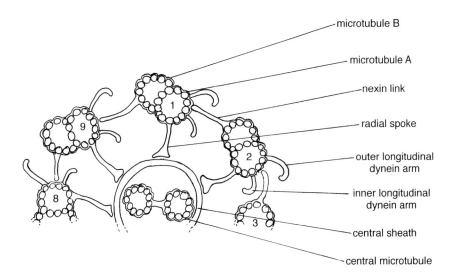

Fig. 4.13. Doublets of the middle and principal piece of the stallion spermatozoon numbered as shown. The relationship between the A and B microtubules is illustrated, along with the positioning of the two longitudinal arms, inner and outer, (pointing towards the next doublet), the radial spoke (attaching each doublet to the central pair) and a nexin link attaching each doublet to its neighbour around the circumference. (Satir, 1974.)

Graham, 1993) (Figs 4.12, 4.14 and 4.15). However, towards the caudal end of the principal piece the dense fibres taper away.

The one major difference between the mid-piece and the principal piece is the loss of the surrounding mitochondria and the gain of the fibrous outer sheath in their place. The fibrous sheath is arranged in two fibrous ribs connecting two longitudinal columns, which run along the ventral and dorsal aspects of the principal piece, overlying the dense fibres three, four and eight. Towards the end of the principal piece the dense fibres taper away. The first fibres to terminate are numbers three and eight, followed by number nine and then numbers one, five and six. All fibres terminate prior to the end piece. As the dense fibres taper away towards the caudal end of the principal piece, the fibrous sheath lies increasingly closer to the axenome. It is likely that the function of the fibrous sheath is to provide support, and yet allow flexibility for the translation of the contraction and sliding of the doublets within the axenome into controlled tail movement (Amman and Graham, 1993).

End-piece

The end-piece is the caudal end of the spermatozoon and is reported to measure 2.79 ± 0.9 µm in length (Dott, 1975; Bielanski and Kaczmarski, 1979)

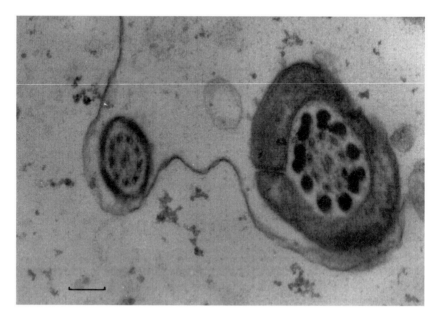

Fig. 4.14. Cross-section through the mid-piece of one spermatozoon (on the right) and the cross-section through the principal piece of another (to the left). Despite the poor focus, the replacement of the outer mitochondria with a fibrous sheath is clearly evident in the principal-piece cross-section compared with that of the mid-piece. (Bar = 0.17 µm.) (Photograph by Jennifer Jones.)

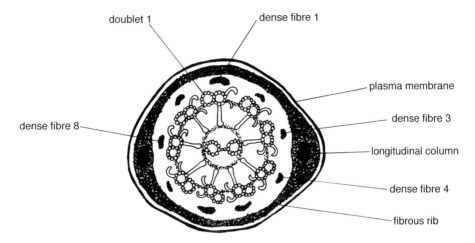

Fig. 4.15. Transverse cross-section through the principal piece of the stallion spermatozoon. Dense fibre and doublet 1 are positioned at 12 o'clock and the remainder are numbered in a clockwise direction. Note the loss of the outer mitochondria helix.

(Fig. 4.16). The nine doublets plus the two central microtubules continue through the first half of the end-piece. They then taper out over a short distance, leaving just the fibrous sheath, which then tapers away, but not necessarily evenly, disappearing on one side before the other. The microtubules of the doublet do not disappear evenly, as the B microtubule terminates prior to the A microtubule (Bielanski and Kaczmarski, 1979; Kimball, 1983).

Plasma membrane

The whole of the spermatozoon is contained within its plasma membrane, which is anchored in specialized areas and forms the outermost component of the spermatozoon. The plasma membrane remains intact, except in the region of the acrosome as a preliminary to fertilization or as a result of senescence or death of the spermatozoon.

The structure of the membrane is consistent throughout, in that it is composed of three layers or zones: lipid bilayer, phospholipid–water interface and glycocalyx (Fig. 4.17).

The lipid bilayer (5 nm in thickness) is subdivided into polar phospholipids, which orientate themselves so that the hydrophilic polar head groups are situated externally, and the hydrophobic fatty acid chains orientated internally towards each other (Amann and Pickett, 1987; Hammerstedt *et al.*, 1990). The major lipids present are phospholipids and cholesterol in a ratio of 0.64:0.36 (Chow *et al.*, 1986; Parks and Lynch, 1992). It is suggested, by some workers, that although this bilayer configuration is the one most commonly observed and an efficient one for maintaining a permeability

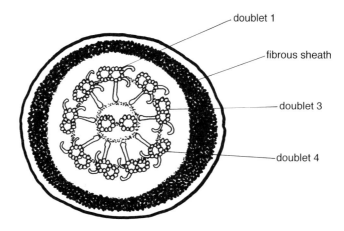

Fig. 4.16. Transverse cross-section through the first half of the end-piece of the stallion spermatozoon. As in Fig. 4.15, doublet 1 is at 12 o'clock and the remainder are numbered in a clockwise fashion. Note the loss of the dense fibres.

barrier, an alternative hexagonal phase II configuration may be present. In this configuration the lipids are arranged in a cylindrical form with their hydrophobic phospholipid heads orientated towards the centre of the

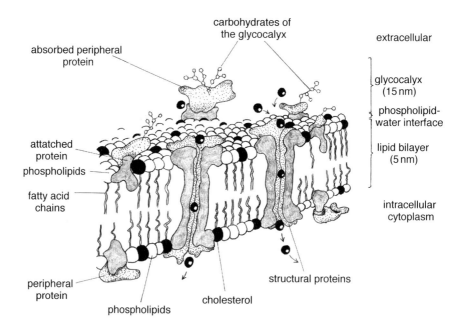

Fig. 4.17. Simplified structure of the plasma membrane of the equine spermatozoon. (Adapted from Amann and Graham, 1993.)

cylinder. It is postulated that this configuration reduces the efficiency of the membrane to act as a barrier but allows membrane fusion (Hammerstedt et al., 1990). The amounts of cholesterol relative to phospholipids determine the fluidity of the membrane. In general, the higher the relative concentration of phospholipids, the more fluid is the membrane. Cholesterol therefore acts, along with integral proteins, as a stabilizer ensuring a normal lamella configuration of the phospholipids and the bilayer. The concentration of cholesterol is known to vary with location within the plasma membrane, being highest at the acrosome region (Lopez and Souza, 1991). It has been shown that in other domestic species the class of phospholipids and the nature of their side chains also vary with their position. The major classes include choline, ethanolamine and sphingomyelin. It is likely that a similar pattern is to be found in equine spermatozoa (Amann and Picket, 1987; Hammerstedt et al., 1990). Indeed, work by Parks and Lynch (1992) demonstrated that, as with boars, bulls and roosters, these were the major phospholipids in the horse, with the addition of phosphoglycerides. Glycolipids in all four species were minor, making up less than 10% of the total polar lipids. The peak phase transition temperature for phospholipids in horses is 20.7°C, compared with 24.0°C, 25.4°C and 24.5°C for boars, bulls and roosters, respectively. Similarly the horse demonstrated the lowest peak phase transition temperature for glycolipids at 33.4°C, compared with 36.2°C and 42.8°C for boars and bulls (no peak phase transition temperature was recorded for roosters). It is possible that these differences reflect the differing tolerance of spermatozoa to rapid drops in temperature.

Proteins are also found amongst the lipids and make up 50% of the weight of the plasma membrane. These proteins act as either structural (integral) proteins or as attachment points for other peripheral proteins. The structural proteins may also act as channels or pores through which small molecules may pass to the cytoplasm of the spermatozoon; the remaining structural proteins are found between the two bilayers of the membrane. Attachment proteins act as surface receptors for other peripheral proteins from the surrounding medium by means of their negatively charged carbohydrate side chains. Proteins which attach to, or are part of, the membrane are known to participate in spermatozoon–ovum interactions. These proteins differ between species, though their function in binding is similar (Calvete et al., 1994; Dobrinski et al., 1997; Thomas et al., 1997).

In addition, α-1,4-galactosyltransferase has been located on the equine spermatozoon membrane where it mediates the binding of the spermatozoon with the glycoconjugate residues in the zona pellucida of the ovum (Fayrer-Hosken et al., 1991).

The next area of the membrane is the phospholipid–water interface, which is the junction between the hydrophilic polar head groups of the lipid layer and the surrounding medium (largely water) and in which the glycocalyx is found. The glycocalyx is a polysaccharide outer coat of the equine spermatozoon, 15 nm in depth. Its exact function is unclear but it is likely to be involved in antigenicity, cell adhesiveness, specific permeability and ATP activity (Winzler, 1970). It is known that within the glycocalyx there are attachments for peripheral proteins (Hammerstedt *et al.*, 1990). These proteins are likely to be provided by seminal plasma and act as a stabilizing influence on the spermatozoon during its passage through the male, and subsequently the female, tract. They may also be involved in capacitation (Hernandez-Jauregui *et al.*, 1975).

As stated previously, the plasma membrane of the spermatozoon conforms to the above structure regardless of its location. However, as seen, minor differences do occur in the class of phospholipids and their associated fatty acid chains. The function and significance of these differences is unclear. Some variations may be associated with areas of specialization or attachment to the underlying spermatozoon structure. There are three such areas: over the caudal ring of the head (the junction of the head and the neck), the annulus (the junction of the mid-piece and the principal piece) and the principal piece (Eddy, 1988; Bedford and Hoskins, 1990). These areas are characterized by densely packed particles, presumably involved in the attachment. In the caudal end of the head region and the annulus these densely packed particles are arranged in a band around the circumference of the spermatozoon. In the principal piece, however, they are arranged in a zipper fashion along the length of the principal piece, again presumably providing the attachment. It has also been suggested that a further point of attachment may be found overlying the post-acrosomal region, as this is a relatively stable area (Hancock, 1957). Proteins within the bilayer also apparently change: for example, Calvete *et al.* (1994) isolated a zona pellucida protein similar to AWN-1 previously found in boars, termed stallion AWN protein, which was restricted to just the equatorial segment.

The structure of the plasma membrane may, under certain circumstances, be altered or changed – for example, during cooling or freezing. Under such circumstances the stability of the membrane may be affected due to induced changes to the lipids of the bilayer. Such disturbances are potentially irreversible and so lead to spermatozoon metabolic malfunction and death. The effect of cooling on spermatozoon membranes is discussed in more detail in Chapter 7.

4.3.2. Spermatogenesis

Spermatogenesis is the production of spermatozoa and may be divided into three phases: spermatocytogenesis, meiosis and spermiogenesis (Fig. 4.18) (Johnson *et al.*, 1997). The whole process takes place within the seminiferous tubules of the testis and is aided and controlled by Sertoli (nurse) cells. These Sertoli cells form a close association with each developing spermatogonium, providing them with structural support, a communication system to the rest of the body, nutrients required for development and protection from immunological rejection. Sertoli cells, by means of the microtubular components of their plasma membrane, drive the movement of spermatozoa

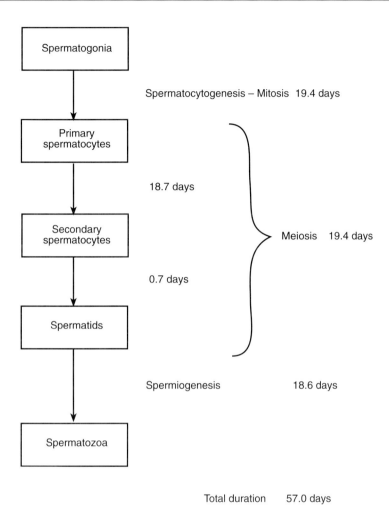

Fig. 4.18. Duration of the various stages of spermatogenesis in the stallion. (Swiestra et al., 1974, 1975; Amann, 1993b; Johnson et al., 1997.)

ready for their release into the lumen of the seminiferous tubule, at spermiation. Spermatogenesis is under endocrine control, which is mediated via the Sertoli cells (section 3.3.5).

The average daily spermatozoa output of a stallion is 5×10^9, with some significant variation, depending upon many of the factors discussed later in this section (Amann et al., 1979b; Johnson and Neaves, 1981; Johnson, 1985, 1991a; Johnson et al., 1991a,b). Table 4.7 gives comparative daily spermatozoa production figures for many mammals.

Significant information on the length of spermatogenesis and its component stages was obtained by Swierstra et al. (1974, 1975) using radioisotope injections into the testicular artery of stallions. This radioisotope was taken up by the germinal cells whilst they synthesized DNA and was identified in the progeny of these germinal cells at set times by removal of the testis and treatment of the testicular tissue with photographic emulsions, showing up the radioisotope as black granules. Using this technique the length of the various stages within spermatogenesis was ascertained.

Spermatocytogenesis

Spermatocytogenesis is the first stage of spermatogenesis and in the stallion takes 19.4 days (Figure 4.18). Spermatocytogenesis is the development of the spermatogonia that arise post-natally from gonocytes in the base of the seminiferous tubules. The number of spermatozoan divisions has not been identified for the horse, but in other mammals the number ranges from 1 to 14 (Courot *et al.*, 1970). In the horse, five different types of spermatogonia are evident through the spermatogenic phase: A_1 cells with small, flattened nuclei; A_2 cells with large nuclei, with centres apparently empty of euchromatin; A_3 cells with the largest nuclei, plus either a single nucleolus or large fragments of nucleoli; B_1 or intermediate cells with a large oval to spherical nucleus containing large chromatin flakes; and B_2 cells with small spherical nuclei, with small chromatic flakes (Fig. 4.19) (Johnson, 1991a).

The division of the A_1 spermatogonia has two main functions: firstly, to produce more stem cell spermatogonia (uncommitted A_1 spermatogonia) by mitosis, to continue the supply of spermatogonia for future spermatozoa production; and secondly, to produce committed A_1 spermatogonia which will go on to produce primary spermatocytes via $A_{1.2}$, $A_{1.4}$, $A_{1.8}$, A_2, A_3, B_1 and B_2 routes and, eventually, spermatozoa (Johnson, 1991a; Johnson *et al.*, 1997).

The necessity to maintain a continual supply of stem cell spermatogonia is self-evident. The rate of mitotic division may be increased in order to repopulate germ cells after testicular damage. Evidence has been presented which suggests that such repair may originate from a specific reserve

Table 4.7. Comparisons of daily spermatozoa production (DSP) and duration of spermatogenesis in a variety of mammals (Swiestra *et al.*, 1974, 1975; Pickett *et al.*, 1989).

Animal	DSP g^{-1} parenchyma ($\times 10^6$)	DSP both testes (10^9)	Duration of spermatogenesis (days)	Weight paired testes (g)
Stallion	16	5.37	57	340
Boar	23	16.2	39	720
Bull	11	6.7	61	685
Ram	21	9.5	47	500
Dog	12	0.37	61	34
Rabbit	25	0.16	48	6.4

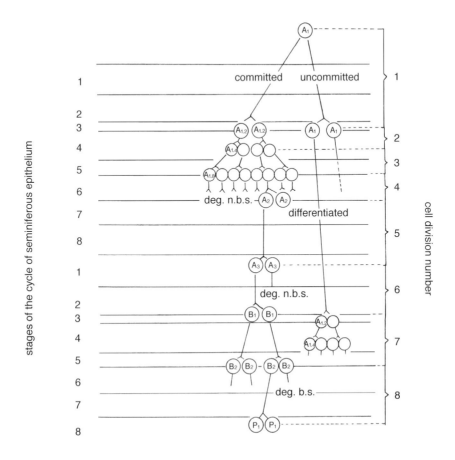

Fig. 4.19. The various steps in spermatocytogenesis in the stallion, assuming eight spermatozoal divisions (as indicated on the right) occurring through eight stages of the cycle of the seminiferous epithelium (as indicated on the left). The entry of a single A_1 spermatogonium results, on average, in two (P_1) primary spermatocytes entering the meiosis phase. Many spermatogonia degenerate during spermatocytogenesis. (Adapted from Pickett *et al.*, 1989.)

population of spermatogonia, individual cells of which may be viable for 60 days or more (Huckins, 1971, 1978).

A_1 spermatogonia destined to produce spermatocytes, once committed to this line of development, but still undifferentiated, divide to give a pair of $A_{1.2}$ spermatogonia which stay together as a pair connected by an intercellular bridge (Johnson, 1991a). These bridged $A_{1.2}$ cells continue to divide, by mitosis, to form a row of four $A_{1.4}$ and finally a row of eight $A_{1.8}$ spermatogonia. At this stage each $A_{1.8}$ spermatogonium differentiates and has the potential to divide by mitosis to form two A_2 spermatogonia. These then, again by mitosis, give rise to A_3, B_1 and B_2 differentiated spermatogonia. Finally the B_2 spermatogonia divide to form two preleptotene primary spermatocytes which enter the first division of meiosis. The entering of A_1 spermatogonia into the undifferentiated cycle of $A_{1.2}$, $A_{1.4}$ and subsequently $A_{1.8}$ can only occur at set times within the cycle of the seminiferous epithelium, details of which are given later. In summary, spermatozoa production takes one cycle of 57 days from spermatogonia through spermatocytes and spermatids to spermatozoa (Johnson et al., 1997). However, spermatogonia cannot commence (jump on to) this cycle randomly; they can only enter every 12.2 days, or at stage 1 of the seminiferous epithelium (Figs 4.19 and 4.22), giving a regular pattern of spermatozoa development in any cross-section of seminiferous tubule (Swiestra et al., 1974, 1975).

If the yield of preleptotene primary spermatocytes (P_1) per spermatogonia entering spermatocytogenesis is assessed, it is evident that the expected 64 primary spermatocytes (P_1) per spermatogonia ($A_{1.2}$) entering spermatocytogenesis does not result; many of the subtypes A_2, A_3, B_1 and B_2 degenerate and do not undergo further mitosis (Fig. 4.19). Indeed Johnson and Tatum (1989) indicated that the yield for the division of A_3 to B_1 and B_1 to B_2 was nearer to one than the expected four.

Season affects the rate of degeneration and the stage at which it occurs. In theory, degeneration can occur at any stage of spermatogenesis, but degeneration of B_2 spermatogonia primarily occurs during the breeding season, whereas during the non-breeding season degeneration primarily occurs at the $A_{1.8}$ and the A_2 stage of development, with some A_3 (Hochereau-de Riviers et al., 1990; Johnson, 1991b). Significant degeneration of B_2 spermatogonia during the breeding season results from an overproduction of A spermatogonia, the resultant spermatogenesis of which cannot be supported by the limited numbers of Sertoli cells (Pickett et al., 1989). Each Sertoli cell is only able to support an average of 2.5 primary spermatocytes and 9–10 spermatids of a given generation (Swierstra et al., 1974; Johnson and Thompson, 1983; Johnson et al., 1994). Overproduction of spermatogonia requiring support that is not available will result in significant degeneration; the loss may be up to 32% of the potential primary spermatocytes. This loss does not occur during the non-breeding season, when fewer A spermatogonia are produced. Therefore, most B_2 spermatogonia that arise do result in two primary spermatocytes. These differences in production of A spermatogonia and the degeneration of later spermatogonia between the breeding season and non-breeding season compensate for each other. As a result the number of primary spermatocytes per gram of testes tissue is similar in the breeding (21) and non-breeding seasons (19) (Johnson and Tatum, 1988, Johnson, 1991b; Amann, 1993b).

During the non-breeding season there is a 23% drop in the number of potential spherical spermatids (A) formed. This is likely to be due to a reduction in the ability of the Sertoli cells to support germinal cells during the non-breeding season, as the ratio of germinal cells to Sertoli cells is reduced at this time (Hochereau-de Rivers et al., 1987; Johnson, 1991a). A general correlation of 0.83 exists between Sertoli cell numbers and daily spermatozoan output, and, more specifically, if Sertoli cell numbers are related to type A spermatogonia, a correlation of 0.81 is reported (Johnson *et al.*, 1994).

Meiosis

Meiosis is the means by which a single diploid cell divides to produce two haploid cells. This process only occurs in the gonads of the male and the female, allowing the production of spermatids and ova that are haploid in nature. The process also allows for the exchange of genetic material between chromosomes in the dividing cell (Fig. 4.20) (Johnson *et al.*, 1997).

In the stallion, meoisis follows spermatocytogenesis. It starts with the preleptotene primary spermatocytes and results in the development of four spermatids (haploid) from each initial primary spermatocyte (diploid). Meiosis may be considered in two stages: the first and second meiotic divisions. The first meiotic division involves the multiplication and exchange of genetic material and results in two diploid secondary spermatoctyes; it is by far the largest phase, taking 18.7 days. The second division, which results in the halving of the genetic material and the production of two haploid spermatids per single secondary spermatocyte, takes 0.7 days.

THE FIRST MEIOTIC DIVISION

The first meiotic division is subdivided into several stages – prophase, metaphase, anaphase and telophase – usually denoted by 1 (e.g. prophase 1) as they are part of the first meiotic division.

Prophase is further subdivided into leptotene (early prophase), zygotene (middle prophase), pachytene, diplotene (late prophase) and diakinesis. It is a more complicated and slower process than the equivalent stage in mitosis and the prophase of meiosis 2, and so takes up the greatest part of the 18.7 days for the first meiotic division. Preleptotene primary spermatocytes immediately begin meiosis by active DNA synthesis. This may take 14–28 h (depending upon the species) and accounts for the relative slowness of prophase 1 (Courot *et al.*, 1970). Chromatin clumps may be found near the nuclear membrane of these preleptotene primary spermatocytes and subsequently disperse to produce the fine chromatin filaments evident in leptotene primary spermatocytes. These filaments may not be evenly distributed and so the nucleus of leptotene primary spermatocytes may appear asymmetrical (Johnson, 1991b). During this leptotene stage the chromosomes elongate and become visible as apparently single structures. However, by this stage, most (if not all) of the DNA has been doubled during the preleptotene phase, evident from the doubling of the total DNA content. These structures are termed univalents (Kimball, 1983).

During zygotene the chromosomes within the cell pair, lengthwise with their homologue, in a process termed synapsis. This pairing of chromosomes allows for the exchange of genetic material between different or non-sister

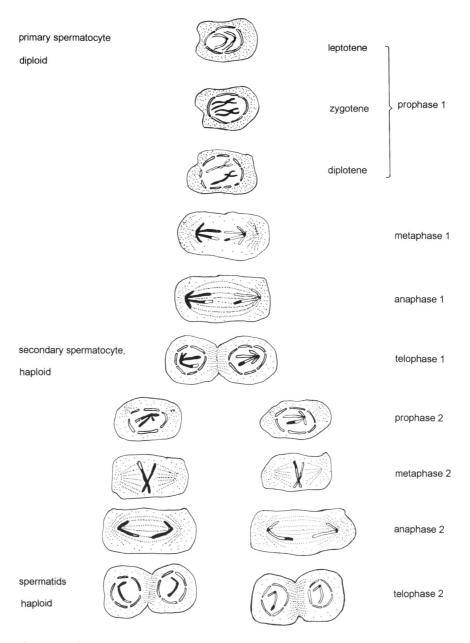

Fig. 4.20. The process of meiosis that results in spermatogonia in the stallion.

chromosomes (that is, of maternal and paternal origin) at attachment sites termed synaptonemal complexes. At this stage the two sex chromosomes, XX or XY, are isolated into a sex vesicle (Burgos *et al.*, 1970).

The pachytene phase follows, commencing with large chromosomes. During this phase each chromosome becomes evident as two chromatids and shortens in length. Pachytene is the longest phase and the exchange of genetic material started in zygotene continues.

The diplotene stage is marked by the pulling apart of the two homologues. At this stage the fact that each homologue is made up of two chromosomes, which are in turn made up of two sister chromatids, becomes evident. The pulling apart is not complete and two areas of attachment remain. Firstly, the attachment of sister chromatids at their centriole persists and, secondly, the attachment of the non-sister chromatids is evident at points termed chiasma. These points mark the position where cross-over of segments of non-sister chromatids has occurred. This cross-over is reciprocal, in that the total genetic material in each non-sister chromatid remains unaltered after the process is complete. There are normally several chiasma in a single bivalent and, as can be imagined, there are numerous permutations from the cross-overs between the four chromatids. The only restrictions placed on cross-overs are that they do not occur between sister chromatids and there is not a simultaneous cross-over of more than two chromatids at a single point at a set time (Kimball, 1983; Johnson, 1991b).

Diakenesis, the final stage of prophase 1, is relatively short lived. During diakenesis the homologous partners continue to separate and the chromatids continue to shorten.

Metaphase 1 marks the disappearance of the nuclear membrane and the appearance of a spindle on which the chromosomes arrange themselves. Each pair of homologues arrange themselves on either side of the equator of the spindle, with their centromeres acting as attachment points. This random arrangement of homologues allows for further genetic variability in the resultant cells. As it is exceedingly rare for all univalents of paternal origin to locate themselves on one side and univalents of maternal origin on the other, a random arrangement occurs (Kimball, 1983; Phillips and Chilton, 1991).

Anaphase 1 heralds the migration of the centromere of each bivalent to either pole (end) of the cell, thus separating the bivalents into univalents again. The centromeres continue to attach each pair of sister chromatids together (Kimball, 1983; Phillips and Chilton, 1991).

Telophase 1 is the continuation of the separation of the bivalents started in anaphase 1, resulting in two separated cells, each having only one member of each of the original homologous pair. The division has not been just the straightforward division of a single bivalent (maternal and paternal) into two univalents, as there has been an exchange of genetic material and chromatid segments along the way. Telophase 1 results in the production of two haploid secondary spermatocytes. These spermatocytes then enter the second meiotic division (Kimball, 1983).

THE SECOND MEIOTIC DIVISION
The second meiotic division is again subdivided into prophase, metaphase, anaphase and telophase, denoted this time by 2 in order to differentiate them from those of the first meiotic division.

Prophase 2 follows a short interphase prior to the secondary meiotic division. At this time the secondary spermatocytes have spherical nuclei with chromatin flakes of varying sizes. Prophase 2 marks the appearance of another spindle within the cells (Kimball, 1983).

As with metaphase 1, metaphase 2 marks the disappearance of the nuclear membrane and the arrangement of the univalents on the equator of the spindle, with the centromeres again acting as attachment points (Kimball, 1983).

Anaphase 2 sees the separation of the centromeres of the sister chromatids and the separation of the chromatids to either pole (Kimball, 1983).

Telophase 2 marks the reformation of the nuclear membrane around the chromatids of each pole and the emergence of two cells from the original single secondary spermatocyte (Kimball, 1983).

On the completion of meiosis, four haploid cells, termed spermatids, emerge from the division of each diploid primary spermatocyte. During meiosis, not only is there duplication of DNA but there is also evidence that RNA synthesis occurs, especially during mid-pachytene of the first meiotic division (Loir, 1972; Slaughter *et al.*, 1989).

Throughout spermatocytogenesis and meoisis, cohorts of developing cells, originating from the same spermatogonia, are interconnected by intercellular bridges. These bridges may function as communication systems between cells, ensuring synchronous development or degeneration. They may also facilitate the production of committed spermatogonia, the differentiation of haploid spermatids and the phagocytosis and digestion of residual bodies remaining after spermiation (Huckins, 1978; Johnson, 1991b).

Spermiogenesis

Spermiogenesis is the final stage of spermatogenesis, lasting 18.6 days, and is the process by which spermatids are differentiated into spermatozoa (Johnson *et al.*, 1997). Spermiogenesis is divided into four phases: the Golgi, cap, acrosome and maturation phases. The spermatids within these phases are denoted as Sa, Sb, Sc and Sd, respectively. The division of the phases is largely based on the development of the acrosome region.

GOLGI PHASE

The acrosome region is an enzyme-containing vesicle and, like other enzyme-containing vesicles in the spermatid, it is developed from the Golgi apparatus. The Golgi responsible for the development of the acrosome region is evident in the Sa spermatid close to the nucleus. The Golgi produces vesicles that fuse to form a large vesicle in close association with the nucleus of a late Sa spermatid. This vesicle also produces membrane-bound enzymes. The resultant large vesicle is termed the acrosome vesicle and now comes in close contact with the nuclear membrane. At the point of attachment an indentation may be seen as if the acrosome vesicle is applying pressure to the nuclear membrane. As more Golgi vesicles fuse with this acrosome vesicle, it enlarges and spreads over the surface of the nuclear membrane. In association with the formation of the acrosome vesicle there is the appearance of an acrosome granule. Late Sa spermatids also show migrating centrioles. The centrioles within the cell cytoplasm migrate to an area near the nuclear membrane from where the distal centriole gives rise to the basal plate, from which develops the axenome of the flagella. The attachment point of this distal centriole and its subsequent axenome to the nucleus is termed the implanation fossa.

The flagellar canal is formed by an inversion of the spermatid plasma membrane from the surface of the cell body to an area termed the annulus just below the implantation fossa. It is through this canal that the flagellum develops and extends towards the lumen. The drawing of the plasma membrane towards the nucleus to form the flagellar canal may be encouraged by positive pressure within the cell from indentation of the nuclear surface at the implantation fossa (Johnson, 1991b). This arrangement allows new growth of the flagellum to be directed away from the body of the spermatid and so facilitates continual flagellum growth and elongation. The canal may also facilitate flagellation (vibrating) of spermatids, especially in the maturation phase, and the attachment of the nuclear membrane during the flagellum attachment to the nucleus. The canal may also provide a means of connection between the area of ATP demand, the flagellum, and the rest of the cell. Mitochondria will, therefore, be drawn along the canal towards this area of high ATP demand. They will then be isolated in what becomes the middle piece after the annulus has become fixed. Finally, the canal may act to prevent the attachment of the randomly moving mitochondria to the axenome or outer dense fibres during their development (Kimball, 1983; Johnson, 1991a).

CAP PHASE

Spermatids in the cap phase are termed Sb. During this relatively short phase, the Golgi apparatus becomes less distinct and moves away from the nucleus. The flagellum continues to develop and elongate. Up to this stage the nucleoli of the equine spermatids are spherical with both coarse and fine granules situated more or less centrally in the cell. However, as the cap extends further over the nucleus to reach its final coverage of two-thirds of the surface, the nucleus begins to change shape and elongate (Ortavant *et al.*, 1977; Johnson *et al.*, 1990).

ACROSOME PHASE

Spermatids in the acrosome phase are termed Sc. It is during the acrosome phase that significant elongation of the nucleus occurs (Johnson *et al.*, 1990). In addition, a specialized organelle, unique to spermatids, becomes evident (Fig. 4.21). This organelle, termed the manchette, is only transient. It is composed of a series of microtubules attached to each other by linking arms, and is isolated, enfolding the caudal or lower end of the now elongating

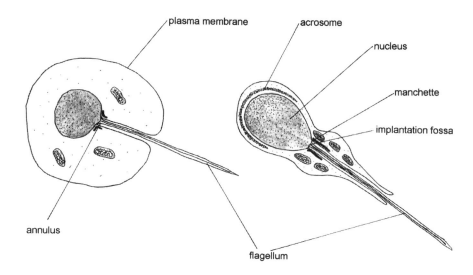

Fig. 4.21. A simplified diagram to illustrate the final stages of spermiogenesis, and the development of the flagella.

nucleus near the developing flagellum. It then extends down to enclose the upper part of the developing flagellum. The manchette persists after elongation and remnants of it are evident in the residual body left behind after spermiation. There should be no evidence of it in mature spermatozoa, except in cases where parts of the residual body are left on mature spermatozoa due to incomplete maturation (Goodrowe and Heath, 1984; Johnson, 1991b).

MATURATION PHASE

Spermatids in the maturation phase are termed Sd. This phase is the final stage of spermiogenesis and also spermatogenesis, resulting in spermatid development. During this phase the manchette continues to migrate caudally, supporting the flagellar canal. The annulus migrates to its permanent site at the junction of the mid-piece and the principal piece, drawing down caudally with it the plasma membrane of the flagellar canal. As a result the flagellar canal shortens and subsequently disappears. At this time, and apparently very quickly, the mitochondria migrate and arrange themselves around the flagellum as the annulus migrates down (Johnson *et al.*, 1990) (Fig. 4.21).

Evidence of the speed of this migration is indicated by the fact that it is very rare to identify a spermatid in which the annulus has migrated but the mitochondria are not yet arranged around the mid-piece. Also, in the cross-section of a Sertoli cell (the centre of which is the characteristic site of maturation phase spermatids), some spermatids may show completion of the maturation phase whereas others, of the same cohort, show indications that the phase has only just been initiated.

During these phases, development and changes within the flagellum are evident. The means by which the characteristic nine dense fibres and fibrous sheath of the spermatozoa are formed is unclear. However, it is likely that they develop in a similar fashion to the development of cell cilia and flagella in general – that is, from the microtubules of the distal centriole which develop on to the columns of the capitellum situated at the implantation fossa and hence extend down the flagellum (Kimball, 1983). The fibrous sheath may develop from a spindle-shaped body, which has been identified in developing human flagella (Holstein and Roosen-Runge, 1981), or longitudinal columns found at the distal end of the principal piece in rats (Irons and Clermont, 1982; Johnson, 1991b).

At the end of the maturation phase a significant amount of cytoplasm is left attached to the spermatid. This excess cytoplasm is held within residual bodies and attached to the spermatid by a cytoplasmic stalk. It is subsequently phagocytosed by the supporting Sertoli cells. These spermatids are released from their residual bodies and Sertoli cells into the lumen of the seminiferous tubule. Remnants of this excess cytoplasm still remain on the spermatid as the cytoplasmic droplet, which is evident in all spermatozoa that have not yet matured in the epididymis.

Spermiation

During the process of spermiogenesis, a spermatid undergoes morphological changes from a spherical cell with a spherical nucleus, to an elongated cell with an elongated streamlined head containing a condensed nucleus and a specialist area of penetrative enzymes, plus a tail required for the movement of the cell. In addition, during this process, nucleotide synthesis takes place, probably for DNA repair, along with a change in chromatin proteins, and RNA synthesis to allow the production of specific proteins for unique spermatid structures such as the manchette (Loir, 1972; Johnson, 1991b).

Spermiation is the final release of a mature spermatid, now termed a spermatozoon, into the lumen of the seminiferous tubule. From here it passes to the rete testes and on to the epididymis for final maturation.

Seminiferous epithelium cycle

If a cross-section of seminiferous tubules is taken, a series of layers or changes in spermatozoa development is evident (Le Bland and Clermont, 1952; Johnson *et al.*, 1990). Working from the lamina propria of the seminiferous tubules towards the lumen, these layers are delineated by stages of development of the spermatozoa within the layers (Fig. 4.22). Four to five layers are evident, each having spermatozoa at a set stage of development. The time interval between each layer or stage at a set point is 12.2 days of the spermatogenic cycle.

For example, in a set cross-section of seminiferous tubule epithelium (Fig. 4.22), there may be A and $A_{1,2}$ spermatogonia in the layer nearest the lamina propria. Above them and nearer the lumen there will be spermatozoa

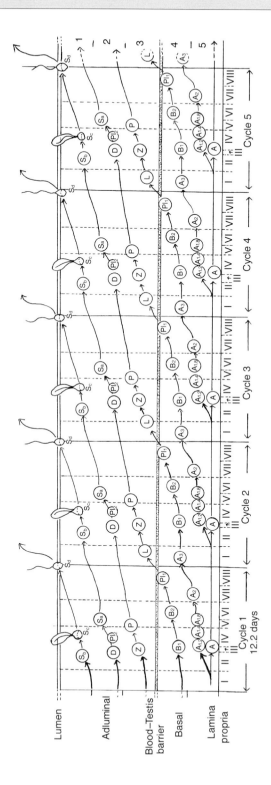

Fig. 4.22. The cycle of the seminiferous epithelium at a given point with the seminiferous tubule of the testis of the stallion. Five cycles of 12.2 days are illustrated, subdivided into eight stages or cellular associations. The lengths of these associations (in days) are as follows: I – 2.1; II – 1.8; III – 0.4; IV – 1.9; V – 0.9; VI – 1.6; VII – 1.5; VIII – 1.9. The diagram shows the progressive development from spermatogonium (A) near the lamina propria to spermatozoon (Sd) released into the lumen. Spermatogonia A_1, A_2, A_3, B_1 and B_2; first meiotic division primary spermatocytes Pl_1 (preleptotene), L (leptotene), Z (zygotene), P (pachytene) and D (diplotene); second meiotic division primary spermatocytes Pl_2 (preleptotene) the remaining stages of the second meiotic division are not shown; S_a, S_b, S_c and S_d progressively more mature spermatids. The relative time taken for each of the stages is indicated by the distance between them on the diagram. The development of one spermatogonium (A) to one spermatozoon (Sd) takes 4.7 cycles, which at 12.2 days per cycle results in 57.3 days in total. (Adapted from Amann, 1993b.)

12.2 days further advanced in development, that is B_1 spermatogonia. In the third layer, progressing further again towards the lumen of the seminiferous tubule, there will be pachytene primary spermatocytes (zygotene Z); in the fourth layer, secondary spermatocytes (Pl_2); and finally in the fifth layer, nearest the lumen, acrosome spermatids (Sc). No stages other than those given in the example will be seen. However, a section taken from slightly further along the seminiferous tubule may yield in the first layer A_2 spermatogonia against the lamina propria. If so, the second layer would be B_2 spermatogonia/preleptotene primary spermatocytes (Pl_1), the third layer would be pachytene spermatocytes (P), the fourth layer Golgi spermatids (Sa) and finally the fifth layer acrosome spermatids Sc nearing maturity. Again, the order and stage of spermatozoa development in the five cycles is fixed, no other stages being seen.

As spermatozoa mature and develop through these cycles they are found initially imbedded within the Sertoli cells. As they develop they become increasingly separated from their Sertoli cells until they are released as mature spermatids. The cycle of seminiferous epithelium is 12.2 days in length and is the cycle of spermatozoa development seen within the epithelium of the seminiferous tubules. Hence, at a set point in the seminiferous tubules, gametes can enter as committed $A_{1,2}$ spermatogonia every 12.2 days. Similarly, they leave as mature spermatids every 12.2 days at that set point in the tubule (Swierstra et al., 1974, 1975; Johnson et al., 1978a; Johnson, 1991a).

As a further complication each 12.2-day cycle of seminiferous epithelium is further divided into a series of cellular associations (Fig. 4.22). The exact number of cellular associations varies with the criteria used to identify each group of developing spermatozoa cells; for example, the criteria might be morphological changes of the germ cells (Roosen-Runge and Giesel, 1950) or the development of the spermatid acrosome (Le Bland and Clemont, 1952). Using these rather arbitrary criteria, the following number of cellular associations (or stages) per cycle of seminiferous epithelium have been suggested: 14 or eight stages in the bull (Hochereau-de Riviers, 1963; Berndtson and Desjardins, 1974); eight in the ram (Ortavant et al., 1977); and eight in the horse (Swierstra et al., 1974; Johnson, 1991a). The suggestion of eight stages for the horse was arrived at by Swierstra et al. (1974) using morphological changes in germinal cells as the criteria. These eight stages are illustrated in Fig. 4.22, and so in the two previous examples given to illustrate the cycle of the seminiferous epithelium the first example was of cellular association stage IV and the second example was of stage VII.

The relative length of each stage may be arrived at by assessing the relative frequency in appearance of these various stages of spermatid development in a random cross-section sample taken from a normal testis. For example, 16.9% of cross-sections show characteristics of stage I: that is, no mature spermatozoa lining the lumen, an outer (fourth) layer of Golgi spermatids (Sa), a third layer of pachytene primary spermatocytes (P), a second layer of leptotene primary spermatocytes (L) and a first layer, nearest the lamina propria, of A_3 primary spermatocytes. The duration of stage I is calculated as 16.9% of 12.2 days, which is 2.1 days. The durations of the other stages are similarly calculated and are also given in Fig. 4.22.

Factors affecting spermatogenesis

There are numerous factors that might affect daily spermatozoan production (DSP) from an individual stallion. These include season, age, testicular size, frequency of use, breed, environmental factors (including temperature and radiation, also nutrition and toxic chemicals or drugs), physical and hormonal abnormalities, disease and infection. These will be discussed in turn, the initial few in some detail. Significant discussion of the latter factors is really beyond the scope of this book and more suited to a text on stallion infertility.

SEASON

As was discussed in section 3.3.2 and as illustrated in Figs 3.19–3.23, season has a significant effect on the reproductive efficiency of the stallion, in particular DSP and libido, as well as the volume of the ejaculate (Pickett and Shiner, 1994).

By collecting semen samples from a group of stallions for 1 year on a weekly basis, the changes in seminal characteristics with season were demonstrated by Pickett and Voss (1972) and Pickett *et al.* (1976). The largest differences observed were in total seminal volume produced, which ranged from 104 ml in June to 45 ml in January. Some of this difference in total seminal volume can be accounted for by a significant increase in the volume of gel produced during the breeding season. The gel-free fraction volume ranged from 45 ml in February to 81 ml in June. However, some of the difference in seminal volume was due to an increase in DSP, indicated by the change in the number of spermatozoa produced per ejaculate. This ranged from 10×10^9 in January to 22×10^9 in July. This effect of season is further supported in work by Dowsett and Pattie (1987), who also assessed the percentage of dead spermatozoa at various times of the year. Their work indicated a slight decline in the percentage of dead spermatozoa with season (18.5% in spring compared with 21.4% in winter). This effect of season was further supported by more recent work using morphometric analysis of spermatids with spherical nuclei in the testes of stallions slaughtered at varying times of the year (Pickett *et al.*, 1989). The hormonal mechanisms by which this effect is achieved have been discussed in section 3.3.2. It is apparent that in addition to a seasonal effect on spermatogenesis, there is a significant effect of season on the number of Leidig and Sertoli cells (Johnson and Thompson, 1983; Johnson, 1986a; Johnson and Tatum, 1989; Johnson *et al.*, 1991a) (Table 4.8).

The close association and involvement of Sertoli cells in spermatozoan development would suggest that the effect of season on spermatogenesis is, at least in part, mediated via the number of Sertoli cells available to support the developing spermatozoa and not only the number of developing spermatozoa that a single Sertoli cell can support. Season is also associated with a significant

Table 4.8. Effect of season upon Sertoli cell and Leidig cell populations within the stallion's testis and upon number of germ cell types accommodated by a single Sertoli cell (Johnson and Thompson, 1983; Johnson and Tatum, 1989).

	Season		
Item	Non-breeding	Breeding	Significance (P)
No. of Sertoli cells per testis (10^6)	2.6 ± 0.2	3.6 ± 0.2	< 0.01
No. of Leidig cells per testis (10^6)	1.7	2.9	–
Germ cell types (no. accommodated/Sertoli cell)			
Type A spermatogonia	1.1 ± 0.1	1.5 ± 0.1	< 0.01
Preleptotene and leptotene plus zygotene primary spematocytes	2.9 ± 0.4	2.9 ± 0.2	NS
Pachytene plus diplotene primary spermatoctyes	2.6 ± 0.2	3.0 ± 0.3	NS
Spermatids with round nuclei	8.1 ± 0.8	10.9 ± 0.8	< 0.05
Spermatids with elongated nuclei	8.0 ± 0.8	10.2 ± 0.7	< 0.05
All germ cell types combined	22.8 ± 2.1	28.5 ± 1.7	< 0.05

increase in the number of type A spermatogonia, which, even though there is still significant degeneration of germ cells between type B spermatogonia and primary spermatocytes, is large enough to result in an overall increase in spermatozoan production (Johnson, 1985; Johnson and Tatum, 1989; Johnson et al., 1997). It is known that during the breeding season there is a small but significant increase in the number of type A spermatogonia produced. These in turn need supporting by Sertoli cells. As the number of Sertoli cells increases with the breeding season, they are available to support the resultant type B spermatogonia. As a result, an increase in spermatogonal divisions early in spermatogenesis is seen (Johnson, 1991b). Even during the breeding season there is a significant degeneration of type B spermatogonia, due to the lack of Sertoli cells to support them all (Johnson, 1985). During the non-breeding season this lack of support is even more evident and so is the resultant effect on spermatogenesis, though the effect is reduced somewhat by the reduction in type A spermatogonia production at this time.

This change in Sertoli cell number is indicated by a change in testes weight. In addition, an increase in diameter of seminiferous tubules is evident during the breeding season (Berndtson et al., 1983). It can be noted here that the older the stallion, the greater is the effect of season on the efficiency of spermatozoan production. The greatest effect on actual DSP is evident in stallions over the age range 6–12 years, with a decline of over 50% in DSP between the breeding and non-breeding season being apparent. However, the decline in the efficiency of spermatozoan production (number of spermatozoa produced per gram of testicular tissue) appears less when testicular weight is taken into account (Table 4.9).

Table 4.9. Reduction in efficiency of spermatozoan production (number of spermatozoa produced g^{-1} of testicular tissue) seen between breeding and non-breeding season in stallions of varying ages (Johnson and Thompson, 1983).

Age (years)	Decline in efficiency of spermatozoan production with season (%)
2–3	21
4–5	12
6–12	19
13–20	34

AGE

The effect of age on DSP was partly considered in section 3.3.1, where it was concluded that DSP increased from puberty (17–22 months) until it reached a plateau at 4–5 years of age, thereafter remaining largely unchanged until 20 years of age (Nishikawa, 1959b; Johnson and Neaves, 1981; Thompson and Honey, 1984; Jones and Berndtson, 1986; Berndtson and Jones, 1989; Pickett and Shiner, 1994; Dowsett and Knott, 1996). There is conflicting evidence regarding the effect of age greater than 20 years on reproductive function: some work indicates a decline in DSP and a possible increase in the percentage of morphologically abnormal spermatozoa with age over 20 years; other work does not support these findings (Van der Holst, 1975; Johnson and Thompson, 1983; Johnson *et al.*, 1991a,b; Amann, 1993b). Dowsett and Pattie (1987) indicated a significant decline in spermatozoan concentration in semen from stallions over the age of 14 years compared with those 3–13 years old (83.24×10^6 ml^{-1} compared with 147.78×10^6 ml^{-1}), along with a decline in the total number of spermatozoa produced (3.25×10^9 compared with 5.05×10^9) and a significant increase in the percentage of dead spermatozoa (22.8% compared with 18.8%). Stallions of 1 and 2 years of age showed significantly lower results than all ages in all these semen categories. Dowsett and Knott (1996) demonstrated a decline in semen quality with age below 3 years and over 11 years.

This possible decline in spermatogenesis with old age is supported by evidence that the number of Sertoli cells per gram of testicular tissue does decline with age, from 39 in 2–3-year-old stallions in the breeding season, to 29 in 4–5-year-old stallions and 20 in 13–20-year-old stallions (Pickett *et al.*, 1989). However, this is disputed by more recent work which indicates that there is no significant relationship between age (4–20-year-old stallions) and Sertoli cell numbers (Johnson *et al.*, 1991a). Jones and Berndtson (1986) demonstrated a decline in Sertoli cell number by up to 50% between 2-year-old and 20-year-old stallions. They reported that this was compensated for by the significant rise in the number of developing germ cells accommodated by each Sertoli cell. Any decline in spermatozoan production per gram of testicular tissue may also be partly compensated for by the increase in testis size and weight with age (Thompson *et al.*, 1979a; Johnson and Neaves, 1981) (Tables 3.2 and 4.10).

Table 4.10. Effect of age on scrotal width (Thompson et al., 1979a).

Age (years)	Average scrotal width (mm)
2–3	96[a]
4–6	100[b]
> 7	109[c]

All three means differ significantly ($P < 0.05$)

DSP is controlled not only by spermatogenesis but also by spermatozoan reserves. Spermatozoa are primarily stored in the tail of the epididymis, and it is these spermatozoa, rather than those within the body and the head of the epididymis, that are ejaculated. The greater the stallion's reserves, the greater is his ability to cover mares consistently and successfully. The number of spermatozoa held within the tail of the epididymis is affected by two major factors: the frequency of use, and stallion age. Amann et al. (1979b) investigated the effect of age and ejaculation frequency on spermatozoan reserves. Using various ejaculation frequencies on different ages of stallion and subsequent castration and examination of spermatozoan reserves, it was concluded that the spermatozoan reserves of sexually rested stallions (and, therefore, the number of spermatozoan produced per ejaculate) increased with advancing age. There was an increase in spermatozoan reserves in all three areas of the epididymis, the most significant being in the body (4.2×10^9 spermatozoa in 2–4-year-old stallions compared with 9.5×10^9 in 10–16-year-olds). The reserves within the tail region also significantly increase (18.7×10^9 in 2–4-year-olds compared with 28.0×10^9 in 10–16-year-olds). This gave a total increase in spermatozoan reserves of approximately 30% with advancing age (28.5×10^9 in 2–4-year-olds, 40.8×10^9 in 5–9-year-olds, 45.4×10^9 in 10–16-year-olds). Age is also known to affect other seminal characteristics, namely motility, pH, gel volume and total volume of semen, as well as concentration and total number of spermatozoa (Squires et al., 1979a) (Table 4.2). Voss et al. (1988) indicated that semen volume, spermatozoan concentration and scrotal width sharply increase to 5 years of age and then continue to increase more gradually to a peak at 12 years of age.

TESTICULAR SIZE

Table 3.2 illustrated the effect that testicular weight has on output of spermatozoa and how this is related to age. Testicular weight is correlated with testicular size, which can be more easily measured, either by calipers or by ultrasonic scanning (Rath and Brass, 1988; Love et al., 1991). No adverse effect of using linear scanning has been reported (Rath and Brass, 1988).

Using these methods, a linear relationship between testicular size and daily spermatozoan output (DSO) is evident (Fig. 4.23) (Love et al., 1991). A correlation between age and scrotal width (0.62) has also been reported (Pickett et al., 1989) along with scrotal width and spermatozoan motility

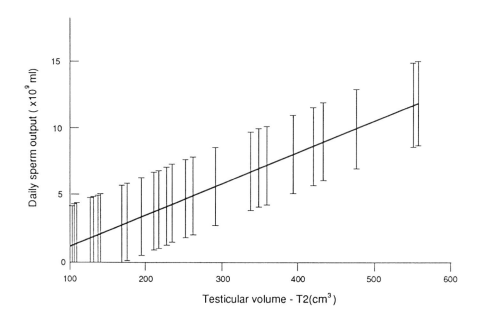

Fig. 4.23. Relationship between predicted daily spermatozoan output and testicular volume for 26 stallions. (Love et al., 1991.)

(0.34) (Sigler and Kiracofe, 1988). Such a relationship would be expected, as the amount of a product depends upon the amount of tissue producing it. Experiments conducted by Thompson et al. (1979a) produced various correlation coefficients between various size parameters of the testes and DSP and DSO. The best correlation (0.75) existed between total scrotal width (measured across the two testes from the point of greatest curvature of one testis to the point of greatest curvature of the other) and DSP. Testicular size is, therefore, a very good indicator of the spermatozoa-producing capacity of that stallion and may be advocated as a parameter when selecting for reproductive potential (Pickett and Shiner, 1994). Testicular size in bulls is highly heritable (65% or more) and is very likely also to be heritable in stallions. The correlation between testicular width and DSO at 0.55 is not as good, as factors additional to testicular size affect output of spermatozoa at ejaculation (for example, spermatozoa reserves, ejaculation frequency, etc).

The size of the testis indicates the volume of parenchyma within. The correlation between parenchyma weight and Sertoli cell number is reported to be 0.85 (Johnson et al., 1994). Hence an increase in testicular size indicates an increase in the number of Sertoli cells available to support developing germinal cells. Indeed, the correlation between parenchyma weight and DSP is 0.89 and between the number of Sertoli cells and DSP is 0.85 (Johnson et al., 1994). This correlation between testicular size and DSP has been exploited in the development of an index for testis shape (Kenney index) which is reported to have a 0.62 correlation with DSP (Silva et al., 1996). The size of the testis also includes the tunica albuginea and it must be remembered that in older stallions the tunica albuginea tends to be thicker, and so the increase in testicular size with age is not entirely due to an increase in parenchyma (Table 4.11) (Johnson and Neaves, 1981).

In order for the correlation between testicular size and DSP to hold, it is necessary that as testicular size increases the volume of functional parenchyma also increases. Therefore, the functional ability of the parenchyma is important. Testicular function may also be indicated by testicular consistency, which can be assessed manually. Evidence of hardened areas within the testes indicates non-functional testicular tissue and, therefore, a decline in DSP. Abnormalities in the epididymal area may indicate a decrease in the ability of the epididymis to store spermatozoa, and so a reduction in the stallion's potential DSO. The presence of large testes is no guarantee that spermatozoa will be produced, nor that the spermatozoa produced will have acceptable motility and morphology. A stallion with average-sized testes may prove to be azoospermic (Pickett et al., 1977). Testicular size alone should not be used as an indicator of DSO, but it may prove to be a useful guide when used in association with a general semen evaluation.

FREQUENCY OF USE

Figures 3.19–3.23 (from Pickett and Voss, 1972) clearly illustrate the effect that two ejaculates, collected 1 h apart, have on various seminal and spermatozoan parameters. In all cases, regardless of the time of year, the second ejaculate resulted in a lower quality sample, produced more reluctantly. As a general rule the greater the use made of a stallion (that is, the more frequently ejaculation occurs), the lower is the number of spermatozoa produced per ejaculate (Dowsett and Pattie, 1987; Pickett and Shiner, 1994). However, no real effect is seen at a frequency of less than one ejaculation per 2 days (Pickett et al., 1975c). It is also apparent that a stallion which produces a second ejaculate, within 1 h of the first, containing less than 50% of the total number of spermatozoa obtained in the initial ejaculate may have a subfertility problem (Pickett and Voss, 1972; Parlevliet et al., 1994). The relationship between the same semen parameters achieved in the first and second ejaculate within 1 day may be used as a good indicator of a stallion's potential performance (Amann, 1993b).

Table 4.11. Effect of age on the weight of the tunica albuginea as a percentage of testicular weight (Johnson and Neaves, 1981).

Age (years)	Weight of the tunica albuginea as a percentage of testicular weight
2–3	10%
4–8	9%
13–20	14%

When the frequency of a stallion's use is assessed within a week, no significant effect of ejaculation frequency on semen volume was observed. Ejaculation frequencies of once per week, three times per week or six times per week had no significant effect upon total seminal volume or gel-free seminal volume (Pickett *et al.*, 1975c). However, there is a tendency for the volume of the gel fraction to decrease with increasing frequency of use. When spermatozoan concentration was analysed in the same group of stallions, then a significant effect of collection frequency was observed between six times per week (142×10^6 ml^{-1}) and three times per week (248×10^6 ml^{-1}). No difference was observed when increasing the frequency of use from once per week (288×10^6 ml^{-1}) to three times per week (248×10^6 ml^{-1}). These differences in concentration were reflected in differences in total output of spermatozoa (once per week, 11.4×10^9; three times per week, 11.7×10^9; and six times per week, 5.9×10^9). When the total spermatozoan output per week was assessed, then again frequency of use had an effect. No difference was evident between a frequency of six times per week (35.3×10^9) and three times per week (35.2×10^9). On the other hand, a significant difference did exist between both these two frequencies and that of once per week (11.4×10^9).

It is apparent that stallions ejaculating three times a week have reached their limit of weekly spermatozoan production. Increasing the frequency of use to daily (six times per week) had no effect on the total number of spermatozoa produced per week; the total is simply divided between more ejaculates (Pickett *et al.*, 1975c). There is, therefore, no additional advantage for AI purposes in collecting daily as opposed to every other day. At a frequency of three times per week, spermatozoan reserves are being used as quickly as they are being replenished. Increasing the frequency to six times per week, or even to five times in 24 h at hourly intervals, has no effect on the rate of passage of spermatozoa through the head and body of the epididymis (Amann *et al.*, 1979b).

The popular belief that increasing the frequency of use of a stallion results in a decrease in fertility rates due to the ejaculation of immature spermatozoa is apparently incorrect. It is the decrease in the number of spermatozoa ejaculated, due to depletion of spermatozoan reserves, that results in depressed fertility rates. Pickett *et al.* (1975c) assessed the percentage of progressively motile spermatozoa in samples obtained at a frequency of once per week, three times per week and six times per week and failed to demonstrate any differences in the quality of the spermatozoa produced over the three collection regimes. In addition, stallion libido was not affected by the three frequencies of collection.

Further work was carried out by Pickett *et al.* (1975c) to investigate if there was any advantage, for the purpose of AI, of collecting twice a week (Tuesday and Friday) as opposed to four times per week (twice on Tuesday and twice on Friday). Collection four times a week resulted in a lower concentration per ejaculate (196×10^6 ml^{-1} compared with 272×10^6 ml^{-1}). However, when the total number of spermatozoa produced per week was calculated, the regime of four collections per week resulted in a greater total number of spermatozoa (35.1 × 10^9) than twice-a-week collections (26.2 × 10^9). In the four-times-per-week collection regime the second ejaculate, when compared with the first, consistently produced a lower spermatozoan concentration (145 × 10^6 ml^{-1} compared with 222 × 10^6 ml^{-1}), a lower volume of the gel fraction (49 ml compared with 58 ml) and a lower total number of spermatozoa per ejaculate (5.5 × 10^9 compared with 12.0 × 10^9). Again the regime of four collections per week yielded the same total number of spermatozoa per week (35.1 × 10^9) as the regimes of three times and six times per week discussed previously.

If frequency of collection is increased to twice per day compared with once a day, there is, as one would expect, no change in the total number of spermatozoa produced, only in the number produced per ejaculate (Pickett *et al.*, 1985). This again indicates that there is a finite number of spermatozoa available per day for ejaculation and that increasing the ejaculation frequency has no effect on the rate of spermatogenesis or upon the rate of passage and, therefore, the period of time that spermatozoa spend within the epididymis.

Finally, if the number of ejaculates within 1 day is assessed, it can be seen that stallions of different ages perform differently under high frequencies of service. Squires *et al.* (1979a) investigated the effect of five successive ejaculates on spermatozoan output in different-aged stallions after a period of 12 days sexual rest. Stallions within the age range of 2–3 years suffered more than those of ages 4–6 and 9–16 years. The biggest effect was observed between the first and second ejaculate; subsequent ejaculates had less of an effect. If the total number of spermatozoa over the five ejaculates is assessed, then the 9–16-year-old stallions produced the most at 22.3 × 10^9; 4–6-year-olds produced 18.0 × 10^9; and the 2–3-year-olds 8.9 × 10^9. By the fifth ejaculate the 2–3-year-olds were only producing 0.5 × 10^9 spermatozoa per ejaculate, and of these stallions over 50% were producing samples containing fewer than 200 × 10^6 spermatozoa. The stallions aged 4–6 and 9–16 years were still producing in excess of 1.0 × 10^9 spermatozoa per sample.

It is evident that, even after a prolonged period of rest, young stallions cannot cope with a high frequency of covering and still maintain their fertility rates. Older stallions, however, are able to cope with the occasional very heavy covering load without a detrimental effect on their fertility, provided that they have had a period of sexual rest. This is another indication that young stallions, though perfectly capable of covering mares successfully, cannot be expected to manage a full season.

Dinger and Noiles (1986) suggested that an estimation of DSO (which is related to the available gonadal spermatozoan reserves, which in turn is related to frequency of use) could be: total number of spermatozoa in the first ejaculate of a sexually rested stallion × 27.5%. This estimation of DSO had a correlation of 0.92–0.95 with the actual DSO.

BREED

There are significant differences in the semen characteristics for different breeds of stallion (see Table 4.3). In general, Arab stallions outperformed all the other breeds and performed the best in the specific direct comparisons carried out by Dowsett and Pattie (1987), showing significantly higher total output of spermatozoa. This work also demonstrated that Arabs showed the lowest percentage of dead spermatozoa and the lowest volume of gel. Work by Pickett *et al.* (1988a,b) indicated that Paint horses have a lower total number of spermatozoa per ejaculate (8.5×10^9) than Thoroughbred stallions (13.0×10^9) and it was postulated that this was related to the difference in scrotal width between the two breeds (102 mm compared with 111 mm). In this work Appaloosa stallions were also assessed for scrotal width, which was reported to be the lowest of the three breeds investigated at 85 mm. The percentage of normal spermatozoa also differed between the breeds, the most noteworthy being 23% for Arabs and 42% for Appaloosa stallions.

The possibility of further breed effects was investigated by Oba *et al.* (1993). This work demonstrated no differences in ejaculate volume, spermatozoan motility, spermatozoan concentration and pH, between six breeds compared, which included Arab, Appaloosa and Thoroughbred. However, a difference in spermatozoan survival rates was indicated, with the Thoroughbred outperforming all the others. It has been suggested that differences in the activity of certain enzymes, such as glutamic pyruvic oxaloacetic transaminase or alkaline phosphatase, may be apparent between different breeds (Serban *et al.*, 1982) and may account in part for some of the breed differences observed.

ENVIRONMENTAL INFLUENCES

Many environmental factors have been associated with changes in spermatogenesis. These include light, temperature, radiation, nutrition and toxic elements. The subject of light has been dealt with in some detail in section 3.3.2 and so will not be considered again. The remaining factors will be discussed in turn.

Temperature. Temperature has a significant effect on spermatogenesis. The scrotum of the stallion acts to protect the testes but also functions to regulate testicular temperature. Normal testicular temperature is 36°C, which is 3°C below normal body temperature. Areas with a temperature as low as 30.5°C have been identified in the normal testes (Pickett *et al.*, 1989). Elevation of testicular temperature may be a result of a rise in body temperature due to fever or an increase in the environmental temperature. Both will result in an elevation of the temperature of the germinal epithelium and hence have a direct effect on spermatogenesis (Johnson *et al.*, 1997). The elevation in temperature need only be for 2 h if temperatures of 40.5°C are reached. It has been suggested that summer temperatures in hot climates may be responsible for a depression in spermatogenesis in horses (Amann, 1993b), boars (Cameron and Blackshaw, 1980) and bulls (Bearden and Fuquay, 1992). The germinal cells particularly at risk are pachytene primary spermatocytes, but B spermatogonia and spherical spermatids can also be affected (Mazzari *et al.*, 1968; Waites and Ortavant, 1968; Harrison, 1975). As the spermatozoa mainly affected are those early on in spermatogenesis, the effect of no more than a transient increase in scrotal temperature may not be evident until 50 days or so later (Blanchard et al., 1996).

Work by Friedman et al. (1991a) indicated that inducing elevated testicular temperatures for 48 h resulted in testicular impairment in stallions for 60 days. The first effect observed was on the total number of progressively motile spermatozoa, evident as early as 5–6 days. This was followed by an effect on total spermatozoa per ejaculate, the percentage of normal spermatozoa and the concentration of spermatozoa evident between 10 and 50 days. Spermatozoan abnormalities became apparent between 15 and 50 days, with detached heads appearing primarily at 10–15 days and 45–50 days, underdeveloped spermatozoa (that is, with cytoplasmic droplets) at 15–50 days and, finally, early round germinal cells at 35–45 days. Further work by Friedman et al. (1991b) suggested that a longer recovery time, of up to 73 days, was required. This work also compared the duration of elevated testicular temperature (37°C over 24 or 48 h) and demonstrated that the degree of change observed was affected by the duration of elevated temperature. Provided the original rise in testicular temperature is only transient, recovery should occur and normal output of spermatozoa should be restored within 2 weeks of the decline.

Prolonged elevated temperatures may also affect spermatozoa within the epididymis, resulting most commonly in detached heads or tails. In such a case a drop in fertility would be evident within 3–4 days, with subsequent severe depression in 50 days (Austin et al., 1961; Waites, 1968; Amann, 1993b). The effect need not necessarily be bilateral; unilateral effects have been reported in stallions that have been recumbent for a while due to illness, resulting in only one testis experiencing elevated temperatures.

It is also possible that the effect of elevated environmental temperatures on reproductive efficiency is mediated via an endocrine route. Studies in rams and bulls have demonstrated an inverse relationship between ambient temperatures and circulating testosterone levels (Gomes et al., 1971; Rhynes and Ewing, 1973). However, Friedman et al. (1995) failed to demonstrate such a link in horses.

Depression in testicular temperature may also be detrimental to spermatozoan production but limited information specific to the horse is available (Johnson, 1991a,b). Sertoli cells seem to be susceptible to depressed temperatures, showing a significant decline in secretory capacity at 32°C compared with 37°C (Hagenas et al., 1978). This is likely to have a corresponding detrimental effect on the germinal cells. Low environmental temperatures may have an additional indirect effect via an increased energy demand for maintenance, with dietary energy being diverted away from reproductive function during times of high maintenance demand (Kemp et al., 1988).

Most of the work carried out to investigate the effect of temperature on spermatogenesis has been done on farm livestock other than the horse. Little direct information on the horse is available but it is likely that the same applies in the horse as that which occurs in species investigated. Circumstantial evidence goes some way to support this extrapolation.

Radiation. As with all tissue, germinal cells are susceptible to radiation (Jainudeen and Hafez, 1993; Johnson *et al.*, 1997), specifically to gamma radiation and X-rays, which affect primarily spermatogonia rather than spermatocytes or spermatids (Oakberg, 1960; Meistrich, 1989). The radiation levels used in many of these studies were far in excess of any that may be found naturally in the environment or during normal management practices. Today, the one area where a valuable stallion may be exposed to radiation is during security checks where levels in the range of 0.5–1.0 µSv are used. England and Keane (1996) indicated that such levels, and even levels as high as 10 µSv, have no detrimental effect on the motility, morphology or longevity of spermatozoa.

Nutrition. Very limited information is available on the specific effect of nutrition on spermatogenesis, especially with specific reference to the horse. In general, it is considered that low and excessive body condition are both detrimental to reproductive efficiency. This is likely to be through an effect on libido: a generally depressed attitude and libido are associated with low or excessive body fat.

Energy deficiency, especially for prolonged periods, has been associated with declining libido and depressed circulating testosterone levels (Jainudeen and Hafez, 1993). It has also been demonstrated that restricted energy diets in boars and rams can reduce the number of spermatozoa produced per ejaculate (Leatham, 1970; Parker and Thwaites, 1972; Kemp and den Hartog, 1989; Murray *et al.*, 1990). No direct evidence of such an effect is available for stallions, though it is likely that there will be a similar effect. Under modern management, it is more likely that excessive dietary intakes, rather than deficiency, will be a problem. The effect of excessive energy intakes is unclear in stallions but is likely to lead to increased testicular insulation, thus elevating testicular temperatures as well as depressing libido due to lethargy and physical weight.

Protein deficiency is reported to have a detrimental effect on reproductive efficiency but not to the same extent as with low energy intakes (Jainudeen and Hafez, 1993).

Deficiencies in vitamins A and E and selenium have been reported to be associated with a reduction in spermatogenesis in other farm animals. Supplementation with either vitamin A or E, above minimum recommended requirements, is reported to have no effect on semen quality in stallions (Rich *et al.*, 1984; Ralston *et al.*, 1986). It has been suggested that there may be a link between iron deficiency and low libido and poor semen quality in bulls. The supplementation of copper, cobalt, zinc or manganese has been reported to be beneficial to semen quality and fertility rates (Jainudeen and Hafez, 1993). Danek and Wisniewski (1992) produced somewhat inconclusive evidence indicating a possible detrimental effect of zinc deficiency on spermatozoan concentration and numbers per ejaculate in stallions. This conclusion is supported by work by Danek *et al.* (1996a) which suggested that dietary calcium excess is associated with depressed seminal plasma zinc concentrations and lower total number, concentration and percentage of viable spermatozoa.

It is apparent that any detrimental effect of nutrition is more evident in young pubertal animals than in older mature animals (Jainudeen and Hafez, 1993; Brown, 1994). The mechanism by which dietary intake influences reproductive function is unclear, but it is increasingly evident that it involves both an indirect and a direct route (Jainudeen and Hafez, 1993), with the indirect route involving nutritional mediation of LH and testosterone release. Brown (1994) proved that low energy intake in rams and bulls resulted in lower LH release. With reference to section 3.3.3, the critical role of LH in the control of spermatogenesis is apparent. The direct route is unclear, but it is evident that high energy diets can have a beneficial effect on Leidig cell size and function, circulating testosterone levels and spermatozoan production in the absence of significant changes in gonadotrophin release (Martin and Walken-Brown, 1995). Again, this work has been carried out in farm livestock other than the horse, though it is likely that similar patterns will be evident in the stallion.

Toxic chemicals and drugs. Like many other factors affecting spermatogenesis, the effects of toxic agents may not be manifest until 60 days after the event. These effects may be directly on spermatozoan production or indirectly on epididymal function or accessory gland secretions. Spermatogonia are susceptible primarily due to their relative proximity to the paternal system. More developmentally advanced germ cells are increasingly isolated from the paternal system, in terms of distance and the presence of Sertoli cells, which protects them from immunological and toxic challenge (Johnson and Thompson, 1983; Johnson, 1991a; Amann, 1993a). At the level of the testes the toxins may act upon the Sertoli cells, Leidig cells or directly upon the germinal cells, and so affect spermatogenesis.

The Sertoli cells are the most likely area to be affected by toxins, reducing their ability to respond to FSH (Chapin and Foster, 1989; Amann, 1993a) or their ability to nurture the developing germinal cells. As a result, those germinal cells with the highest demand (for example, spermatocytes and spermatids) suffer the most and undergo degeneration.

Toxins may also affect the Leidig cells (Kleeman *et al.*, 1990; Klinefelter *et al.*, 1991). Steroids can also act detrimentally on Leidig cell function, as can inhibitors of androgen or oestrogen synthesis, reducing or confusing the hormonal messages received by the Leidig cells and, therefore, their reactions.

Germinal cells, being highly active and dividing, are very susceptible to environmental factors, especially toxins. Drugs used to treat the rapidly dividing cells of tumours, along with fungicides such as methyl 2-benzimidazolecarbomate (carbendazin) and pesticides or fumigants such as dibromochloropropane and ethyl dibromide, adversely affect spermatogenesis (Amann, 1993c).

The major toxin affecting the hypothalamic–pituitary–testes axis is testosterone used as a pharmacological agent to improve, or address, natural hormone deficiencies in the stallion. Administration of testosterone or its artificial analogues is the most commonly used agent in stallions with naturally low circulating testosterone levels. The hormonal control of stallion reproduction is a finely balanced system of positive and negative feedbacks, resulting in a delicate homeostatic balance. Disruption of this balance is an automatic consequence of the administration of exogenous hormones. Administration of androgens to improve stallion libido and muscle development does have the desired effect but also impinges upon the fine balance of the hypothalamic–pituitary–testes axis, acting as an additional negative feedback on the function of the axis similar to that of naturally circulating testosterone. As such, GnRH and subsequently LH and probably FSH levels are depressed, resulting in reduced testosterone production by the Leidig cells, and so the levels of testosterone bathing and acting upon the Sertoli cells are reduced, along with spermatogenesis. In fact, levels of testosterone circulating in the interstitial fluids of the seminiferous tubules in such stallions are reported to be equal to that in the peripheral blood, rather than the normal level of 20–100 times greater (Berndtson *et al.*, 1979; Squires *et al.*, 1981a, 1982; Blanchard *et al.*, 1983; Pickett *et al.*, 1989). Administration of altrenogest (progestagen) to stallions is also reported to have a detrimental effect on reproductive function, resulting in lower libido, reduced total scrotal width, reduced daily output of spermatozoa and percentage of normal spermatozoa, and depressed plasma concentrations of LH and testosterone (Squires *et al.*, 1997).

Anabolic steroids (drugs structurally similar to testosterone and testosterone proprionate) have a similar detrimental effect on spermatogenesis (Johnson *et al.*, 1997; Koskinen *et al.*, 1997). These drugs are in common use, though they are banned substances as far as many equine competition authorities and governing bodies are concerned. They include stanozolol, boldenone undecylenate and nandrolone decanoate and they all have detrimental effects on spermatogenesis similar to those associated with testosterone proprionate (Squires *et al.*, 1982; Blanchard *et al.*, 1983). It is suggested that these detrimental effects may be reversible, but evidence from human research suggests that the use of anabolic steriods may result in up to a 40% decline in testes size (Koskinen *et al.*, 1997).

Administration of human chorionic gonadotrophin (hCG) and GnRH have been used in an attempt to improve reproductive function, by stimulating an increase in endogenous testosterone. Such treatment has met with limited success, due to its interference with the natural balance of the hypothalamic–pituitary–testes axis (Amann and Ganjam, 1981; Zwain *et al.*, 1989; Blue *et al.*, 1991; Boyle *et al.*, 1991; Seamens *et al.*, 1991; Roser and Hughes, 1992; Amann, 1993b; Roser, 1995; Dowsett *et al.*, 1996).

Administration of other drugs may have an indirect effect on spermatogenesis by depression in appetite or the occurrence of diarrhoea. For this reason, the use of anthelmintics is avoided by some practitioners during the height of the breeding season. However, preliminary work suggests that there are no detrimental effects on spermatozoan quality or quantity or testicular size from short-term (4 weeks) treatment with $1\,g\,day^{-1}$ phenylbutazone (bute) (McDonnell *et al.*, 1992).

EXERCISE
The effect of exercise on seminal characteristics and reproductive hormones has been investigated but no effect has been found (Dinger *et al.*, 1986; Lange *et al.*, 1997).

HORMONAL ABNORMALITIES
As would be expected, stallions with low circulating testosterone levels have depressed DSO as well as decreasing libido. As mentioned previously, testosterone, hCG and GnRH therapy have been used with mixed success to address this problem. The lack of success may well be due to the fact that depressed pituitary function is the cause of infertility in only 1% of cases (Boyle *et al.*, 1991; Roger and Hughes, 1991). Abnormal hormone levels may be associated with hypothyroidism, resulting in delayed puberty, smaller testes, decreased spermatozoan production and decreased libido. Feminization of the genitalia may also be observed. It has been postulated that changes in thyroid function may be the cause of stallion summer infertility associated with elevated environmental temperatures (Brachen and Wagner, 1983).

PHYSICAL ABNORMALITIES
The major physical abnormalities that affect spermatogenesis are those that involve the testis and its function. It is beyond the scope of this book to discuss the numerous conditions in detail and so a summary of the major problems will be given. Firstly, the classification of a 'normal' stallion should be considered. It has been reported that up to one in five stallions can be considered to be abnormal in some aspect of the testes. The range is vast, from the minor to the life-threatening condition. In general, testes abnormalities can be divided into congenital (inherited) or acquired (disease and accidental damage).

Cryptorchidism. A cryptorchid stallion, or a rig, is an animal in which either one or both of the testes have failed to descend into the scrotal sack. The passage of the testes from a position just ventral to the kidneys should occur, as a gradual process, *in utero* or during the first few months of life (Fig. 4.24). A cryptorchid stallion may be further classified as illustrated in Figs 4.25–4.27 (Cox, 1993a,b; Davies Morel, 1993).

The failure of testes to descend may be temporary (most will descend within 3 years of birth) or permanent. Evidence suggests that more than 75% of cases involve retention of the right testis (Bishop *et al.*, 1964, 1966). Circumstantial evidence and work reported by Cox (1993a) would suggest that the incidence of cryptorchidism is higher in ponies. The retention of one or both testes results in a significant decline in testicular weight even if it does subsequently descend. The size may be reduced by up to 20-fold in the

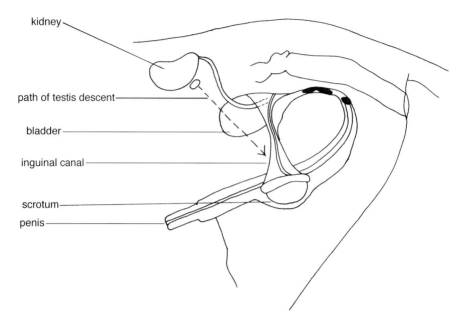

Fig. 4.24. The normal passage of descent of the testes in the stallion.

abdominally retained testis; the reduction in size of the inguinally retained testis is not as great, but a difference of up to sevenfold has been reported (Bishop *et al.*, 1964).

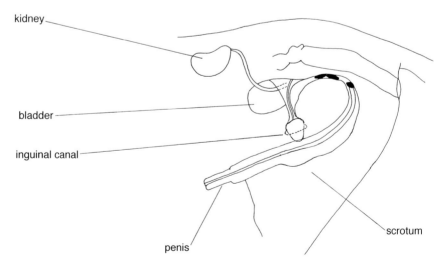

Fig. 4.25. An inguinal cryptorchid stallion is characterized by the testes having only partly descended and remaining associated with the inguinal ring. In a unilateral abdominal cryptorchid, one testis has failed to descend; in a bilateral cryptorchid, both remain in the region of the inguinal ring.

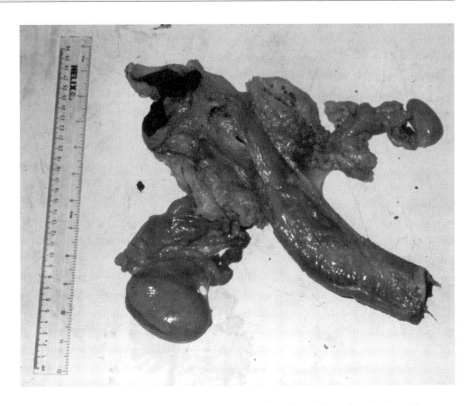

Fig. 4.26. The testes from an inguinal cryptorchid stallion. Note the relative difference in size between the retained testis (upper right) and the normally descended testis (lower left).

The morphological appearance of retained testes varies with their position. The nearer they are to the scrotum, the more morphologically normal they appear. Abdominally retained testes are flabby in nature, with parenchyma consisting of some seminiferous tubules plus interstitial tissue held within a loose connective tissue. Spermatogenic activity is absent and germ cells are rarely developed beyond primary spermatocytes. As the stallion gets older, the seminiferous tubules appear as isolated groups and the proportion of fibrous tissue increases. In general, the testes appear similar to those of a normal 4-month-old foal (Arighi *et al.*, 1987; Cox, 1993a). Testes which have partly descended, though failed in the last descent into the scrotal sack, show greater testicular development, with most of the testis being made up of seminiferous tubules and fewer interstitial cells. Sertoli cells are evident, as are spermatogonia-like cells. Indeed testes lying in the inguinal region may produce viable spermatogonia and so fertilization is possible with spermatozoa from such testes. The testes in such cases resemble those of a 12-month-old foal (Arighi *et al.*, 1987).

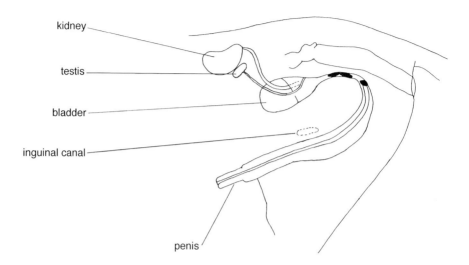

Fig. 4.27. An abdominal cryptorchid stallion is characterized by the testes lying up within the body cavity. Again, failure of descent may be seen in either one (unilateral) or both (bilateral) testes.

Hernias. Stallion hernias may be classified in a number of ways (Figs 4.28 and 4.29). All have the potential to affect spermatozoan production due to an elevation in testicular temperature from the close proximity of the herniated part of the gastrointestinal tract, mesentry or omentum (Cox, 1988).

Testicular hernias may be aquired, usually due to accident or strain (Varner and Schumacher, 1991), or are congenital, due to inherited abnormality (Schneider *et al.*, 1982; Cox, 1993a). They may be indirect or direct, depending upon whether the tunica vaginalis has been ruptured (Ashdown, 1963; Van der Velden, 1988a,b; Cox, 1993a). The most common form of hernia is the indirect, especially in young foals, where large inguinal rings are the prime cause (Wright, 1963). Spontaneous recovery normally occurs within 3–6 months and no long-term detrimental effects have been reported (Varner and Schumacher, 1991).

Testicular hernias may be further classified as inguinal or scrotal, depending on the extent of herniation. Inguinal hernias result from intestinal tissue passing solely through the internal or deep inguinal ring (Fig. 4.28). Scrotal hernias result from further herniation, where the intestine extends beyond the superficial or external inguinal ring (Varner and Schumacher, 1991) (Fig. 4.29).

Apart from the mortal risk of intestinal strangulation, the biggest problem associated with testicular hernias is the effect on testicular function due to elevated testicular temperature from the close proximity of the intestine. As discussed previously, an increase in testicular temperature has a direct effect on function (earlier in this section). If the condition persists, testicular

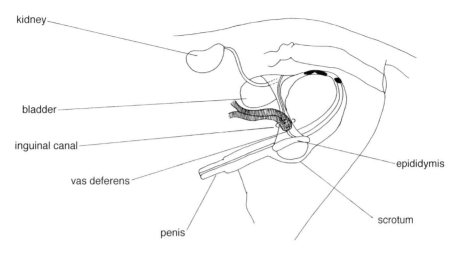

Fig. 4.28. An inguinal hernia in the stallion in which a loop of intestine folds through the inguinal ring.

hypoplasia or atrophy may result, along with the possibility (especially in the case of aquired hernias) of the development of adhesions between the intestine and the overlying tissue and the development of a fibrous sack (Varner and Schumacher, 1991; Cox, 1993a).

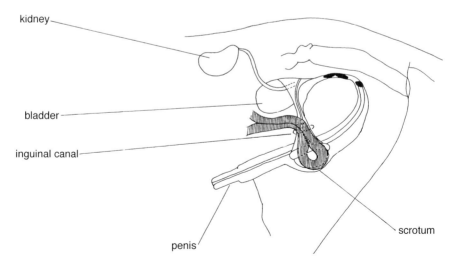

Fig. 4.29. A scrotal hernia in the stallion is a more extreme case of inguinal hernia: a loop of intestine has entered the scrotum and there is significant danger of complete ligation of the intestine.

Testicular hypoplasia or degeneration. Testicular hypoplasia or degeneration is often a secondary condition, resulting from another primary abnormality (often cryptorchidism or herniation). Both hypoplasia and degeneration are terms given to an underdeveloped and, therefore, underfunctioning organ. In general, hypoplasia is the term given to a condition present from birth. In the case of testicular hypoplasia, the testes, for some reason, have never developed beyond an immature stage, resulting in inadequate development of primitive germinal cells in the fetus along with retardation of the normal accelerated germinal cell development at puberty (Roberts, 1986a; Varner and Schumacher, 1991; Varner *et al.*, 1991b). Its causes are many, including cryptorchidism and hernias, but also malnutrition, endocrine malfunction, infections, irradiation or toxins, and it may also be an inherited condition (Roberts, 1986).

The extent of the problem varies considerably. In mild cases, the testes appear normal (though possibly slightly small) during external examination; however, on examination of the testicular tissue it is apparent that most seminiferous tubules exhibit spermatogenesis up to primary spermatocytes, but interspersed between these are completely hypoplastic tubules. In more severe cases, testes are significantly smaller than normal and, if the condition is advanced, the testes may have become hard due to the overdevelopment of connective tissue; on examination of the tissue it is apparent that most of the seminiferous tubules are hypoplastic (Ladd, 1985). Spermatozoan production depends on the severity of the condition, varying from slight impairment to azoospermic samples. Any spermatozoa that are ejaculated have a higher incidence of abnormalities. In such cases the libido of the stallion is often not affected (Varner and Schumacher, 1991; Varner *et al.*, 1991b).

Testicular degeneration refers to the condition of underdeveloped testes which has developed post birth. That is, testicular development did originally occur to some extent but some subsequent problem has resulted in a degeneration of the tissue. The germinal epithelium of the testis is highly sensitive to extrinsic factors, as discussed previously. As such, testicular degeneration is a major cause of infertility in the stallion. Unlike hypoplasia, degeneration is an acquired condition, but stallions suffering from hypoplasia do have an increased chance of also suffering from degeneration and reducing still further their chances of producing viable spermatozoa (Varner and Schumacher, 1991). Degeneration may be temporary or permanent, its severity depending upon the severity and duration of the causative agent. It may be unilateral (the cause being localized in origin) or bilateral (a systemic cause) (McEntee, 1970). The condition is also evident as a shrinking of the testes, often showing small epididymides with a reduced number of spermatozoa within (Watson *et al.*, 1994). Spermatozoa counts are depressed (Roberts, 1986a) and a decline in output of spermatozoa is observed with an increase in the percentage of morphologically abnormal spermatozoa (Friedman *et al.*, 1991a; Blanchard and Varner, 1993). Microscopically, degeneration is evident primarily in midgeneration spermatogonia. Undifferentiated germinal cells are relatively resistant, as are mature spermatozoa. Once the causative agent has been removed, the stem cell population provides a pool from which a new generation of spermatogonia can develop (Huckins, 1971, 1978).

The causes of testicular degeneration are many and varied, the prime ones being elevated testicular temperature, hydrocoel, scrotal haemorrhage, increased scrotal insulation due to scrotal oedema, scrotal dermatitis and cryptorchidism (McEntee, 1970; Varner and Schumacher, 1991; Blanchard and Varner, 1993). Minor affectors include hormonal disturbances, radiation, malnutrition, production of antispermatozoa antibodies, toxins, tumours, obstructions of the vas deferens, testicular torsion and old age (Rossdale and Ricketts, 1980; Pickett *et al.*, 1989; Varner and Schmacher, 1991; Varner *et al.*, 1991b).

In most cases, testicular degeneration is reversible, provided that the duration of the problem is limited and the causative condition can be alleviated. Infective and traumatic degeneration is more likely to be permanent (Burns and Douglas, 1985; Blanchard and Varner, 1993).

Testicular torsion. The extent to which the testes of a stallion may twist or turn is variable, as is the resultant effect. Torsion occurs most commonly in younger stallions. The twist may occur through an angle of up to 360°, a condition difficult to detect immediately, as the testes, on cursory examination, would appear to be positioned correctly. More commonly, the angle of twist is 180°, resulting in the epididymis being in a cranial or anterior presentation at palpation. This condition is evident during testes descent (Hurtgen, 1987). A minor torsion may be transient and may present just a little pain and a slight decrease in spermatozoan concentration of ejaculates. Such torsions may correct themselves (Threlfall *et al.*, 1990). Major torsion can result in symptoms similar to orchitis (discussed later in this section), including acute colic pain, scrotal swelling and obstruction of the blood supply, which if present in a chronic case may lead to degeneration (Kenney, 1975; Threlfall *et al.*, 1990). There is some dispute as to the effect of the condition on semen quality. It is evident that if degeneration does result, then semen quality will suffer. However, Hurtgen (1987) stated that the condition *per se* does not affect libido and semen quality.

Testicular neoplasms. Testicular neoplasms are rare in horses, though their exact incidence rate is difficult to ascertain as the majority of stallions are gelded at a young age. Most neoplasms can be divided into germinal neoplasms and non-germinal neoplasms (Caron *et al.*, 1985). Germinal neoplasms include epididymal, dermoid cysts (especially prevalent in cryptorchids), teratomas (Fig. 4.30) and seminomas (malignant germinal cells of seminiferous tubules). Non-germinal neoplasms include Sertoli and Leidig cell tumours (Bostock and Owen, 1975; Morse and Whitmore, 1986; Schumacher and Varner, 1993). Such conditions normally occur in older stallions and, in the case of seminomas, are more prevalent in cryptorchid stallions (Vaillencourt *et al.*, 1979). Often the condition is not associated with pain or elevated temperatures but a firm swelling may be felt in the testicular tissue and the affected testis may be enlarged. Neoplasms are causes of testicular degeneration and, therefore, associated with depressed spermatozoan counts and high incidence of morphological abnormalities (Hurtgen, 1987).

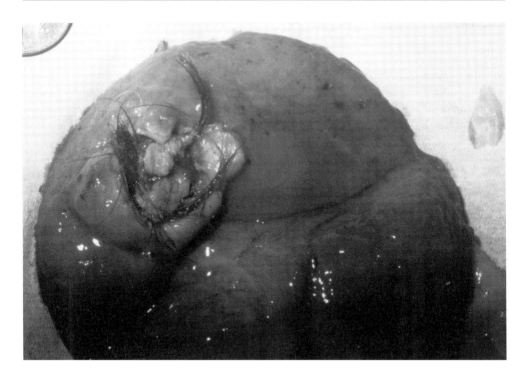

Fig. 4.30. A teratoma, containing mainly hairs, found within the testis of a stallion post mortem.

IMMUNOLOGICAL INFERTILITY

Semen contains many antigens, including those in the seminal plasma and those that are spermatozoan bound. Spermatozoon antigens include spermatozoon-specific antigens, histocompatability antigens, blood-group antigens and other somatic tissue antigens. These antigens may be antigenic within the male system (autoantigenic) or within the female system (isoantigenic).

The autoantigenic effect is prevented from causing an autoimmune response by the Sertoli cells which form the blood–testis barrier within the testes. If this barrier is breached or incompetent then an autoimmune response will result, along with the destruction of spermatozoa (Jones, 1980; Wright, 1980). Such antibodies have been isolated in bull semen, but do not seem to have a direct correlation with breeding ability (Purswell *et al.*, 1983). Such autoimmunity has been induced experimentally in stallions immunized with their own spermatozoa. It has, therefore, been suggested as a cause of idiopathic subfertility in stallions (Teuscher *et al.*, 1994).

TESTICULAR DISEASE OR INFECTION

Infection of the testes may have a systemic or localized cause. Either will result in elevated testicular temperature and associated decline in spermatogenesis. The magnitude of the decline in fertility rates and the time period reflects the severity of the disease and the duration of the problem. Recovery, when it occurs, will be somewhat delayed, as the prime site of effect, as far as spermatogenesis is concerned, is the germinal cells. The spermatogenic cycle being 56 days, this period of time must be allowed after recovery for semen quality to return to normal (Varner *et al.*, 1991b).

Systemic disease, causing orchitis, normally results in bilateral inflammation of the testes and epididymis. Bacterial agents causing such orchitis include *Streptococcus equi* (strangles), *Streptococcus zooepidemicus*, *Klebsiella pneumoniae*, *Actinomyces bovis*, *Pseudomonas mallei* (glanders) and possibly *Salmonella abortus equi*. Viral agents may also be a systemic cause of orchitis: these include equine viral arteritis, equine infectious anaemia and equine influenza (Rossdale and Ricketts, 1980; Ladd, 1985; Roberts, 1986a; De Vries, 1993; Slusher, 1997). Systemic infections cause chronic rather than acute orchitis and as such have more of a chance of causing low-grade testicular degeneration and, with it, permanently depressed semen quality.

Localized infections may be caused via a wound, often to the scrotum, but may also be caused by infection via the inguinal canal (Varner and Schumacher, 1991). Such infections tend to cause acute orchitis (De Vries, 1993) which may be unilateral or bilateral and presents itself initially as soft, flabby, swollen testes. If the condition persists, chronic orchitis may result. Semen quality will be poor, with a decline in spermatozoan concentrations and an increased incidence of abnormalities (Hurtgen, 1987). The major infective agents associated with localized orchitis are *Staphylococcus* species, *Escherichia coli*, *Streptococcus zooepidemicus* and *Streptococcus equi*. In cases of acute orchitis, rises in testicular temperature are also a potential hazard (Blanchard and Varner, 1993).

Orchitis may also be caused by trauma or parasites. The parasite most often associated is *Strongylus edentatus* larvae (Smith, 1973). These can migrate into the testicular tissue, causing orchitis or obstruction of the testicular artery within the pampiniform plexus. This will have an additional detrimental effect upon the efficiency of the counter-current heat exchange mechanism (Roberts, 1986a; Varner *et al.*, 1993).

Finally, orchitis may be caused as a result of damage to Sertoli cells and hence the blood–testes barrier. This will result in an autoimmune response set up by the stallion's system (Zhang *et al.*, 1990b; Papa *et al.*, 1990).

4.3.3. Metabolism and capacitation

Having investigated the structure, development and maturation of spermatozoa, it is now necessary to consider their metabolism, function and movement. Detailed understanding of these areas should allow the development of appropriate conditions for the storage and transfer of spermatozoa with minimal interruption to their normal function and fertilizing capacity.

Metabolism

Testicular spermatozoa differ in several ways from spermatozoa found within the cauda epididymis (that is, post maturation). Many of these differences are related to changes in metabolism and/or biochemical reactions within the spermatozoa. Testicular spermatozoa are largely immotile, though they may show limited motility after incubation *in vitro*. Testicular spermatozoa are also incapable of fertilization (Setchell *et al.*, 1969; Voglmayr, 1975; Voglmayr *et al.*, 1978).

Upon examination it can be seen that the composition of testicular spermatozoa differs from that of caudal epididymal and ejaculated spermatozoa, especially in relation to their lipid content. Testicular spermatozoa have higher concentrations of phospholipids (Setchell, 1991). These phospholipids have been identified in bull and ram spermatozoa as phosphatidylcholine (63%), phosphatidylethanolamine (15%) and lysophosphatidylcholine (10%) (Parks *et al.*, 1987). In the case of testicular spermatozoa, the predominant fatty acid within the phospholipids is palmitic acid (16C saturated fatty acid), whereas phospholipids in caudal epididymal spermatozoa contain predominantly docosahexanoic acid (22C unsaturated fatty acid with six double bonds) (Parks and Hammerstedt, 1985). A significant decline in the cholesterol content of the membranes of cauda epididymal spermatozoa and testicular spermatozoa has also been reported. It has been suggested that this change may be necessary for spermatozoon transit and storage within the epididymus (Lopez and Souza, 1991).

If testicular spermatozoa are incubated *in vitro* in a phosphate-buffered saline resembling serum, they metabolize a high percentage of available glucose to carbon dioxide, amino and carboxylic acids plus inositol. On the other hand, if cauda epididymal spermatozoa are similarly treated they convert most of the glucose to lactate (Setchell *et al.*, 1969; Voglmayr, 1975). It is apparent that some significant changes occur in the metabolic and biochemical reactions within the spermatozoa between their release from testicular tissue and their presence in the caudal epididymis; that is, during maturation. Spermatozoa emerging from the cauda epididymis have, therefore, already undergone and completed the vast majority of their metabolic changes. The components required for spermatozoan function and metabolism are in place, having been developed during spermatogenesis. The metabolic processes within mature spermatozoa are, however, limited to those providing energy for movement, initiation of glycolysis and other catabolic processes, the maintenance of ionic balance and cell function. Their ability to divide and repair cell damage has been lost (Hiipakka and Hammerstedt, 1978; Amann and Graham, 1993).

For subsequent activity the prime energy source available to spermatozoa are carbohydrates from the extracellular substrate. Stallion spermatozoa readily metabolize monosaccharide glucose but have a limited ability to utilize fructose. They also have a limited and variable ability to utilize other sugars or more complex carbohydrates as energy sources (Mann, 1964a). In contrast to stallion spermatozoa, bull and ram spermatozoa are able to utilize fructose and sorbitol as energy sources (Mann, 1964a), and will rapidly use all the monosaccharide energy sources available in seminal plasma within 15–20 min at 37°C. This does not occur so rapidly with stallion spermatozoa (Amann and Graham, 1993).

As can be seen in Fig. 4.31, the use of extracellular glucose results in the production of two ATP molecules via anaerobic metabolism and 36 by aerobic metabolism, making 38 ATP molecules available for the maintenance and movement of the spermatozoon. In order for the energy sources to be utilized, they have to be transported into the spermatozoon across the plasma

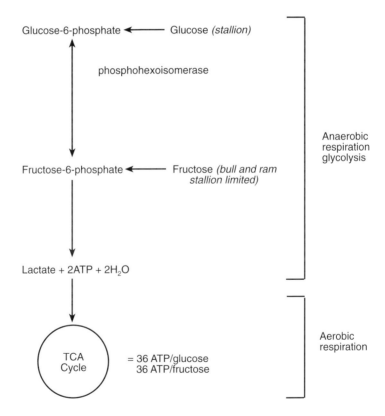

Fig. 4.31. Summary diagram of the use of extracellular glucose in the stallion spermatozoon. Spermatozoa produced by the bull and the ram can also use fructose as a major energy source, fructose entering at the level below glucose. The stallion spermatozoon has only a limited capability to do this. Both glucose and fructose yield identical amounts of energy: 38 ATP molecules (2 ATP from anaerobic respiration and 36 ATP from aerobic respiration). (Mann, 1964a; Amann and Graham, 1993.)

membrane. This is achieved by means of carrier proteins. Little specific information is available on such processes in stallion spermatozoa (Fig. 4.17), but it is apparent from bull spermatozoa that the transport of extracellular glucose across the plasma membrane is a rate-limiting step in the metabolism of sugars (Hiipakka and Hammerstedt, 1978). The process is also ATP dependent and so can be blocked by inhibitors such as cytochalasin B (Peterson *et al.*, 1977). It is not, however, linked to sodium or potassium ion transportation (Hiipakka and Hammerstedt, 1978). Both glucose and fructose carrier proteins have been identified in the bull spermatozoon, each type of carrier protein being largely specific for its own monosaccharide. Evidence exists that glucose carrier proteins do have a very limited ability to transport fructose molecules (Hiipakka and Hammerstedt, 1978). As stallion spermatozoa are largely unable to utilize fructose as an energy source, it is likely that they only possess glucose carrier proteins and that any fructose that is used is due to the limited ability of glucose carrier proteins to transport fructose.

The use of exogenous substrates accounts for 90% of the cell's energy requirements. The remaining 10% is obtained from intracellular sources (Amann and Graham, 1993). Endogenous substances used by spermatozoa include phospholipids (Mann, 1964a,b). Glycogen is not stored within the spermatozoon and so does not provide an endogenous source of energy; spermatozoa lack the appropriate enzymes for its use (Inskeep and Hammerstedt, 1985).

Work with bull spermatozoa indicates that 60% of the total ATP produced is used for motility and 40% for repeated phosphorylation and dephosphorylation of glycolytic intermediates through substrate cycling. Very little ATP is used in transmembrane ion movement and the maintenance of transmembrane ion pumps and other cellular processes. The production rate of endogenous ATP is relatively constant, regardless of the temperature of the surrounding medium. The production of ATP from exogenous sources and hence the total ATP produced is, however, affected by the medium and temperature (Amann and Graham, 1993).

The extent to which aerobic and anaerobic respiration is used to produce energy in spermatozoa varies considerably between species. Bull, ram and human spermatozoa apparently have a much higher dependency upon anaerobic metabolism and a limited ability or need to use aerobic respiration. The stallion spermatozoon is the converse, relying heavily upon aerobic respiration as a source of ATP (Peterson, 1982; Setchell, 1982).

Aerobic respiration produces significant quantities of hydrogen peroxide within the mitochondria. Hydrogen peroxide has an adverse effect on the permeability of the plasma membranes via lipid peroxidation, which in turn disrupts the function of enzymes. This effect on the plasma membrane reduces its structural integrity and the motility and viability of the spermatozoon (Alvarez and Storey, 1984 a,b). This is potentially a significant problem in stallion spermatozoa that rely heavily on aerobic metabolism. However, seminal plasma antioxidants act to reduce the deleterious effect of hydrogen peroxide, by buffering (Amann and Graham, 1993).

This reliance of stallion spermatozoa on aerobic respiration also causes potential problems in spermatozoon storage. Once the oxygen within the seminal plasma, or any semen extender used, has been depleted, spermatozoa rely solely upon anaerobic metabolism for the production of ATP, the end-product of which is lactic acid. As stallion spermatozoa are not able to use anaerobic metabolism at the same rate as bull and ram spermatozoa, they potentially suffer more as a result. This is of particular concern in slow cooling of spermatozoa from body temperature to 5°C for cool storage (Mann, 1964a,b). The build-up of lactic acid also has a detrimental effect through a significant drop in pH. A reduction in pH may decrease metabolism, leading to a drop in ATP production and so a decline in spermatozoan motility. This has been demonstrated in rabbit spermatozoa, which die rapidly at pH levels of less that 5.8 (Mann, 1964a,b). The maintenance of extracellular pH at 6.2–7.8 in the stallion is very important for spermatozoan viability. This maintenance of pH is affected by the many components added as semen extenders and also by temperature, especially freezing, which increases ionic strength as water in the solution crystallizes, so lowering hydrogen ion coefficients and hence increasing the pH (Amann and Graham, 1993).

In addition to biochemical metabolism, other biochemical changes are needed within the spermatozoon. Freshly ejaculated spermatozoa are not able to fertilize an ovum. Biochemical changes, induced naturally by secretions of the mare's genital tract, are required before fertilizing ability is conferred. Two major modifications have to be achieved before fertilization can take place. The first of these is the acrosome reaction – the fusion of the plasma membrane with the outer acrosomal membrane. The second is hyperactivation of the spermatozoa, as a result of which their motility is changed from progressive to non-progressive oscillatory in fashion, the heads of the spermatozoa whipping from side to side in a figure-of-eight configuration that aids passage through the investments of the ovum (Yanagimachi, 1990). These two changes form part of the process termed capacitation.

Capacitation

Prior to ejaculation the stallion spermatozoon acquires (from the epididymis) a glycoprotein coating. This is then added to in the form of proteins by the seminal plasma after ejaculation. These proteins protect the spermatozoon during its journey through the mare's reproductive tract, maintaining membrane integrity and fertilizing ability. Prior to fertilization this protection needs to be removed: capacitation involves the removal of this protective layer, allowing exposure of receptor sites, alteration in transmembrane ion movement and freedom of the flagellum to allow hyperactivity. Capacitation may occur anywhere within the mare's uterus, though more importantly it is thought to occur specifically in the isthmus region of the fallopian tubes, resulting in the release of spermatozoa from the oviduct epithelial cells, making them available for fertilization (Thomas *et al.*, 1995a). Capacitation is reported to take 6 h in the mare's uterus (Baranska and Tischner, 1995).

Changes occur over the whole surface of the spermatozoon, facilitating the binding of its head to the zona pellucida of the ovum, the acrosome reaction, and the passage of extracellular monosaccharides into the mid-piece, and hence energy output by mitochondria and hyperactivity of the tail region (Yanagimachi, 1990; Amann and Graham, 1993).

In addition to changes in the protective covering of the spermatozoon, changes in the plasma membrane are also apparent, including the efflux of cholesterol from the plasma membrane (Langlais and Roberts, 1985). This cholesterol is passed out into the mare's fallopian tube via specific proteins (Amann and Graham, 1993). As cholesterol has an integral part to play in stabilizing membranes, its efflux has a detrimental effect on the stability of the membranes from which it is removed. As a result, the bilayer of the membrane becomes reorganized, enzymes within become exposed and the receptor characteristics are altered. This destabilization also affects the lipids within the membrane, altering its permeability to ions, particularly calcium, and so the membranes fuse. The acrosome reaction occurs as a result of an increase in the concentration of calcium ions, which induces vesiculation of the plasma membrane. Hence cholesterol and its resultant destabilization of the plasma membrane has an effect on induction of the acrosome reaction.

The acrosome reaction results in the release of acrosin enzyme from the vesicles, which is required for fusion with the ovum plasma membrane, plus the exposure of ovum-binding sites on the inner acrosome membrane. Successful completion of the acrosome reaction can be determined by indirect immunofluorescence, using a monoclonal antibody which recognizes an integral acrosomal membrane. Lack of immunofluorescence indicates an incomplete acrosomal membrane, i.e. one that has undergone the acrosome reaction (Zhang *et al.*, 1990a).

All these changes take on average 18–20 h in stallion spermatozoa incubated *in vitro* (Brackett *et al.*, 1982). The exact time *in vivo* is unclear. The importance of the relative timing of events is evident: a premature acrosome reaction would mean a waste of acrosin, as its area of need is in the vicinity of the ovum. Similarly, premature hyperactivation of the flagellum would reduce the chances of enough energy being left to allow the spermatozoon to penetrate the zona pellucida and initiate fertilization.

The need for capacitation would on the face of it pose a potential problem when considering AI. However, contact with the secretions of the mare's tract is ensured after insemination. The need for capacitation does pose an extra hurdle for *in vitro* fertilization and other such techniques which bypass direct contact between the spermatozoa and the mare's tract prior to fertilization. With such techniques, capacitation has to be induced *in vitro*. Several capacitation activators have been investigated, but the use of uterine tubal epithelial cells or extracts is the most successful. Capacitation like membrane changes, may also be achieved using calcium ionophore A23187 alone (Ellington *et al.*, 1993; Slonina *et al.*, 1995; Christensen *et al.*, 1996; Hochi *et al.*, 1996), or with caffeine (Blue *et al.*, 1989; Campo *et al.*, 1990), cysophosphatidylserine or liposomes of phosphatidylcholine (Blue *et al.*, 1989) or heparin or hypotaurine. Considerable variation exists in the ability of spermatozoa to penetrate horse oocytes after induced capacitation treatment (Zhang et al., 1991). Work by Samper et al. (1989) suggested that the usual requirements for calcium ions for capacitation can be substituted for by elevating the pH.

Capacitation also involves the hyperactivation of spermatozoan tails. This is accompanied by a higher energy or ATP demand. Indeed the metabolism of spermatozoa significantly increases prior to hyperactivated motion; an increase of as much as fourfold has been reported in rabbits (Rogers and Brentwood, 1982). The hyperactivation of the tails is likely to be essential to ensure that the passage of the spermatozoa is not impeded by large epithelial cells in the vicinity of the ovum.

4.3.4. Function of the spermatozoon

The prime function of a spermatozoon is to pass on its genetic make-up to subsequent generations by fertilizing an ovum. The process of fertilization has been alluded to throughout the text, but detailed consideration of this area is beyond the scope of this book. However, it is apparent that the spermatozoon's success in achieving fertilization depends upon the full and efficient functioning of its component parts at the appropriate times, during its passage through the mare's reproductive tract and on to penetration of the ovum. The function of the spermatozoon's three component parts are distinct: the head as conveyer of genetic material; the mid-piece as provider of ATP for power and metabolic activity; and the principal piece for propulsion by means of its contractile microfilaments.

The various component parts are characterized by differences in their membranes. For example, the ratio of cholesterol to phospholipid in the membrane of the acrosome region is lower than that for other membranes. Similarly, the arrangement and concentration of membrane proteins varies (Parks et al., 1987). The efficiency of these component parts in carrying out their functions is reflected in the integrity of their membranes. Therefore, the analysis of specific components will indicate their functional efficiency. Techniques such as staining (Cross and Miezel, 1989) and the use of monoclonal antibodies (Feuchter et al., 1981; Chakraborty et al., 1985; Blach et al., 1988; Amann and Graham, 1993) can be used to assess the acrosome membrane. Fluorescent stains for DNA (Pinkel et al., 1985) can be used to assess the genetic component of the head region. Staining, using eosin–nigrosin stain, can be used to assess the integrity of the whole plasma membrane (Mayer et al., 1951; Swanson and Bearden, 1951). The hypo-osmotic swelling test can be used to assess the principal piece plasma membrane, as can filtration using Sephadex or glass wool (Jeyendson et al., 1984; Samper et al., 1991). Multiple assays can be used quite successfully to assess spermatozoon viability and efficiency via assessment of the membranes or its component parts.

4.3.5. Movement of the spermatozoon

Movement of the spermatozoon is achieved by the regular beating of the flagellum (tail or principal piece), thus propelling the spermatozoon through its surroundings. Its movement is tempered by its medium: the greater the viscosity, the greater the resistance to movement and the more energy is required to achieve forward propulsion (Rikmenspoel, 1984). As secretions vary within the mare's tract, spermatozoa are affected by their position. This is especially evident when in contact with oviductal cells (Thomas et al., 1994a,b). The mechanisms for the movement of the principal piece show many similarities to the contraction of skeletal muscle (Kimball, 1983).

As discussed previously (section 4.3.1), the principal piece of the spermatozoon consists of nine doublet microtubules surrounding a central pair of microtubules. Movement of the principal piece is achieved by shear forces generated between neighbouring doublets, thus making one doublet slide against another over a short distance. Each microtubule consists of repeating units of tubulin protein. One of the main tubulin proteins is dynein, which forms a major arm permanently attached to microtubule A and with the ability to attach to microtubule B of the neighbouring doublet (Fig. 4.13) (Satir, 1974, 1984; Kimball, 1983; Bedford and Hoskins, 1990). These dynein bridges, like myosin in skeletal muscle, are the ATPase. Thus, in the presence of ATP, the dynein bridges detach from neighbouring microtubule B and shorten (Fig. 4.32).

This free and shortened dynein bridge tilts at an angle of about 40° and elongates, then re-attaches to microtubule B at a new site lower down. The bridge contracts, sliding microtubule B along microtubule A until the bridge is parallel. This final stage is a return to the resting state and is, therefore, the only one that does not require the presence of ATP. By this means microtubule B slides progressively along microtubule A towards the axenome.

This movement of the microtubule B against A does not in itself explain the bending action that is characteristic of spermatozoa. This is achieved by alternate contractions of doublets 4–8 with doublets 9–3, providing (for example) a shortening or contraction on the concave side (doublets 9–3) that causes bending of the principal piece concurrent with a relaxation on the convex side (doublets 4–8). A reversal in the doublets contracting and relaxing would give a bend in the principal piece in the opposite direction, and hence give the regular beating from side to side that is characteristic of flagellum movement. This is explained further in Fig. 4.33.

As detailed in section 4.3.1, the major microtubules (A) continue further down into the top of the end-piece than do the more minor microtubules (B). If a cross-section is taken near the tip of the end-piece in a straight spermatozoon, the typical 9-and-2 doublet pattern can be seen (Fig. 4.33). If a similar cross-section was to be taken from a bent spermatozoon, the B microtubules of the doublets on the upper convex side (9, 1, 2 and 3 in Fig. 4.33) would not be seen and only those on the lower concave side (4–8) would be apparent, the arms in the convex side having slid down the A microtubules as a result of the contraction process previously described. The ones on the concave side are

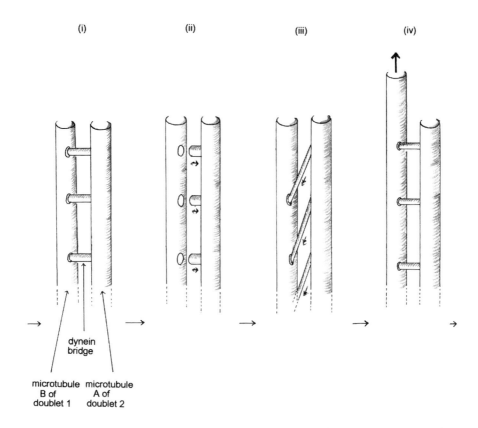

Fig. 4.32. The sliding of the doublets within the axenome of the spermatozoan midpiece. The detachment and attachment of the dynein bridge allow the sliding of one doublet relative to the other. The addition of ATP (i) causes shortening of the dynein bridges and their detachment from the neighbouring B microtubule (ii). The dynein bridges tilt to an angle of 40° and reattach to points lower down the B microtubule (iii). The dynein bridges then shorten and force the microtubule B to slide relative to microtubule A (iv). (Satir, 1984.)

apparent as they slide up the A microtubules. Conversely, if a cross-section was taken from a spermatozoon bent in the other direction there would be a swap-over of the B microtubules visible (9, 1, 2 and 3 in Fig. 4.33), the B microtubules on doublets 4–8 not being apparent. The above hypothesis is verified by electron microscopic examination of spermatozoa and cilia (Satir, 1974; Kimball, 1983). The regular rhythmic side-to-side beating of the spermatozoon's principal piece can be explained by this hypothesis. However, spermatozoa also show helical beating by the principal piece as they rotate or roll. This could be accounted for by the sequential contraction of (for example) doublets 9 to 3, rather than the synchronized contraction of doublets 3–6 followed by the synchronized contraction of doublets 4–8.

As discussed, movement of the spermatozoon principal piece is primarily

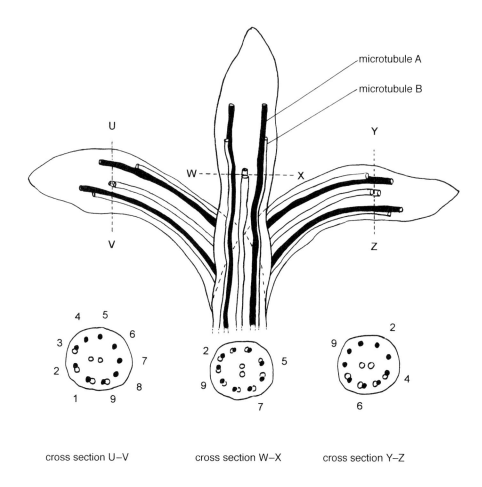

Fig. 4.33. The spermatozoan principal piece. Longitudinal and transverse sections are given to illustrate the sliding microtubule movements which result in the characteristic wave-like movement of a spermatozoon's flagellum. Bending of the flagellum should force the filaments on the concave side to extend the furthest towards the tip. A cross-section taken a little way back from the tip should reveal one set of microtubules (numbers 4–8) during one stroke and the opposite set (numbers 9–3) during the opposite stroke. (Satir, 1974, 1984.)

driven by the availability of ATP, originating from the mid-piece. However, there is also strong evidence that, as with skeletal muscle fibres, calcium ions play a major role (Inskeep and Hammerstedt, 1985; Kimball, 1983) and that adenosine, bicarbonate and the protein carboxymethylation enzyme system are also involved. Details of these systems regulating and controlling movement of spermatozoa have not yet been fully investigated and are unclear.

The pattern of movement, as indicated above, does vary. The advent of

Table 4.12. Characterization of spermatozoan movement is now possible due to computer-assisted image analysis: the movement may be described by several means, as shown (average values for some of these parameters for spermatozoa diluted in 10% sucrose 3% bovine serum albumin at 37°C are also given) (Amann and Graham, 1993).

Characteristic	Description	Value at 37°C
Total motile sperm	(% ≥ 5 μm s^{-1})	68
Motile sperm	(% ≥ 20 μm s^{-1})	67
Progressively motile sperm	(%)	58
Curvilinear velocity (VCL) (μm s^{-1})	The centroid-to-centroid path, and velocity along that path	114
Average path velocity (VAP) (μm s^{-1})	The smoothed centroid-to-centroid path as calculated by one of several smoothing algorithms, and velocity along that path	69
Straight line velocity (VSL) (μm s^{-1})	The linear line between the first and last centroid, and velocity along that path	51
Linearity	VSL/VCL	47
Lateral head displacement	Mean distance of the centroids from the average path	8.4

computer-assisted image analysis has allowed close investigation of these patterns. Details of the equipment used are given in Chapter 6. The use of this analysis allows the movement of spermatozoa to be described using the terminology given in Table 4.12.

The movement of spermatozoa may be interspersed with periods of rest or immobility. Therefore, when viewing a sample, it should be remembered that immotile spermatozoa are not necessarily dead, but may be in a period of rest. Spermatozoa also swim at varying velocities and normally in a progressive manner, which might involve rotational movement along the long axis – that is, helical movement. They also swim in a curvilinear fashion (Figure 4.34) or demonstrate circular motility.

Average values for the various aspects of spermatozoon motion are given in Table 4.12. These are significantly affected by temperature, the viscosity of the medium in which the spermatozoa are suspended and the depth of the preparation (Amann and Hammerstedt, 1980; Amann *et al.*, 1988).

Assessment of motility and its relevance to the viability of spermatozoa is discussed in detail in Chapter 6.

4.4. Conclusion

In conclusion, it is apparent that the current state of knowledge concerning the anatomy and physiology of equine spermatozoa is limited. Much of the information published is based upon assumptions extrapolated from work on other mammalian spermatozoa. While it is highly likely that the main characteristics and functional components of equine spermatozoa are similar to

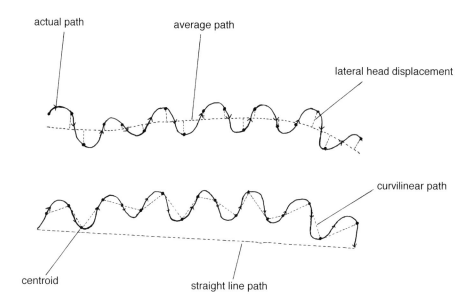

Fig. 4.34. The characterization of spermatozoon movement, now possible due to computer-assisted image analysis. The pattern or trace of the centroid (mid point) of a spermatozoon head is given. (Tessler and Oldsclarke, 1985; Amann and Graham, 1993.)

those of other mammalian spermatozoa, new knowledge increasingly suggests that equine spermatozoa do differ in some important aspects. It is also apparent that a whole host of factors can affect the production of spermatozoa. It is, therefore, very difficult to identify the cause of a poor semen sample, especially as the causes often do not work in isolation and may be only transient in nature. Before many of the hurdles to successful artificial insemination can be overcome, a more complete knowledge of equine spermatozoa is required, especially with regard to their metabolism and response to their environment.

5 Semen Collection

5.1. Introduction

The history of the use and development of artificial insemination has already been considered in Chapter 2, which detailed some of the early methods of semen collection in horses. These methods included the use of sponges, condoms, dismount samples and the early prototype artificial vaginas. This chapter will concentrate on the techniques currently in use for the collection of semen.

5.2. The Artificial Vagina

Semen needs to be collected in an environment which closely mimics that of the mare's vagina, but which eliminates the potentially detrimental effect of natural secretions. The aim is to encourage ejaculation and ensure that the sample is collected under conditions as near as possible to those found with natural service, hence minimizing any collection effect on the sample and on subsequent fertilization rates. These aims may be achieved by using a sterile tube, surrounded by a water jacket that provides a means of temperature control. Ejaculation occurs into the lumen of the tube (which is normally lined with a disposable liner) and is collected into a warmed collecting vessel attached to the distal end of the tube. Figure 5.1 illustrates the major features required of an artificial vagina (AV).

Today there are three major types of AV that satisfy these criteria in slightly different ways. In addition, there are numerous minor variations on the basic themes. The three basic models are the Cambridge, the Nishikawa and the Missouri AVs.

5.2.1. The Cambridge and Colorado models

The Cambridge AV (Fig. 5.2) is essentially similar to the Colorado or Lane model (Lane Manufacturing Co., Denver, Colorado) and the CSU model

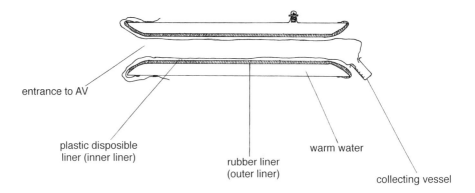

Fig. 5.1. The major features required of an artificial vagina (AV).

(Animal Reproductive Systems, Chiro, California). The Cambridge model is primarily used in the UK and the rest of Europe, whereas the Colorado or Lane model and the CSU AV find greater popularity in the USA.

Fig. 5.2. The Cambridge model of AV, similar to the Colorado or Lane model and the CSU model.

The Cambridge model is constructed of a stiff outer casing (of varying lengths and normally made of plastic, though metal may be used) which is enclosed firstly in an outer rubber lining (Fig. 5.3) to provide protection. This outer lining is tightly clamped to the plastic tube by means of two large jubilee clips attached at either end. Secondly, there is an inner rubber lining (Fig. 5.3), which is slightly longer than the outer lining but smaller in diameter. The inner liner is placed inside the lumen of the AV and stretched over both ends of the casing and outer liner, and also secured by, the two jubilee clips. The inner liner is a tight fit and stretching it is an art. It is advisable to secure one end and then stretch the other end over the casing, trying to ensure that the inner liner is smooth throughout the length of the AV. Any wrinkles in the liner may cause irritation and the AV may therefore be rejected by a fussy stallion, with low libido, especially during the transition period between the breeding and non-breeding season. This second liner and the plastic casing provide the inner and outer parts of the water jacket that enable the internal temperature of the AV to be controlled. It is normal practice to use a third sterile plastic or polyethylene disposable inner liner, which is the one that comes into direct contact with the stallion's penis and into which ejaculation occurs. These liners tend to be much looser in their fitting and do sometimes cause irritation to stallions; in such circumstances, ejaculation directly into the smooth inner rubber liner may be used, omitting the use of the disposable third inner liner.

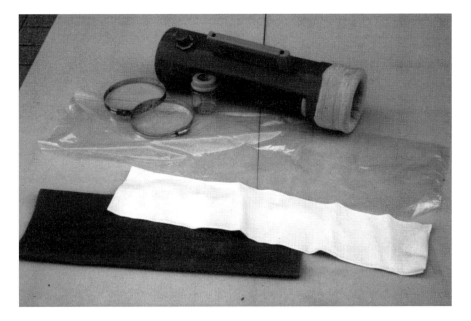

Fig. 5.3. A dismantled Cambridge AV, illustrating the component parts including the outer and inner linings.

The Colorado or Lane model bears a significant resemblance to the Cambridge model, the major differences being that the Colorado model (which is made of plastic) and the Lane model (which is made of aluminium) are designed to be used primarily, but not exclusively, without the use of the additional plastic disposable inner liner. These models also tend to be longer than the Cambridge, which is available in varying lengths for different sized stallions. Some variations on this model include the slight flaring of the AV casing, at the end into which the stallion thrusts. The CSU model is largely a development of the Colorado model, the major modification being a reduction in its total weight when ready for use – a significant advantage over the other two models, which tend to be heavy when filled with water and ready for use. The CSU also has specially designed disposable liners.

All three of these models have the advantage of good heat-retaining capabilities over relatively long periods and they find their greatest popularity in colder climates. Heat retention is conferred by the large volumes of water they contain, but such volumes of water have the major disadvantage of adding considerable weight to the AV when ready for use. A fully assembled Cambridge AV ready for use may weigh up to 10 kg. This weight means that they are heavy to hold and cumbersome to manoeuvre. A further disadvantage is that some variation in the lumen temperature may occur with the movement of the water within the water jacket as the AV is manoeuvred. The tendency is for the AV to be held at a slight angle during collection, with the distal (collecting vessel) end lower, this tendency being further exacerbated by the weight of the AV. As a result the semen tends to be ejaculated into and pass over the warmer end of the AV prior to its collection in the collecting vessel. If the temperature of the water in the AV is too high, there is a greater risk of heat damage to the spermatozoa as they pass over this area on their way to the collecting bottle. These models also have the disadvantage of being a similar size at the distal and proximal ends. This does not give the resistance at the distal end required by some stallions – a requirement that is reported to be particularly evident in Thoroughbreds (J.M. Parlevliet, The Netherlands, 1998, personal communication).

With all these models, the collecting vessel is attached to the tapered end of the inner liner, be it plastic or rubber. The collecting vessel may have a built-in nylon filter, in the shape of a pocket or bag, that fits over the neck of the bottle before it is attached to the inner liner and protrudes into the collecting vessel. A simple piece of sterile gauze may be used in a similar fashion to provide a cheaper alternative to the nylon filters. Either of these filters allows the semen gel fraction, plus any detritus, to be filtered off during the collection process. This eliminates any detrimental effects that such detritus may have on the semen sample, and reduces the time in the laboratory, as removal of the gel fraction is required before evaluation can take place.

The collecting vessel needs to be kept at a constant 38°C in order to protect the spermatozoa from cold shock. This can be achieved by a simple insulating sleeve or cone placed over the collecting bottle and the distal end of the AV. The insulating cone should be pre-heated to 38°C. This can achieved by simply placing it in an incubator, or more precise methods of heating the cone using a rheostat may be used (Pickett, 1993a).

5.2.2. The Nishikawa model

The Nishikawa or Japanese AV (Fujihira Industry Co. Ltd, 11 Hongo 6 Chome, Bunkyo-ku, Tokyo, Japan) was developed in Japan, from where its use has become more widespread; however, its availability is now limited (Fig. 5.4). It consists primarily of a small, rigid aluminium case with a single inner rubber liner and is shorter in length than other types of AV available. As a result it is lightweight and thus easy to manoeuvre. It is also easy to clean and assemble, having only one rubber liner. The collecting vessel is attached directly to the aluminium casing, and as it is short this allows the ejaculate to pass directly into the collecting vessel. This reduces the time that spermatozoa are in contact with the rubber liner and thus the potential detrimental effects of such contact and any contamination risk (Kenney and Varner, 1986). As with the Cambridge AV, the temperature of the collecting vessel may be maintained at 38°C by means of an insulating cone or jacket.

Its major disadvantage is that, due to having only one inner liner, any small pin pricks or faults will result in water from the water jacket passing directly into the lumen of the AV, contaminating the collected sample. There is also a tendency for water to leak from around the valve and also at the ends of the liner if the seals are not tight, again running the risk of water contamination of the sample.

The Nishikawa AV is smaller in diameter as well as length when compared with other models of AV. Although the shorter length has advantages, as semen

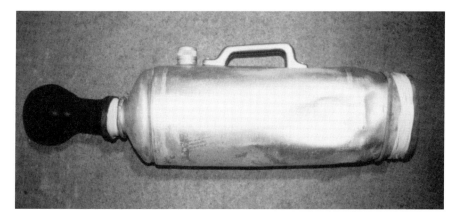

Fig. 5.4. The Nishikawa model of AV. (Photograph by Julie Baumber and Victor Medina.)

is deposited directly into the collecting vessel, the smaller lumen diameter may pose problems for larger stallions. A small lumen does enhance glans penis pressure and, therefore, stimulation; but as the stallion cannot thrust through this AV, stimulation of the proximal portion of the shaft of the penis is reduced. Some models have an additional foam ring, which is placed at the distal end of the AV in order to alleviate this problem and provide additional stimulation to the glans penis (Fig. 5.5). However, the use of a Polish-style open-ended AV (section 5.2.4) demonstrates that for the majority of stallions glans penis stimulation is not required for ejaculation, contrary to previously held beliefs (Tischner *et al.*, 1974; Yates and Whitacre, 1988; S. Revell, Wales, 1998, personal communication).

5.2.3. The Missouri model

The Missouri AV (Nasco, Ft Atkinson, Wisconsin) is the most popular model in use in the USA today and is available in different sizes (Figs 5.6 and 5.7). This model differs from the previous ones in that it is largely made of rubber and has no rigid barrel and so is lightweight and manoeuvrable. The water jacket for the temperature control of the AV is a sealed chamber between the walls of the double-walled rubber liner. The whole unit is then enclosed in a

Fig. 5.5. The Nishikawa model of AV, with foam ring that may be placed at the distal end of the AV in order to provide additional stimulation to the glans penis. (Photograph by Julie Baumber and Victor Medina.)

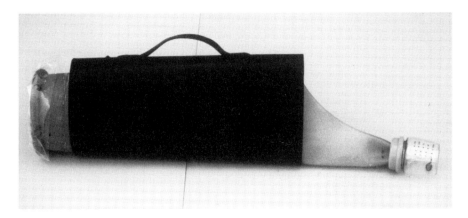

Fig. 5.6. The Missouri AV.

leather carrying case. This model, therefore, has the advantage of a low risk of water contamination. It is also lightweight and easy to manoeuvre, assemble and clean. An air valve is placed in the entrance to the double rubber liner, allowing the use of air or water to fill the chamber. Air under pressure increases the internal pressure of the AV, without adding weight (as occurs when water is used). Some water needs to be used to heat the lumen but the ratio of water to air can be adjusted, depending on the size of the

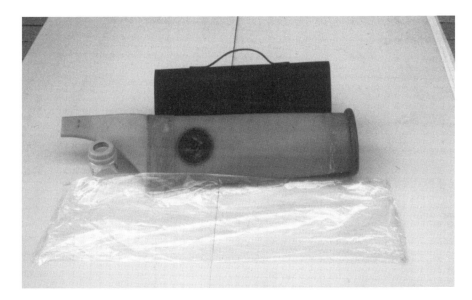

Fig. 5.7. The dismantled Missouri AV, illustrating the component parts and the liner used.

stallion, environmental temperature, etc. In addition, the double liner does not reach to the end of the AV and so ejaculation normally occurs beyond the end of the heated portion of the liner. Inadvertent elevation of the temperature within the water jacket will not, therefore, result in heat-related injury to the spermatozoa collected, though obviously heated-related injury to the stallion may still occur under such conditions.

The Missouri model also has the advantage that the glans penis is accessible at the end of the leather carrying-case through the liner, which at this position is only of single thickness, so that extra stimulation of the glans penis is possible (Yates and Whitacre, 1988). However, as mentioned previously, the necessity for such stimulation is disputed (Tishner *et al.*, 1974; Yates and Whitacre, 1988).

5.2.4. Alternative models

The Hannover AV

The Hannover AV (Figs 5.8 and 5.9) is very similar in design to the Cambridge and CSU models, but lighter in weight and easier to manipulate. The liners used for the CSU model may also be used with the Hannover AV. This AV is particularly popular in The Netherlands and Germany (B.A. Ball, California, 1998, personal communication; J.M. Parlevliet, The Netherlands, 1998, personal communication).

The Roanoke model

The Roanoke AV (Roanoke Laboratories, Roanoke, Virginia) is a much less popular type. It comprises a short plastic casing with a single rubber liner. The assembly of the casing and rubber liner is similar to that seen in the Nishikawa model, hence the Roanoke runs the similar relatively high risk of

Fig. 5.8. The Hannover AV. (Photograph by Julie Baumber and Victor Medina.)

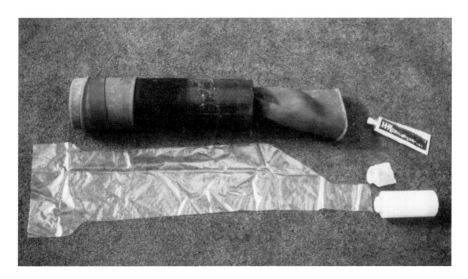

Fig. 5.9. A dismantled Hannover AV, illustrating the component parts and liners. (Photograph by Julie Baumber and Victor Medina.)

water leakage and contamination of the sample collected. As with the Nishikawa, the collecting bottle is attached directly to the outer casing. This provides a light and highly manoeuvrable AV; it is also smaller in size than the Nishikawa and is, therefore, used with small stallions or pony stallions (Yates and Whitacre, 1988).

The open-ended AV

Open-ended AVs, sometimes referred to as the Polish or Cracow AV (Fig. 5.10), are popular in Eastern Europe, especially in Poland, Czechoslovakia and Finland, and are used primarily, but not exclusively, for research, enabling the collection of fractionated ejaculates (Heiskanen *et al.*, 1994b; Langkammer, 1994; Samper, 1995b). The style is similar to that of the Cambridge AV, with a double liner for protection and the formation of a water jacket. However, the open-ended model is shorter, allowing the glans penis to pass out through the distal end of the AV and facilitating the collection of the fractions of ejaculate required. The collection bottle is unattached and semen is collected by placing the bottle near the glans penis to collect the fraction required. It has been suggested that the use of open-ended AVs reduces bacterial contamination of the semen sample collected (Clement *et al.*, 1995). The major disadvantages of this type of AV are the risk of losing the ejaculate, the relative difficulty in filling the jacket with water, and the risk of water leakage (J.M. Parlevliet, The Netherlands, 1998, personal communication).

Fig. 5.10. The open-ended or Polish AV and liner. A plastic glove has been placed through the AV to illustrate the open end. At the bottom of the photograph is a condom, which may also be used for semen collection. (Photograph by Julie Baumber and Victor Medina.)

5.2.5. Preparation and maintenance

All AV components that come into contact with ejaculated semen must be thoroughly clean and disinfected prior to use, including all liners, collecting vessels, glassware, etc. These components must also be non-spermicidal in nature, as must the products used to clean them. Sterile, non-toxic, disposable liners and collecting vessels are readily available for most AVs. These are convenient to use and ensure that there is no transmission of disease between animals and no chemical contamination of samples collected. Many workers favour the use of disposable components.

Cleaning

Rubber liners can be cleaned by rinsing with warm water, preceded by the optional use of a detergent such as ×7 laboratory soap. This is followed either by soaking for 20 min or by just rinsing or swabbing with 70% isopropyl alcohol (J.M. Parlevliet, The Netherlands, 1998, personal communication). Alternatively, sterilization by gas accompanied by appropriate aeration time may be used. Some difficulties have been reported with prolonged immersion of rubber liners in alcohol: the alcohol is reported to be absorbed into the rubber and can be very difficult to rinse off, and a maximum immersion time of 20 min has, therefore, been recommended (Love, 1992). It is of paramount importance that no chemical residue from soaps and disinfectants is left after cleaning, and so the use of the majority of soaps or disinfectants during the washing process is not advised. Some non-residue soaps are available and may be appropriate, but they should be used with care and all items need to be thoroughly rinsed, ideally with warm tap water initially, followed by distilled or deionized water. After cleaning, the rubber liners should be dried in a dust-free environment, such as a dust-free warming cabinet. If the liners are to be stored for any length of time, they should be dusted with talcum powder and kept in a refrigerator or freezer to prevent damage to the rubber from oxidation or dry rot (Love, 1992; J.M. Parlevliet, The Netherlands, 1998, personal communication). For short periods they can be left in a clean, dry, dust-free environment.

Choice of liner material

The choice between rubber, polythene or plastic liners is really one of personal preference and convenience. The use of plastic liners is reported to have no adverse effect on semen quality (Silva *et al.*, 1990); indeed there is some evidence to suggest that the use of plastic liners rather than rubber liners results in a significantly higher percentage of live spermatozoa and percentage progressive motility (Colborn *et al.*, 1990). Similar work by Merilan and Loch (1987) suggested that spermatozoan motility over time (that is, longevity) was better in semen samples collected with an AV with plastic liners compared with rubber ones. The apparent toxic effect of rubber is reported to be greater at higher pH levels and elevated temperatures (Colburn *et al.*, 1990), but other researchers have failed to identify a similar toxic effect (Silva *et al.*, 1990). In addition to a comparison between types of liners, Merlin and Loch (1987) compared two types of AV: the Colorado and the Missouri. No differences in semen quality between the two types were apparent. Some stallions object to the use of plastic AV liners that have a tendency to wrinkle and, therefore, cause discomfort when used. This effect may be minimized by careful insertion of the liner and liberal use of lubricant prior to collection. Non-disposable rubber liners are available as an alternative and their use does not result in the problems of wrinkling associated with plastic liners. However, they are more difficult to clean and, with some AVs, preclude re-use with different stallions in close succession as they have to be dismantled for cleaning.

Assembly

Careful preparation of the AV is essential if a truly representative semen sample is to be obtained. It is of paramount importance that all equipment coming into contact with the semen sample should be non-toxic, non-spermicidal and dry. The AV should be assembled according to the manufacturer's instructions and in a clean and dry environment. Assembly should be done well in advance of the collection as the process can be fiddly and time consuming (especially with the Cambridge AV and similar types). The assembled AV should then remain in a clean, dry environment until imminent use. The water jacket of the assembled AV should be filled with warm to hot water, and time should be allowed for the full mixing and equilibration of the water and equipment temperature. The normal temperature of the lumen of the AV should be 44–48°C, but individual stallions may have a preference for cooler or warmer AVs. This can only be ascertained with experience but lumen temperatures of over 48°C should be avoided, due to the potential heat damage to the spermatozoa and to the stallion (Cooper, 1979; Hillman *et al.*, 1980). This is of particular importance with the Cambridge model as the movement of water within the water jacket often results in the distal end of the AV, over which the semen passes, being at a higher temperature than the entry end of the lumen.

Some damage to spermatozoa is reported to occur at any temperature above 43°C. This presents a dilemma, as for most stallions an AV temperature of 44–48°C provides additional stimulation for ejaculation. The Nishikawa and Missouri models have an advantage as they are shorter, with ejaculation occurring beyond the warm water jacket area, avoiding spermatozoa coming into contact with areas of the AV that are at higher temperatures. However, spermatozoa are equally susceptible to cold shock (Amann and Pickett, 1987; Watson, 1990; Kayser *et al.*, 1992). In this respect the Cambridge AVs have an advantage with their better heat retention properties, and so are more consistent in providing a warm environment for spermatozoa collection in countries with cooler climates. As mentioned previously, an additional feature to help heat retention and reduce the rate of cooling of the collecting vessel is the use of an insulating cone, placed over the distal end of the AV (Fig. 5.11). This device also protects collected semen from damage due to ultraviolet light.

Filters

Some AVs are fitted with in-line filters to remove the gel immediately upon collection (Fig. 5.12) and these are particularly useful with stallions who habitually produce a large volume of gel. Work by Amann *et al.* (1983) indicated the advantage of using nylon filters of pore size 37 µm over those with alternative pore sizes (20, 30, 53 and 74 µm) and over traditional polyester filters as well as stainless steel filters. The use of 37 µm nylon filters significantly reduced the number of spermatozoa that were held within the filter itself and also within the gel retained by the filter. It has been reported that up to 30% of ejaculated spermatozoa are lost in the collecting equipment, of which 47% are held within the conventional polyester filters, 30% within the gel, 13% by adhesion to the rubber AV liner and 10% in the collecting vessel (Pickett *et al.*, 1974b).

Water jacket

The temperature of the water used to fill the water jacket should depend upon the type of AV, the environmental conditions and the immediacy of

Fig. 5.11. An insulating cone to help heat retention of the AV and prevent the cooling of the collection vessel. Such a cone will also protect collected semen from damage due to ultraviolet light.

collection. Whatever the situation, consistent use of very hot water should be avoided as this may damage the rubber of the liners. In general, for immediate use, the Missouri and the Nishikawa AVs can be filled with water at 45–50°C as they are lighter and have only a single rubber liner to be warmed (Fig. 5.13). The Cambridge model also has a single liner but, due to its larger size, it requires temperatures of up to 65°C in order to heat it thoroughly prior to use. Reusable inner rubber liners as seen in the Colorado

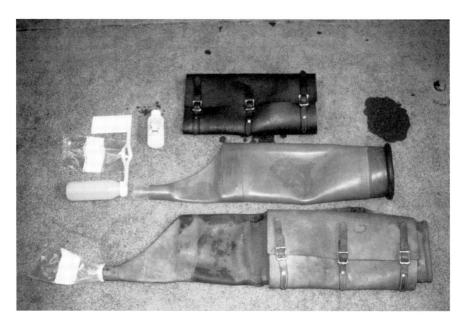

Fig. 5.12. Two different sized Missouri AVs. To the left of the smaller one is a collection bottle with filter that may be used to filter semen directly on collection. (Photograph by Julie Baumber and Victor Medina.)

model, rather than disposable plastic liners, require higher water temperatures of 50–55°C in order to ensure that both inner liners are thoroughly warmed prior to use. The temperature of water required clearly depends upon the heat retention properties of the AV, which in turn are largely related to the volume of water used. With practice, a fine balance between temperature and volume can be achieved, to ensure the correct lumen temperature and that the AV is not too heavy for convenient use.

The volume of water used also has a significant bearing on the pressure within the AV. This internal pressure is quite critical in stimulating ejaculation. It is now believed that stimulating pressure on the glans penis is not normally required for successful ejaculation (Tischner *et al.*, 1974; Yates and Whitacre, 1988); however, stimulation of the base of the penis is required and so must be provided by the AV. The pressure within the AV should vary with the size of stallion in order to ensure that the lumen is the correct size to give adequate stimulation. Excessive pressure not only makes it difficult for the stallion to penetrate the AV; it may also inhibit full erection and hence ejaculation. Excessive volumes of water within an AV also make it particularly heavy to hold (Varner, 1986). When a particular AV is in regular use with a specific stallion, once optimum pressure has been determined then the volume of water used can be measured and this used as a future guide. Alternatively, the filled AV can be weighed so that it can be refilled with a

Fig. 5.13. Filling a Missouri AV with warm water (45–50°C) prior to immediate use. (Photograph by Julie Baumber and Victor Medina.)

similar quantity of water on successive occasions. In models such as the Missouri, air may be used to increase the pressure within the lumen. The use of air in this way has the advantage of having a minimal effect on the weight of the AV. The Missouri, as with other short AVs, allows access to the penis and hence the application of extra pressure.

Lubrication

Immediately prior to use, the internal surface of the AV must be lubricated. This is an attempt to mimic the natural vaginal environment as well as reducing the risk of penile friction damage. The lubricant should be based on water rather than petroleum, as petroleum-based lubricants may have an adverse effect on the rubber liners. Unfortunately, though most obstetric lubricants are water-based, they may have spermicidal properties due to the preservatives they contain. It has been suggested that one of the best lubricants is 10% HR jelly (Fromann and Amann, 1983). Lampinen and Kattila (1994) investigated the effects of eight lubricants on spermatozoan viability: their results suggested that some lubricants such as Vetopgel, Bovivetgel and K-Y jelly reduced spermatozoan motility, whereas others such as mineral oil and Superlibe had no apparent adverse affects.

Whatever lubricant is used, most should be placed around the neck of the AV into which the stallion penetrates. Some lubricant can be placed further down the lumen as well, but the mere action of penetration should ensure that a significant amount of the lubricant placed in the neck of the AV gets distributed along its internal length. Some practitioners prefer to lubricate the AV prior to filling with water, whereas others routinely lubricate the fully assembled AV immediately prior to use. The only proviso is that a short period of time is allowed for the lubricant temperature to be equilibrated with the internal temperature of the AV.

If the AV is to be lubricated some time before use, then it is as well to leave a protective sleeve within the AV to prevent contamination and the drying out of water-soluble lubricants. This can be achieved by inserting a hand into a plastic long-sleeved AI glove and placing a quantity of lubricant on the palm of the glove. The gloved hand is then used to lubricate the lumen of the AV and the hand is then withdrawn, leaving the glove behind (Fig. 5.14). The glove can then be removed immediately prior to use. This is not an easy procedure to carry out with plastic liners such as those in the Cambridge type. Hence lubrication immediately prior to use is best practised with these types of AV. Warmed lubricant can be used in an attempt to increase or maintain the lumen temperature of the AV, though this is not normally required. The smoothness of the proximal end of the AV may be improved by using non-spermicidal glue or acid-free vaseline (J.M. Parlevliet, The Netherlands, 1998, personal communication).

5.3. The Condom

An alternative to the AV is the use of the condom (Fig. 5.10), a large rubber sheath placed over the stallion's penis when fully erect. The condom needs to be large enough to allow adequate room for the collection of the ejaculated semen. After liberal lubrication of the mare's vulva and vagina and of the condom, the stallion is allowed to mount and penetrate the mare as per natural service. The sample is collected within the condom, which is removed immediately upon dismount. The use of condoms does pose some problems, the major one being the contamination of the sample with debris, detritus, epithelial cells and possibly bacteria from the penis as it remains in relatively close contact with the sample for a prolonged period until the condom is removed. There is also the risk of the condom becoming

Fig. 5.14. A Missouri AV lubricated prior to use. In order to prevent the lubricant from drying out, a long plastic AI glove can be left in the lumen until immediately prior to use. (Photograph by Julie Baumber and Victor Medina.)

dislodged or damaged during covering, leading to the accidental covering of the mare. Finally, many stallions using condoms require additional manual stimulation prior to ejaculation.

The condom is not a preferred method of semen collection but it provides an environment that mimics more closely that of the vagina during natural service and so it is useful for stallions that are reluctant or unable to ejaculate into an AV.

5.4. Preparation of the Stallion for Semen Collection

The preparation of the stallion prior to the use of the AV should, in summary, be the same as for natural covering, and in accordance with the guidelines set out for minimal contamination breeding (Kenney *et al.*, 1975; Horse Race Betting Levy Board, 1997). For the export of semen, there are additional criteria and tests that the stallion must fulfil prior to collection. The exact regulations depend upon the country of destination for the semen. The guidelines for the importation of semen into EC countries, both from countries outside the EC and for the movement of semen from country to country within the EC, are governed by Directive 92/65/EEC and more specifically by Decisions 94/63/EC, 95/176/EC, 95/307/EC and 96/539/EC.

Other countries (for example, the USA, South Africa and Australia) follow similar, but often stricter guidelines.

Once all the preparatory measures have taken place and the stallion is ready, he should be brought to the collection area appropriately tacked up. The exact method is a personal choice though, as with natural covering, a stallion should possess both breeding and exercise tack so that when he is brought in he is aware of what is expected of him. Regardless of the specifics of his tack, it should be such that it allows sufficient restraint to prevent accidents but enough leeway to allow him to tease the mare with minimal interference.

As is normal practice for covering in hand, the stallion should be led to a teasing area and be allowed to tease an oestrous mare, in order to stimulate erection (Fig. 5.15). Prolonged teasing is reported to affect semen quality mainly via an increase in the volume and proportion of the gel fraction, which is of little use in AI (Pickett and Voss, 1972).

Once erect, the stallion's penis should be thoroughly washed to remove bacteria, debris and smegma (epithelial cell debris) (Fig. 5.16). Recent work sheds doubt on the success of washing in reducing bacterial counts in semen collected (Clement *et al.*, 1995), but washing does normally take place. Historically, soaps and disinfectants have been used, but, as discussed, such solutions are highly likely to be spermicidal and should be avoided. Washing with warm water (40–45°C) has been shown to be just as effective as washing

Fig. 5.15. Teasing a stallion in preparation for semen collection. (Photograph by Julie Baumber and Victor Medina.)

Fig. 5.16. Washing the stallion prior to collection. (Photograph by Julie Baumber and Victor Medina.)

with soaps or disinfectants (Jones *et al.*, 1984). Inclusion of 5% chlorhexidine in the washing water has been advocated as an effective means of eliminating bacteria but without a detrimental affect on spermatozoan quality (Betsch *et al.*, 1991). Washing of the penis also gives the handler the opportunity to look for penile damage and/or lesions.

As with natural service, excessive washing of the stallion's genitals is not advocated as this will remove not only potentially harmful bacteria but also the natural microflora, as a result of which there is likely to be colonization of the area by opportunistic bacteria, such as *Pseudomonas*. For this reason, many centres only wash stallions occasionally, or in cases of considerable contamination.

Some stallions consistently give semen samples with high bacterial counts, despite routine washing as recommended above. This can be circumvented by the addition of antibiotics to the semen sample as a prophylactic measure, or careful washing with antibacterial scrub may be used, provided the area is carefully and thoroughly rinsed prior to use. However, regular use of such antibacterial soaps is not recommended, for the reasons previously discussed.

Regardless of the solutions used for penile washing, the handler carrying out the procedure must wear disposable plastic gloves, and careful rinsing of the area after washing is essential. Some practitioners recommend drying the penis after washing; others do not, claiming that the process of drying leads to an additional risk of contamination (Varner and Schumacher, 1991; Pickett,

1993a). It is essential that any material used to dry the penis is thoroughly clean and does not pose the risk of recontamination to the area. If the penis is not to be dried, then care must be taken to ensure that there is no water contamination of the sample collected.

Before washing the penis it is as well to wash the surrounding area – for example, the scrotum and the inner aspect of the hindlegs. The washing procedure must be carried out immediately prior to collection, especially if the environment is dusty and the washed area is not to be dried. In such circumstances there is a high risk that environmental dust will recontaminate the area. If this occurs, the washing procedure must be repeated.

5.5. The Collecting Area

Any area that is suitable for natural covering may also be used for semen collection. However, it is of particular importance in semen collection that the surface drains well. A surface that may become slippery or muddy as a result of water leaking from the AV or of teaser urination is not desirable. Similarly a dusty surface, like wood-chip or dry earth or peat, may cause contamination of the stallion's penis and, therefore, of the semen sample collected. A relatively cheap, perfectly serviceable and increasingly popular surface is provided by rubber flooring, which has the additional advantage of being fully washable. The durability of rubber matting is quite good. Synthetic surfaces used more commonly for riding arenas last longer, though they tend to be more expensive and can cause contamination of the penile area if the stallion is particularly excitable. For ease of management, the collecting area should ideally be in a quiet corner of the yard and under cover. An effect of the collecting area on semen quality has been demonstrated by Ionata *et al.* (1991b), who showed that the collection of semen from stallions in an enclosed undercover shed resulted in a higher volume of gel, when compared with ejaculates collected from stallions in an outside environment. This difference is, however, of limited importance in the collection of semen for AI. The collecting area should have easy access, ideally with a door at either end to allow the teaser mare and stallion to enter and exit via different routes. It should be large enough to allow the stallion room to move around without feeling restricted, but not so large that if he did accidentally get loose he would be difficult to catch. An area of 20 m $\times$ 12 m is perfectly adequate. The area may also have a trying board at one end, behind which a teaser mare may be placed, or alternatively a dummy mare may be positioned at one end for the stallion to mount at collection.

5.6. Sexual Stimulation of the Stallion and Encouragement to Ejaculate

In order for the stallion to ejaculate successfully, adequate sexual stimulation is required. Evidence suggests that the quality of the semen sample collected may be affected by the method and length of sexual stimulation. Ionata *et al.* (1991b) demonstrated that teasing a stallion for 20 min prior to a second ejaculation resulted in a significant increase in the total volume of semen collected, mostly due to a greater volume of the gel-free fraction, and a reduction in the number of mounts required per ejaculation, though the concentration of spermatozoa remained lower than from the first ejaculate. The normal method of sexual stimulation is to present the stallion with a mare in oestrus, termed a teaser or jump mare. He will then be encouraged to mount this mare and the resultant semen sample is collected into an AV rather than ejaculated into the mare (Fig. 5.17).

This method of sexual stimulation and encouragement of the stallion to mount is widely practised, especially with young or inexperienced stallions. It has the obvious drawback that it requires a constant supply of mares in oestrus. The alternative to a jump mare is the dummy or phantom mare, a construction of wood, plastic or leather, which is shaped to mimic a real mare. With time, many stallions can be trained to ejaculate using a dummy mare, negating the need for a constant supply of mares in oestrus. The use of either a teaser mare or a dummy is reported to have no effect on the sexual behaviour of the stallion when returned to a natural covering programme (McDonnell and Love, 1990).

Other methods of sexual stimulation and semen collection are occasionally used, and will be discussed in the following sections.

Fig. 5.17. The use of a jump mare provides sexual stimulation and encourages ejaculation (photograph by Angela Stanfield).

5.6.1. The jump mare and her preparation for use

If a jump mare is to be used, it is most important that she is well into oestrus, is docile in temperament and stands well to the stallion. In many countries the health status of such mares must be equivalent to that of the stallion and they must be housed away from stock of lower status. The mare's temperament is particularly important if the stallion is inexperienced at semen collection or is a young animal. A well-placed kick by a mare may cause considerable psychological as well as physical damage to a stallion. This will at best put him out of action for a while but may result in him permanently objecting to the semen collection procedure, rendering him of no future use in an AI programme.

As indicated, jump mares need to be in oestrus at the time of collection. This can be achieved naturally by waiting until the mare comes into season, which is an adequate system where the AI programme is low-key, and only the occasional sample is required (from a single stallion) which is then to be frozen for future use. If the AI programme is larger, with more mares and with the use of fresh and chilled semen, then timing of collection needs to be more controlled and specific. If naturally occurring oestrus is to be relied upon, then a significant number of mares is required to ensure that a mare in oestrus is available as and when a sample is required. Such a system is still erratic, and also expensive in terms of keep. Instead, manipulation of oestrus in a few chosen mares is the normal practice. There are many methods by which the timing of oestrus in a mare can be controlled, and two of the most common are the use of progesterone and/or prostaglandins ($PGF_{2\alpha}$) or their artificial analogues, which are discussed briefly below. Further details of the methods available to synchronize oestrus and ovulation are given in Chapter 8 (section 8.3.1).

The use of progesterone or its analogues in the timing of oestrus works on the principle of imitating the mare's natural dioestrous or luteal phase. This is achieved by mimicking the natural progesterone production during the luteal phase, by the administration of exogenous progestagens. Termination of this induced luteal state, achieved by the cessation of treatment, acts like the end of the natural luteal phase and so induces the changes in the mare's endogenous hormones that are a prerequisite for oestrus and accompanying ovulation. The use of exogenous prostaglandin is similarly an attempt to mimic the mare's natural hormone changes. In the absence of pregnancy, prostaglandin is naturally produced at a set time (12–14 days) post ovulation. As such, it both marks and causes the termination of the luteal phase and the commencement of the endogenous hormone changes associated with oestrus and ovulation. Administration of exogenous prostaglandin, provided it is within certain time limits within the luteal phase, allows the termination of this phase to be controlled, and with it the timing of oestrus and ovulation. These two hormone treatments may be used either alone or in a variety of combinations (Ginther, 1992; Allen and Cooper, 1993; Davies Morel, 1993; Irvine, 1993; Squires, 1993a,b).

Progesterone supplementation may be achieved in several ways, but most commonly via the use of the oral progestagen, altrenogest (Regumate). A dosage of 0.044 mg kg^{-1} for 15 days will result in oestrus in 3.4 (± 1.9) days after cessation of treatment, accompanied by ovulation 5.4 (± 2.2) days later (Squires et al., 1983). Alternatively, progesterone supplementation may be via intramuscular injection of one of the progestagen analogues (Loy and Swann, 1966; Squires et al., 1979b) or by progestagen vaginal sponges (Palmer, 1979). It may be used alone or, more successfully, with a PGF$_{2\alpha}$ injection 24 h after the cessation of progesterone supplementation (Palmer, 1979; Draincourt and Palmer, 1982). Other workers are currently examining the use of intervaginal devices in mares, along the lines of those available for cattle (S. Revell, Wales, 1997, personal communication). In general, progesterone supplementation, administered as a daily dose of 100 mg day^{-1} for 15 days, successfully suppresses oestrus in mares. Oestrus commences at 3–6 days following withdrawal, with ovulation occurring at 5–15 days after cessation of treatment (Loy and Swann, 1966; Van Niekerk et al., 1973; Stevens et al., 1979; Squires et al., 1983).

Progesterone treatment of shorter duration is also successful if used in conjunction with prostaglandin treatment. For example, treatment for 8 days with the progesterone analogue, altrenogest, followed by a single injection of PGF$_{2\alpha}$ (Equimate) on the last day of progesterone treatment, results in oestrus 3–5 days post PGF$_{2\alpha}$ injection (Hughes and Loy, 1978; Palmer, 1979). Some protocols include the additional use of human chorionic gonadotrophin by injection. Human chorionic gonadotrophin (hCG) is a human placental gonadotrophin with properties similar to luteinizing (LH) and follicle-stimulating hormone (FSH). It enhances and supplements the natural release of gonadotrophins, which drive follicular development and, more specifically, ovulation. Its use at day 6 post treatment with progesterone alone, or a progesterone–PGF$_{2\alpha}$ combination protocol, is reported to help to alleviate the problems of variability in synchrony between ovulation and oestrus (Palmer, 1976, 1979). When used on day 8 post PGF$_{2\alpha}$ injection, oestrus synchronization rates of 90% have been reported (Holtan et al., 1977). Further developments and refinements of these basic protocols have been investigated, including a combination treatment of progestagens and oestradiol (Loy et al., 1981; Lofstedt, 1997; R. Pryce-Jones and C. McMurchie, Llandysul, 1997, personal communication) and the use of gonadotrophin-releasing hormone (GnRH) in place of hCG (Meinert et al., 1993; Jochle and Trigg, 1994; Mumford et al., 1995).

In addition to its use in combination treatments with progestagens, PGF$_{2\alpha}$ may be used alone. If the stage of the mare's cycle is known, then a single injection of PGF$_{2\alpha}$ or its analogue between days 6 and 15 of the cycle will result in oestrus 2–5 days post injection, and in mares destined to be covered it will allow fixed-time insemination at 48 and 96 h post injection (Loy et al., 1979). In the majority of circumstances the exact stage of the mare's cycle is unknown, and in such mares a double injection of PGF2α 14–15 days apart will result in oestrus in 77.8–92.0% of mares within 6 days of the second injection (Hyland and Bristol, 1979). Further refinement of this protocol, with the addition of an injection of hCG on day 7 (7 days after the first PGF$_{2\alpha}$)

and on day 21 (7 days after the second PGF$_{2\alpha}$ injection), is reported to result in up to 95% of mares ovulating on either day 22 or day 23 (Allen *et al.*, 1974; Palmer and Jousett, 1975; Voss, 1993).

Alternatively, a mare suffering from cystic ovaries or ovarian granulosa cell tumours may be used as a jump mare. As a result of her condition, her system is continually dominated by the oestrogens being produced by the persistent follicles undergoing prolonged maintenance. Such a condition often manifests itself as nymphomanic behaviour, with mares continually showing oestrus. Such mares have the advantage that oestrous behaviour is not associated with ovulation and hence accidental conception is not a risk. However, the condition is quite rare and affected mares tend to be more unpredictable in nature and should be treated with care.

A further alternative is the use of an ovariectomized mare. Many such mares display adequate receptivity to the stallion to allow their use without the administration of exogenous hormones. However, if oestrous behaviour needs to be enhanced, a single injection of 0.12–2 mg of oestradiol cypionate may be administered (Varner and Schumacher, 1991; Love, 1992).

Once the jump mare has been selected and is ready for use, she should be prepared as for in hand covering and, as with the stallion, this should ideally be within the guidelines for minimal contamination breeding (Kenney *et al.*, 1975; Horse Race Betting Levy Board, 1997). Although no direct contact between the stallion's penis and the mare should occur, accidental contact is not that easy to prevent. A mare prepared as for minimal contamination breeding will present no risk to the stallion should accidental contact occur.

A mare ready for use should normally have her tail bandaged, her perineal area washed, her shoes removed and her hindfeet fitted with covering boots. She should be presented bridled and possibly twitched. In America and Australia the use of hobbles is often advocated, especially if the stallion is of particular value (Fig. 5.18). This ensures the safety of the stallion, though it could be argued that if a mare is likely to kick the stallion, so necessitating the use of hobbles, she is not truly in oestrus and is an inappropriate candidate for use as a jump mare. The use of hobbles is not routinely practised in Europe. However, wither guards and biting rolls are used, especially with stallions that have a reputation for biting or savaging a mare during covering.

5.6.2. The dummy mare

A good alternative to a live jump mare is the use of a dummy or phantom mare (Kenney and Cooper, 1974; Richardson and Wenkhoff, 1976) (Fig. 5.19). Dummy mares were originally designed for stallions who were unable, for some reason, to mount a live jump mare. However, their use has now spread to become routine in semen collection (Kenney and Cooper, 1974; Richardson and Wencoff, 1976). Most stallions, given time and the initial encouragement of an oestrous mare in the vicinity, can be persuaded to mount a dummy mare and ejaculate into an AV. Adjustment by stallions used to natural service may

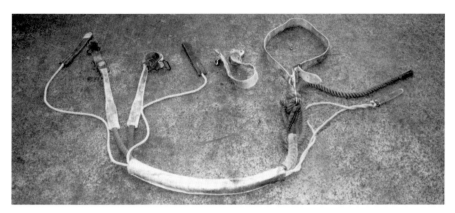

Fig. 5.18. The use of hobbles to restrain a mare used for collection is often advocated in the USA and Australia. (Photograph by Julie Baumber and Victor Medina.)

take some time, but most novice stallions not previously used for natural covering readily learn to accept the dummy mare for collection (Conboy, 1992). Even with stallions experienced in natural service there may, in time, be no need for the presence of an oestrous mare, as the stallion gets conditioned to the use of the dummy which, in itself, provides adequate sexual stimulation.

A dummy mare, in its simplest form, can be a large padded roll or barrel on four wooden legs. Many variations on, or additions to, this basic theme

Fig. 5.19. A dummy or phantom mare may be used as a mount for a stallion during semen collection. Note the hole into which the AV can be placed.

are available. There are a few requirements that should be met with regard to ease of use and safety.

It is essential that the dummy is sturdy and able to hold the weight of the thrusting stallion. Instability will result in potential danger to the stallion and the handlers and may result in the stallion feeling insecure and so objecting to ejaculating while mounted on the dummy. The dummy mares that are available with four legs (one at each corner) tend to be more stable. Those with a single or two central legs allow less scope for the stallion to get entangled in the supports, and are increasingly the more popular design (Fig. 5.19).

The dummy should be covered with a durable and ideally padded material which is non-abrasive and easy to clean (cow's hide may be used as a good non-abrasive material). Some dummy mares also have contours to mimic those of the mare, plus a withers area, or alternatively a neck guard or biting roll can be used to give the stallion material to grasp during collection (Figs 5.20 and 21). Some dummies have provision for an AV to be placed inside them, mimicking the natural position of the mare's vagina and so facilitating the collection process (Fig. 5.19) (Volkmann, 1987). The major advantage of the use of such dummies is that they allow collection to be carried out by a single person, provided the stallion is well trained and familiar with the process, rather than the normal requirement for two. A disadvantage is that such an arrangement makes it difficult to manipulate the AV and provide the manual stimulation required by some stallions.

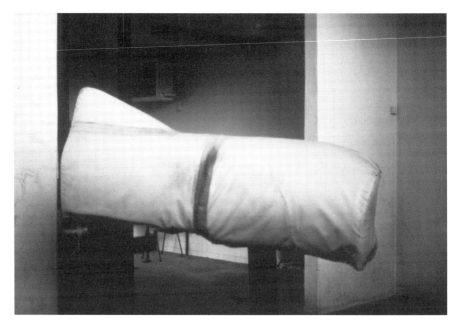

Fig. 5.20. A dummy mare may be designed with contours and a withers area to mimic the natural shape of the mare.

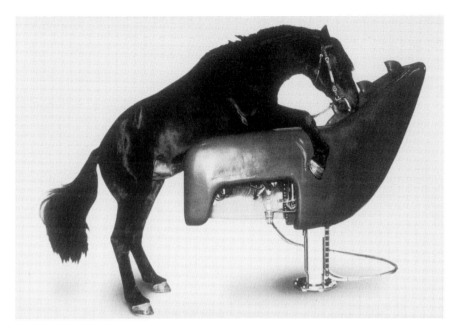

Fig. 5.21. The most sophisticated systems incorporate a complete collection and computer analysis system within the dummy mare. (Equidame phantom, Haico Oy, Loimaa, Finland.)

Many dummy mares are also provided with adjustable legs or supports, which can be altered hydraulically or manually to match the height of the stallion (J.M. Parlevliet, 1992, The Netherlands, personal communication) or even form part of a complete collecting and evaluating system (Fig. 5.21). However sophisticated the dummy is, it must be placed in an area free of obstacles, with plenty of space to allow the stallion to mount, and with the handlers in a safe position. It must also allow an oestrous teaser mare to be placed either alongside or in front of the dummy to encourage reticent or novice stallions.

In the main, the use of a dummy mare provides a safer method of collection for both the handlers and the stallion. It also reduces the number of personnel required, possibly down to a single person (as indicated above). Although, in general, the use of a dummy mare is safer for all concerned, it is not without risk to the stallion. During natural covering, or with the use of a jump mare, the mare provides a shock-absorbing effect when the stallion covers her, allowing him to move forward a step or two as he mounts. Obviously the fixed dummy cannot act as a shock absorber and so there is the risk of penile damage, especially in overzealous stallions that charge the dummy at mounting, or that thrust vigorously.

5.7. Alternative Methods of Semen Collection

The use of mounting and ejaculation into an AV is by far the most popular method for the collection of semen from a stallion. Other methods have been attempted, with limited success, but they may have their uses in stallions that are physically or psychologically unable or unwilling to mount a jump mare or a dummy and ejaculate into an AV. They may also be used to assist stallions with erectile dysfunction (McDonnell *et al.*, 1991; Love *et al.*, 1992).

5.7.1. Manual stimulation

Manual stimulation may be employed to provide extra (or the sole means of) stimulation and encouragement to ejaculate. This may be provided by placing a warm towel (45–50°C) at the base of the shaft of the penis or on the glans penis. Manual stimulation is primarily of use in stallions unable to mount a teaser mare or dummy due to injury or fear. Such stallions require a teaser mare in the vicinity and are collected from with all four feet on the ground. The semen ejaculated may be collected into an AV as per the normal collection procedure or even into a clean, sterile plastic bag placed over the penis (McDonnell and Love, 1990; Love, 1992). A stallion may get so used to this method of collection that he may never need to mount a mare or dummy again, regardless of whether he recovers, thus reducing the risks and easing the process. Indeed, training a stallion to ejaculate by manual stimulation is reported to take no longer than training for conventional collection using a dummy or teaser mare. Manual stimulation applied (as before) by means of a warm towel on the shaft or base of the penis may also be used to encourage libido and mounting of the dummy or teaser mare prior to more conventional semen collection (Crump and Crump, 1989), or to encourage ejaculation in stallions that are slow to ejaculate into an AV once mounted on a teaser mare or dummy. This form of extra stimulation is more commonly required by the larger type of stallion, for which the standard sized AV is rather short. Stallions suffering from non-specific neurological defects have also been reported to respond well to such extra manual stimulation (Love, 1992).

Semen collected at ejaculation resulting from manual stimulation is reported to contain spermatozoa of comparable characteristics to those produced by natural ejaculation (McDonnell and Turner, 1994). The limited work available suggests that there is no detrimental effect, as far as semen quality is concerned, of ejaculation via manual stimulation (Crump and Crump, 1989). It is also reported that stallions so used retain their normal sexual behaviour when returned to natural service (McDonnell and Love, 1990).

5.7.2. Minimal restraint

Stallions differ significantly in their response to handling and restraint during covering and semen collection. Some stallions object to even the slightest restraint and are hard to collect from by conventional means. It is possible to collect semen from an unrestrained stallion, using an AV, though it is potentially very dangerous. This method is used by the University of Wales, Institute of Rural Studies for the collection of semen from semi-wild Welsh Mountain Section A stallions (under 13.2 hands high, or 135 cm). A specially designed collecting and penning area is used for collection (Fig. 5.22). This system works well with stallions that are young, small in size and lack confidence. However, it requires handlers who are alert and quick to ensure that ejaculation occurs into the AV. In the Institute of Rural Studies a relatively large and reliable teaser mare is used, making accidental covering less likely.

Evidence would suggest that the presence or absence of restraint does affect semen quality. Ionata *et al.* (1991a) reported that restraint of stallions resulted in an increased number of spermatozoa per ejaculate (9.1×10^9) compared with unrestrained stallions (8.3×10^9). Restrained stallions are also reported to give seminal plasma of a lower pH.

5.7.3. Collection without mounting

Semen may be collected using an AV but in the absence of a teaser mare or dummy for mounting. Thirteen stallions out of a group of 18 used by Schumacher and Riddell (1986) successfully ejaculated into an AV placed over

Fig. 5.22. The specially designed collection area used by the University of Wales Aberystwyth, Institute of Rural Studies to collect from unrestrained, pony stallions.

the penis after sexual excitation had been induced by the presence of a mare or gelding.

A further method of collecting with minimal restraint and without the need for mounting of a mare is described by Love (1992). This work reports the previous use of a specially designed teasing board, with a teaser mare and the stallion placed on either side. The board was high enough to prevent the stallion reaching the mare but had a hole cut in the middle of it, allowing the stallion to tease the mare through the hole. The hole was positioned low enough for the stallion's head to be near the ground as he reached through to tease the mare. In this position, it is then possible to place an AV over his penis and collect semen as he thrusts forward through the hole, lunging his chest against the lower part of the board. Stallions are reported to accept this method of collection quite readily once the first collection has been successful. A similar arrangement has been described by S. Revell (1998, Wales, personal communication) in which a tyre is mounted in a solid partition and used to collect semen from pony stallions without the need for mounting.

Pharmacological ejaculation

Pharmacologically induced ejaculation has been reported to be successful in the stallion, using xylazine infusion while standing the stallion in a quiet environment (McDonnell and Love, 1991). Xylazine acts as an α-adrenergic stimulator. Other α-adrenergic drugs used with some success in inducing erection followed by ejaculation include imipramine and chlomipramine (McDonnell and Turner, 1994). Additional stimulation using xylazine has been reported to be required for consistent ejaculation following erection (McDonnell and Odian, 1994; Turner *et al.*, 1995). Using such treatment, Card *et al.* (1997) reported conception in four out of five mares inseminated. These α-adrenergic drugs are also known to have an effect on the central nervous system, involving the norepinephrine, dopamine and serotonin systems, each of which plays a role in male sexual arousal, erection and ejaculation (Turner *et al.*, 1995).

Pharmacologically induced ejaculation has been used in clinically normal horses as well as in stallions with ejaculatory dysfunction, aorto-iliac thrombosis and musculoskeletal injury (McDonnell *et al.*, 1987; McDonnell and Love, 1991; Turner *et al.*, 1995). Semen collected pharmacologically using imipramine and xylazine is reported to be of comparable quality to that obtained by natural ejaculation. The sample is normally lower in total volume but higher in concentration, with a reduced gel volume and pH and an increase in the total number of spermatozoa, making the sample particularly useful for freezing (McDonnell and Odian, 1994; Turner *et al.*, 1995). Pharmacological ejaculation using imipramine alone is reported to result in an ejaculate even more like that produced by natural ejaculation. No effect on the total number of spermatozoa, volume of gel fraction or pH has been reported, unlike when xylazine is used. This may suggest an adverse affect of xylazine on accessory gland function (McDonnell and Turner, 1994).

Electro-ejaculation

Electro-ejaculation is the electrical stimulation of the musculature surrounding the vas deferens, urethra, accessory glands and the base of the penis. This stimulation is provided by means of a probe placed in the rectum of the stallion. Electro-ejaculation is a feasible method of collecting semen from bulls and rams, though its use is illegal in some countries such as The Netherlands (J.M. Parlevliet, The Netherlands, 1998, personal communication). It is normally reserved for use with animals unable to mount and can be used without sexual stimulation. It would, therefore, seem to be a feasible method of collecting from stallions with some of the physical and psychological problems mentioned previously. Its use in the horse has been demonstrated but it is not practised, due to the relatively high voltages required in order for an effect to be achieved. The need for high voltages results in general discomfort to the stallion, causing him to strain. As a result, a general anaesthetic is required (Roberts, 1986a; Hafez, 1993a). The semen sample collected by this method tends to be of lower quality, with a lower spermatozoan concentration. Samples collected by electro-ejaculation also run a higher risk of urine contamination, due to the general stimulation of muscles in the penile, accessory gland and bladder area. Urination prior to collection helps to alleviate this problem but does not eliminate it (Roberts, 1986c; Boyle, 1992; Oristaglio Turner *et al.*, 1995).

5.8. Collection Procedure

Several methods of semen collection have been discussed in the previous sections, the majority of which involve the use of an AV. The procedures for the correct use of an AV are varied and depend on the type of AV being used and individual stallion preference. Regardless of the specific techniques that are chosen, the following principles apply to the collection procedure.

The aim when using an AV is to mimic the natural vagina and it is important that, among other things, the positioning of the AV during collection is appropriate. After teasing (Fig. 5.23) the stallion should be allowed to approach the jump mare or dummy. The collection procedure, whether using a jump mare (Fig. 5.23) or a dummy (Fig. 5.24), will be very similar, except that fewer handlers are required in the case of the dummy as only the unpredictable behaviour of the stallion needs to be catered for. As with natural covering, all handlers (and the collector) should be positioned on the same side of the stallion (Fig. 5.23). This is normally the nearside, as it is the side most stallions are used to, but it depends on the preference of the collector. The positioning of all handlers on the same side when using a jump mare means that, in the event of an accident or should problems occur with the procedure, both the mare's head and the stallion's head can be pulled to the same side as the collector. This action will ensure that the hindquarters of the mare and stallion are moved away from each other and the collector. This reduces the risk of injury to all parties, especially as a result of lashing out by either the mare or the stallion. Collection from the offside is perfectly

Fig. 5.23. Events involved in semen collection using a jump mare: (a) introduction or teasing of the stallion and mare; (b) AV ready for use; (c) guiding the stallion's penis into the AV and the correct positioning of the handlers all on the same side of the mare and stallion; (d) slightly inclined positioning of the AV to mimic the natural position of the mare's vagina; (e) dismount of the stallion and slow removal of the AV, ensuring all semen is collected; (f) after dismount the stallion should be turned away from the mare to reduce the chance of injury; (g) the collected sample. (Photographs by Julie Baumber and Victor Medina.)

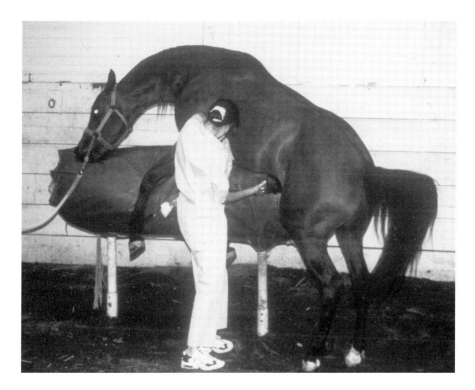

Fig. 5.24. Collection from a stallion using a dummy mare. (Photograph by Julie Baumber and Victor Medina.)

acceptable, provided the handlers are appropriately placed. Indeed, collection from the offside is preferred by some stallions and by many right-handed collectors, especially with the heavier Cambridge type AVs, as it allows the right arm to take the weight of the AV. In Germany, specially designed bridles are used that allow the handler to stand on the nearside and the collector on the offside. The reins of the bridle are so arranged as to allow the handler to pull the horse's head to the right in the event of problems (J.M. Parlevliet, The Netherlands, 1998, personal communication; S. Revell, Wales, 1998, personal communication). Many dummies are fitted with a lumen into which the AV may be placed (Fig. 5.19), so circumventing the problem as to on which side to stand. Once the stallion is drawn and ready to mount the mare, he should be allowed to do so with minimal interference. Excessive interference at any stage, but especially during the period prior to intromission and ejaculation, can discourage or distract the stallion. If the interference is consistently a problem it will result in psychological effects, reducing libido and willingness to use an AV.

Once the stallion has mounted he should be allowed a little time to stabilize himself on the mare or dummy. The collector then steps forward and deflects the penis away from entry into the vagina and into the AV (Fig. 5.23c). The amount of time allowed for a stallion to stabilize himself should not be too long and the collector should be ready to deflect the penis immediately on mounting if there is any danger of the stallion entering the mare or injuring himself on the dummy. Once the AV has been placed over the penis (Fig. 5.23d), the warmth within it plus the thrusts of the stallion should stimulate ejaculation. It may be necessary to reduce the pressure within the AV on entry by allowing some water or air to escape. This is not normally required in stallions used regularly for collection as the required pressure within the AV can be predetermined. However, in new stallions, depending upon the ease with which they thrust and the size of the stallion, a reduction in the AV pressure may be deemed necessary.

During thrusting and subsequent ejaculation, the AV should be slightly inclined at an angle of 25° to the horizontal, equivalent to the natural angle of the mare's vagina, at about mid hip level of the mare (Fig. 5.23d). One hand should hold the handle of the AV while the other supports it from below. The AV should be held in this position, the collector's hips and ribs pushing it against the mare's side, but allowance should be made so that the AV can move slightly as the stallion thrusts. During ejaculation some stallions may grasp the mare or dummy to steady themselves; others may require their legs to be held in place on the mare or dummy, especially if they are small in comparison. This can be done by the mare handler, provided the mare is quiet and happy to stand still.

Successful ejaculation is normally marked by the flagging of the stallion's tail. It can be more accurately confirmed by sliding the hand (which is supporting the AV) along the length of the AV to the base of the penis and feeling for rhythmical contractions of the urethra along the ventral side of the penis. This technique accurately determines whether ejaculation has taken place but the interference may be objected to by some stallions. Immediately ejaculation has been ascertained, the distal end of the AV, with the collecting vessel, should be slowly lowered to aid the collection of semen into the collecting bottle (Fig. 5.23e). During this stage the stallion may dismount. If possible, the AV should remain on the stallion's penis and be slowly lowered to the vertical as his erection subsides. This may not be possible with all stallions but the AV should remain on the penis as long as possible to ensure maximum collection of the ejaculate. Once the AV is removed from the penis, the urethral area may be observed for the presence of a white frothy secretion – a further indication that ejaculation has taken place.

As soon as the AV has been removed and ejaculation confirmed (Fig. 5.23f), the collector should move away from the stallion, holding the AV vertically (Fig. 5.23g). The filler cap should be unscrewed to allow water to drain out and so reduce the pressure within the lumen and encourage any semen deposited towards the entrance of the AV to drain into the collecting vessel. Drainage of the water will also cool the AV and so minimize the risk of overheating of the spermatozoa from excessive contact with the warm lumen environment (J.M. Parlevliet, The Netherlands, 1998, personal communication). Once all the semen has been collected, the collecting vessel should be removed from the AV and taken without delay to an incubator or water bath in the laboratory.

When the stallion has dismounted, the mare should be brought forward and turned towards the handler; the stallion should similarly be turned towards the handler at the same time. This minimizes the risk of injury to both animals and handlers. After collection the stallion does not normally require washing, as his penis should not have come into direct contact with the mare, and so he may be returned directly to his stable or field.

The exact procedure involved in collection may vary from stallion to stallion. Many have certain idiosyncracies and may perform better if these are catered for. As with all stud management, knowledge of the stock being handled helps significantly in obtaining good and consistent results.

5.9. Frequency of Collection

The effect of frequency of collection on semen quality has already been discussed in some detail (section 4.3.2). The frequency of collection that a stallion can tolerate without detrimentally affecting semen quality is highly variable. Evidence would suggest that instead of a single collection per day regularly over the breeding season, double collections (1 h apart) give better results. As might be expected, the first ejaculate in double collections taken on alternate days is reported to produce the best quality semen. However, pooling of these two samples resulted on average in samples with higher spermatozoan motility, a lower incidence of spermatozoan abnormalities and a higher percentage of live spermatozoa before and after freezing than single daily collections. This was at the expense of ejaculate volume, which suffered in the alternate-day collection regime (Arras, 1994). Work by Magistrini *et al.* (1987) showed a similar effect comparing daily collections with collection on alternate days (three times per week). This work noted a decrease in ejaculate volume and in spermatozoan concentration, along with a slight increase in spermatozoan motility when collections were made at the higher frequency. It also assessed the effect that season might have on spermatozoan quality for use with AI. Semen collected during the winter tended to be lower in volume, of both the gel and gel-free fractions, but higher in spermatozoan concentration and with lower spermatozoan motility (Magistrini *et al.*, 1987). For AI, it is generally considered that a stallion may be collected from once per day, thus allowing monitoring of the quantity produced and any adaptation in management. For stallions in less demand, collection three times per week is often practised (J.M. Parlevliet, The Netherlands, 1998, personal communication).

The standard procedure normally used for assessing a stallion for inclusion in an AI programme is either two ejaculates taken 1 h apart, followed 3 days later by a single ejaculate for evaluation, or two ejaculates taken 1 h apart and followed by daily collection of samples for evaluation for 6–7 days (Pickett and Voss, 1972; Kenney, 1975; Sullivan and Pickett, 1975; Swierstra *et al.*, 1975; J.M. Parlevliet, The Netherlands, 1998, personal communication). The second system, though more laborious, also allows the stallion's daily spermatozoan output to be estimated, indicating the number of mares that he can cover or the number of AI doses that can be obtained.

5.10. Training the Stallion for Collection

As with training any animal for a specific task or purpose, the key words are patience, consistent discipline and praise, and safety for both handlers and stock. The same applies to the training of a novice stallion. The act of mounting a mare is a natural instinct, so some stallions adapt very readily to the use of a jump mare, dummy and AV. However, it is most important that the stallion perceives the process as a pleasurable event and is happy to continue. Unpleasant psychological or painful physical associations with the use of the AV may permanently reduce his libido, not only for AV collection but also in natural covering.

With the more novice stallions, it is particularly important that the collecting area is roomy, safe and free from obstacles and has a good non-slip floor. The stallion's shoes should have been removed to minimize the risk of damage to himself, the mare or dummy, or possible entanglement in the mare's halter. Ideally, the stallion will have been allocated a handler who will know him well and in whom he has trust. Confidence in and respect for the handler help significantly in all training.

Initially the presence of a teaser mare, well into oestrus, is required regardless of whether the stallion is to be trained to use a dummy or a jump mare or both. The teaser should be of a quiet disposition, well into oestrus, of an equivalent size to the stallion, accustomed to the process of semen collection and familiar with the collecting area. An older mare dark in colour often proves the most successful. Stallions, especially novices, find lighter coloured or bicolour horses (piebald, skewbald, etc.) a distraction from the job in hand. The novice stallion, suitably bridled with a long, but strong, lead rein, should be led towards the teaser mare. They should be introduced to each other either side of a teasing board. The stallion should then be allowed to tease the mare in the normal fashion for natural service, moving from her head towards her vulva, gauging her response. Conboy (1992) suggested that at this initial stage the ease with which a stallion can be trained for semen collection is apparent from his reaction to the teaser mare. He classified stallions as follows:

- Type 1 – an aggressive teaser;
    - keen, alert and often noisy;
    - may rear and lunge in an attempt to get to the mare;
    - achieves erection in 1–5 min.

- Type 2 – definite interest shown but cautious and appears to lack confidence;
    - smells and nuzzles the mare;
    - generally quiet and reserved;
    - achieves erection in 5–15 min.
- Type 3 – disinterested;
    - refuses to smell or touch the mare;
    - fails to develop an erection.

Conboy's work indicated that type 1 and 2 stallions are relatively easy to train to use an AV. Type 3 stallions may never show any real interest in covering naturally, let alone in using an AV, and are not a realistic consideration for AI.

Once a stallion has shown interest in the mare he should be allowed to tease her at will until he achieves an erection. The balance between confidence and discipline at this stage is very important. As with the introduction of stallions to natural service, the main aim of the first few attempts is to build up the stallion's confidence – discipline should be kept to a minimum at this stage. Nevertheless, he should not be allowed to get away with downright dangerous behaviour. The use of the voice in such circumstances can prove very effective, rather than having to resort to physical reprimand.

When the stallion indicates that he is ready to mount the teaser mare, she should be twitched if required and placed in position to allow him to do so safely, or, if a dummy is to be used, the teaser mare should be placed in front of or alongside the dummy mare. In the case of a jump mare, once he has mounted her and provided he has an erection, his penis should be diverted into the AV and the procedures given previously for collection from experienced stallions followed. If he is to be trained to use a dummy, he should be persuaded by gentle encouragement and guidance, not force, to mount the dummy rather than the teaser mare. When this has been achieved, and provided he has an erection, the same procedure as given above can be followed.

It may take a little longer and more patience to train a novice stallion to mount a dummy, compared with a jump mare, though success rates of 99% are reported for 3-year-old stallions from a variety of breeds from which semen is collected for breeding soundness tests in The Netherlands (J.M. Parlevliet, The Netherlands, 1998, personal communication). However, once he has been successful and associates the process with a pleasurable experience he will become increasingly willing to use the dummy.

As with all training, events do not always proceed according to plan. The more common problems encountered include failure to achieve an erection, or loss of erection during mounting or during diversion of the penis into the AV. However, provided the procedures are kept consistent, and the environment is one of calm, the majority of stallions will eventually be successful. For the first few occasions the prime objective is to get him to use the AV. When this has been achieved successfully on a number of occasions, the finer points of collection may be introduced – for example, washing the penis, and mounting the mare or dummy directly rather than at an angle.

If a stallion fails to ejaculate, then the following should be considered. The temperature of the lumen of the AV should be checked: it may require a slight increase. The pressure within the lumen of the AV should also be checked and may need adjusting down for larger stallions and vice versa for smaller ones. Extra manual stimulation may be required at the base of the penis or at the glans penis. Finally, if disposable liners are used these may become excessively wrinkled with successive attempts. The stallion's penis is very sensitive and any slight changes in the temperature, pressure or internal surface of the AV may have a significant effect on his performance. If, even after these adjustments have been made, mounting or ejaculation does not occur, then a change of mare after a short rest period may encourage ejaculation (J.M. Parlevliet, The Netherlands, 1998, personal communication). If the stallion is being trained to use a dummy, he may benefit from being allowed to mount the teaser mare a couple of times in order to build up his confidence. If the stallion fails to mount or ejaculate with the teaser mare after several attempts, he should be removed and taken back to his box and a further attempt made the following day – ideally with another mare, as some stallions develop preferences or aversions to different mares. It is most important that the stallion continues to find the procedure a pleasurable one. Conboy (1992) further suggested that type 1 and 2 stallions will usually mount a dummy mare within 15 min in the first training session and that 90% of these stallions will ejaculate into the AV by the second day.

Alternative methods of training stallions to use dummy mares have been reported – for example, the use of a blindfold. The stallion is initially trained to mount a teaser mare and ejaculate into an AV. Once he is accustomed to this procedure he is blindfolded and the procedure is repeated; the mare is then replaced by a dummy and the stallion is allowed to mount the dummy blindfolded. The blindfold is removed after ejaculation but before he has dismounted the dummy. Eventually the procedure is repeated with the blindfold entirely removed and the stallion willingly mounting the dummy for collection (Richardson and Wenkoff, 1976).

The type 3 stallion as classified by Conboy (1992) presents a much greater challenge and may in the end prove to be a lost cause. Such stallions may react favourably to the use of 10–20 mg of diazepan, 5–7 min prior to exposure to the teaser mare, or may need to be turned away for a month or so and then reintroduced to the training programme from the beginning at a later date.

Once the novice stallion has been collected from on a number of occasions and has developed his confidence, discipline may be introduced. This will instil manners and reduce the risk of injury to horses and handlers. Some stallions become increasingly keen with time, and need to be encouraged to mount the jump mare or dummy steadily and not be too zealous. They should also learn to accept the washing procedure without fuss and be prepared to enter and leave the collecting area willingly but steadily.

With time and patience a happy medium between libido, self-confidence, discipline, respect and safety can be achieved to the benefit of all.

5.11. Conclusion

There are numerous methods that may be used for the collection of equine semen. The most popular method throughout the world is the use of a dummy or teaser mare along with an artificial vagina. This system is very successful because it closely mimics the natural mating process. Collection of semen from most stallions presents no problems: most seasoned breeding stallions and young inexperienced stallions adapt readily to ejaculation into an AV rather than into the vagina of a mare. The technique is not associated with any subsequent reticence on the behalf of the stallion to exhibit his natural sexual behaviour; hence stallions may be used in both natural covering programmes and AI programmes within the same season. These facts present semen collection as a relatively simple technique, which can be used on the great majority of normal stallions as well as those with veterinary complaints. This significantly reduces the limitations previously imposed upon a stallion's breeding influence, by geography, physical injury or complaint.

6 Semen Evaluation

6.1. Introduction

Semen evaluation is primarily carried out to give an indication of the fertilizing capacity of spermatozoa produced by a certain stallion. This may be done as part of an assessment process prior to purchase, as a routine management practice at the beginning and at set stages throughout the breeding season, or to assess a stallion's suitability for inclusion in an AI programme.

Traditional tests used in the evaluation process are rightly criticized for their lack of objectivity and for the significant variations between evaluators. As a result of new technological advances more standardized and objective measurements, especially for motility and concentration, may be made, though these are not without their critics. A series of set tests that may be carried out on a sample are discussed in detail in the following sections. Whether all the tests are used on a particular sample or just a selected few is a matter of personal choice and of the reason why the sample is being assessed. In general, as with all investigative work, the more tests that are carried out, the better the overall assessment will be and the more accurate the conclusions drawn. Care should be taken in interpreting results from a single evaluation, due to the numerous variables that can affect spermatozoan quality. According to Rousset *et al.* (1987), assessment of 2–7 ejaculates per stallion, depending upon the parameters to be investigated, is adequate to give a repeatability of 0.8 for the mean values obtained. This compares well with the 17 mare breeding seasons that are reported to be necessary in order to measure stallion fertility based upon pregnancy rate per oestrous cycle, with the same accuracy as that based upon semen evaluation. However, there is no test as good as fertilization itself, and so despite considerable advances in technology and the accompanying improvement in accuracy, no test or group of tests can predict the fertility of any given sample precisely.

This chapter will also consider some of the factors specifically affecting semen characteristics. Those specifically affecting production of spermatozoa are dealt with in Chapter 4.

6.2. Semen Evaluation as Part of an AI Programme

As a routine part of an AI programme, all stallions should have a full assessment of a semen sample carried out at the beginning of the season. This enables any problems that might have occurred since the last season to be picked up and allows a standard to be set for a particular stallion for that particular year. The standard can be used for comparison if problems are suspected later in the season. Once a stallion has passed this broad semen evaluation and has been accepted on an AI programme for a particular season, then a further simpler evaluation may be used for assessing (for example) motility, concentration and morphology of each sample collected prior to its extension for insemination or storage. This shorter assessment is adequate to indicate whether the quality of the samples collected is being maintained. If problems are suspected through a change in these semen parameters or from knowledge of the stallion himself, then a full evaluation is advised before the sample is used for AI. When interpreting the results from semen evaluation the past reproductive use of a stallion must be considered, especially within the last month.

Ideally, for an accurate evaluation, samples should be either: one taken after 3 days' sexual rest preceded (by 1 h) by a double collection 1 h apart; or the last collection of a series of seven daily collections taken, preceded (by 1 h) by a double collection 1 h apart; or both collections taken 1 h apart after 1 month's sexual rest (Pickett and Voss, 1972; Kenney, 1975; Sullivan and Pickett, 1975; Swierstra *et al.*, 1975). In most commercial AI programmes such regimes are not economically viable, and single sampling, as indicated above, interpreted with caution, can provide adequate information for most routine AI practices.

6.3. General Semen Handling

In all procedures involving the handling of live semen samples it is imperative that a constant temperature of 37–38°C is maintained throughout. Variation from this temperature will result in cold or heat shock, both of which will significantly affect the viability of the sample and produce erroneous results. The effects of cold and heat shock are discussed in more detail in Chapter 7 (section 7.4.1). When the sample has been brought to the laboratory (ideally immediately post collection or thawing), it must be placed in an incubator at 38°C until all the equipment for evaluation is ready for use. All tests should be carried out on aliquots removed from this sample, after mixing, as and when required, without the need to remove the sample from the incubator, hence ensuring the maintenance of a constant temperature in the original sample.

Before evaluation begins it is essential that all equipment (pipettes, microscope slides, cylinders, instruments, etc.) are clean, dry and pre-warmed to 38°C along with all solutions that are to be used with a live semen sample.

Pre-heating can be done either by an incubator or a water bath or by means of a warmed stage. An incubator or water bath is to be preferred as they also provide protection and a more consistent warming of large objects and fluids. Small portable incubators are particularly useful, providing convenient and clean transport containers for all the equipment (Fig. 6.1). In addition, the microscope should have a warmed stage. If such a microscope is unavailable, adequate warmth can be provided for about 15–30 min by using a thick piece of pre-warmed glass on the stage, on to which the slide is placed. Provided a number of these glass slabs are available, they can be swapped over and returned to the incubator for re-warming as and when required. The thickness of the glass gives it good heat-retaining properties, but if the glass is too thick it can affect the transmission of light through to the slide for illumination.

In general, a semen sample may be left in the incubator at 37–38°C for up to 30 min without a significant detrimental effect on the evaluation results. A delay of more than this runs a risk of spermatozoan mortality and hence erroneous results in the evaluation process. It is best that all tests which rely upon live spermatozoa, such as live:dead ratio and motility, are carried out as speedily as possible to minimize any potential effect of spermatozoan senescence on the results obtained. Evaluations relying on dead spermatozoa, such as morphological assessment, may be carried out last, provided staining or fixing has been carried out when the spermatozoa are in optimum condition (Fig. 6.2).

Fig. 6.1. A small portable incubator is particularly useful for the storage of glassware and semen during the evaluation process.

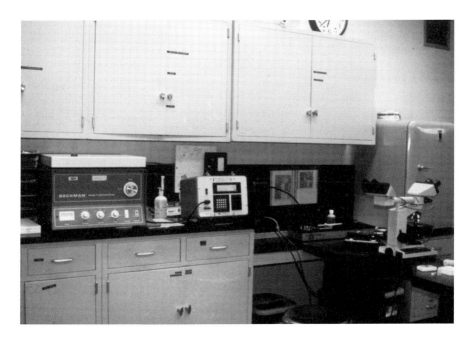

Fig. 6.2. A typical laboratory set up to evaluate semen. The laboratory should be convenient to the collection site and the equipment should include, among other things, incubator, microscope, heated slide warmer and centrifuge. (Photograph by Julie Baumber and Victor Medina.)

6.4. Gross Evaluation

As soon as the semen sample has arrived in the laboratory the first process is to filter the sample to remove the gel fraction, detritus and sloughed epithelial cells (Fig. 6.3). This process may be bypassed if an in-line filter has been placed in the AV liner, in which case the sample will have been filtered on collection. AV liners with non-toxic filters are readily available but tend to increase the cost, especially when using disposable liners. Any filter used must be non-toxic, clean, dry and sterile. It should also be fine enough to allow all the spermatozoa and the majority of the seminal plasma to pass through, but to retain the gel fraction along with any detritus or debris that might have been collected with the sample. Conventionally polyester filters are used, but Amann *et al.* (1983) indicated that nylon filters with a pore size of 37 µm minimize the loss of spermatozoa within the filter and the retained gel. As a temporary measure a double-folded piece of clean, dry, sterile muslin can be placed over the neck of the collecting vessel as it is attached to the liner prior to collection. This acts as an effective filter for both debris and the gel fraction of the semen sample.

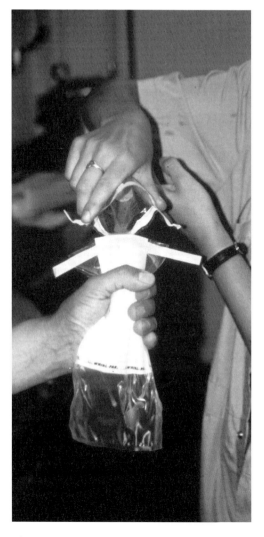

Fig. 6.3. Prior to evaluation, the semen sample requires filtering to remove the gel fraction and any debris. (Photograph by Julie Baumber and Victor Medina.)

6.4.1. Volume

After filtration, the semen is poured into a warm, graduated cylinder (a size of 100 ml will usually suffice) so that the volume of the gel-free fraction of the semen sample can be determined. It is normal practice that the collecting vessel attached to the AV is also graduated so that an indication of the total volume of the ejaculate can be obtained when in-line filters are not used. The volume of gel produced can then be calculated by subtraction.

Factors affecting seminal volume

The normal volume produced by a stallion varies with many factors. As with all evaluation it is ideal to have information on expected values for each stallion to act as a reference point or standard to refer back to if abnormalities are suspected. In general, the total volume of a sample should be between 30 and 250 ml. On average, stallions will produce 100 ml in total, of which 20–40 ml will be gel, leaving 60–80 ml of useful gel-free semen (Dowsett and Pattie, 1982; Pickett *et al.*, 1988b; Fayrer-Hosken and Caudle, 1989; Ricketts, 1993). Significant variations in these volumes are seen which may be due to nothing more than individual stallion variation. However, they may be indicative of abnormalities, a change in the breeding season, age, workload, length of teasing, etc. (Pickett *et al.*, 1988a,b). The availability of a standard for a stallion will help to ascertain the reason for any change. Work by Pickett and Voss (1973), Squires *et al.* (1979a), Pickett *et al.* (1988a,b), and Johnson and Tatum (1989) illustrated well the effect of some of these factors, and a summary of some of their findings is given in Tables 4.2, 4.8 and Fig. 3.19.

THE EFFECT OF SEASON

The effect of season on the volume of seminal plasma has been demonstrated by several workers. Differences of up to 50% have been reported between the height of the breeding season and the non-breeding season. These differences are largely accounted for by changes in the volume of gel produced (Pickett *et al.*, 1976; Pickett and Voss, 1972; Pickett and Shiner, 1994).

THE EFFECT OF WORKLOAD

The effect of workload (that is, ejaculation frequency) has been investigated by several authors (Pickett and Voss, 1972; Kenney, 1975; Pickett *et al.*, 1975c; Sullivan and Pickett, 1975; Swierstra *et al.*, 1975). In summary, it is apparent that the major effect of increasing the ejaculation frequency of a stallion is on the volume of gel fraction produced, with the first ejaculation invariably having a higher gel fraction than subsequent ejaculations (Pickett *et al.*, 1976). When frequency of collection throughout a week was investigated, it was concluded that for the majority of stallions a frequency of collection ranging from one per week to six per week had no significant effect on seminal volume, either total or gel-free (Pickett *et al.*, 1975c). However, a second ejaculation within 24 h did show a significant drop in total volume (Dowsett and Pattie, 1987; Pickett and Shiner, 1994). No significant effect should be seen in the great majority of stallions with a collection frequency of less than one ejaculate per 24 h.

THE EFFECT OF AGE

Age has also been demonstrated to have a significant effect on seminal volume. Squires *et al.* (1979a) showed an increase in seminal volume from 2 to 16 years of age. The increase was particularly evident between the ages of 4–6 and 9–16 years, where volumes were shown to increase with age by 120%. It was also apparent that older stallions produced a higher gel fraction, though the difference was not significant.

THE EFFECT OF TEASING

The length of time that a stallion is teased prior to ejaculation has also been shown to have an effect on the volume of seminal plasma produced. The longer a stallion is allowed to tease an oestrous mare, the greater is the volume of semen produced, largely due to an increase in the gel-free fraction (Ionata *et al.*, 1991b).

THE EFFECT OF CLINICAL CONDITIONS

As will be discussed in later sections of this chapter, significant increases in volume may be the result of contamination with urine (urospermia), blood (haemospermia) or pus (pyospermia). These are normally picked up initially as changes in the appearance of a sample and are discussed in the next section. In addition it has been reported that conditions such as testicular rotation may also be implicated if seminal volumes are lower than might be expected (Pickett *et al.*, 1988a,b).

6.4.2. Appearance

Once the filtered sample is in the measuring cylinder, its colour and consistency can be assessed. This is normally done by eye but requires microscopic examination for confirmation. A good sample should appear milky white in colour, though it may range from watery to creamy, depending upon the spermatozoan concentration within the sample (Kuklin, 1983). The sample should also have a neutral smell and be evenly turbid with no evidence of clots, and should be the consistency of thin single cream (Kuklin, 1983; Frayrer-Hosken and Caudle, 1989).

Factors affecting seminal appearance

Abnormal colours may indicate contamination with urine, blood or pus. Significant increases in the volume of such samples will also be an indication of the extent of the problem.

UROSPERMIA

Urospermia is the term given to the condition resulting in contamination of a semen sample with urine, for which there are several causes (Voss and Pickett, 1975; Slusher, 1997).

Causes. One of the reasons for the condition of urospermia is a disruption of the neural mechanisms involved in ejaculation, which can lead to accidental bladder contraction and hence urine leakage during ejaculation (Rasbech, 1975; Leendertse *et al.*, 1990; Mayhew, 1990). Control of the bladder sphincter and seminal emission is via the α-adrenergic sympathetic nervous system. Hence any interference with, or disruption of, this pathway might be involved in the condition (Voss and McKinnon, 1993). It has also been reported that the condition may be associated with self-mutilation in stallions (Samper, 1995a). In addition, neuropathies which cause bladder paralysis may also result in urospermia. This may be associated with equine herpes virus type I, as a secondary effect, or a result from poisoning with sorghum grass or due to cauda equina neuritis (Varner and Schumacher, 1991; Voss and McKinnon, 1993). Stallions suffering from urospermia may appear normal, with no neurological defects and with adequate libido and mating ability. The condition may be continuous or intermittent in appearance and its occurrence may be unpredictable. Contamination may occur at any time during ejaculation and may be as little as 1 ml or as much as 250 ml or more in volume (Varner and Schmacher, 1991). Evidence suggests that contamination is not likely to be due to just leakage but to an all-or-nothing effect (Nash *et al.*, 1980).

Diagnosis. Gross contamination of a sample is apparent from the semen's yellow to amber colour and characteristic odour of urine. A sediment may also be evident. Microscopically, crystals associated with urine may be observed. In order to diagnose slight contamination, concentrations of urea nitrogen or creatinine may be measured (Danek *et al.*, 1994a), values > 25–30 mg dl^{-1} and > 2 mg dl^{-1}, respectively, being diagnostic of the condition (Hurtgen, 1987; Varner and Schumacher, 1991). Field tests are now available which enable urine contamination of 10% or greater to be diagnosed. Field tests using test strips are also available for urea nitrogen, nitrates and creatinine (Althouse *et al.*, 1989). The test strips diagnose the presence of urine by changing colour (Voss and McKinnon, 1993). Urea nitrogen strips (Azostix, Miles, Inc., Ag-Vet, Shawnee, Kansas) turn from yellow to green to indicate a positive result. Nitrate reagent strips (Multistix, Miles, Inc., Ag-Vet, Shawnee, Kansas) turn from yellow to radiant orange when positive (Voss and McKinnon, 1993). Elevated osmolarity (> 350 mOsm) may also be indicative of urospermia.

The effect of urine contamination on spermatozoan viability. Urine contamination within a semen sample adversely affects the motility of spermatozoa and their capacity to fertilize an ovum (Hurtgen, 1987; Varner and Schumacher, 1991). Urine contamination may cause an elevation in sample pH which reduces spermatozoan viability primarily via an effect on plasma membrane structure. Finally, changes in osmotic pressure are seen due to the introduction of hyperosmotic urine. This will cause water to pass out of the spermatozoa and down the osmotic gradient to the seminal plasma, resulting in desiccation of the spermatozoa and a reduction in viability (Hurtgen, 1987; Samper, 1995a). The severity of the above problems seems to be dose dependent. It is apparent that stallion spermatozoa can tolerate minute amounts of urine without deterioration but the threshold level for contamination has not yet been determined and is unlikely to be a simple single value.

Treatment. Samples with minor urine contamination may be diluted in an appropriate extender immediately upon collection, to try to counteract, or protect the spermatozoa from, the potential urine-related damage. The sample may also be centrifuged after dilution in an appropriate extender; the supernatant is then decanted off, and the urine along with it, and the spermatozoan pellet is re-suspended in urine-free extender prior to insemination (Varner and Schumacher, 1991).

Treatment of the condition is not simple and is often unsuccessful. Treatment may be by alteration in management techniques – for example, by delaying covering or collection until immediately after natural urination. Alternatively, urination may be stimulated by the administration of diuretics, such as frusemide or ephedrine sulphate, immediately prior to covering or collection (Voss and McKinnon, 1993). The effectiveness of such management practices in reducing the problem is disputed (Hurtgen, 1987). Urination may also be encouraged by the presence of faeces, from another stallion or a mare in oestrus, so inducing the stallion's natural territory-marking activity. Some stallions can be trained to urinate upon command, while others urinate in response to feeding, fresh bedding or exercise. As a more extreme measure, bladder catheterization has been used to evacuate urine prior to collection, but repeated use of this technique is not advised due to the risk of urethritis or cystitis (Voss and McKinnon, 1993; Varner and Schumacher, 1991). Pharmacological therapy has been used primarily to close the neck of the bladder, using α-blockers. This treatment apparently works in men suffering from retrograde ejaculation, but this condition is not reported as a significant cause of urospermia in horses and so the effectiveness of the treatment is limited (Varner and Schumacher, 1991; Voss and McKinnon, 1993; Samper, 1995a). Other alternatives, including the use of oxytocin to induce contraction of the bladder sphincter, or bethanecol chloride or flavoxate hydrogen chloride, have been tried but with only limited success (Varner and Schumacher, 1991; Voss and McKinnon, 1993). Significant work is still required before the major causes of the condition can be ascertained and feasible treatments and regimes devised. One of the biggest problems is the unpredictability of the condition, with many stallions showing the condition only intermittently. In general, however, urine is passed during the latter stages of ejaculation. An open-ended AV (section 5.2.4) may be used to identify and collect the various fractions of the ejaculate, so allowing the later jets to be discarded.

HAEMOSPERMIA

Haemospermia is the term given to contamination of a semen sample by blood.

Causes. The causes of haemospermia are varied and more commonly may include lacerations of the exterior of the penis caused, for example, by a tail hair from the mare entering the vagina at covering, or other accidental or puncture wounds of the glans penis area. The condition may also be caused by urethral defects (Schumacher *et al.*, 1995), the damaged areas tending to haemorrhage at erection and ejaculation due to the increased blood pressure.

If it is not treated immediately, the condition is made worse and may predispose to infections or, more specifically, to surface infestation with *Habronema* or *Druschia* larvae (summer sore) (Hurtgen, 1987; Varner and Schumacher, 1991; Voss and McKinnon, 1993). Habronemiasis on the glans penis area may itself also cause haemorrhage. The larvae are carried by house and stable flies, which are attracted to the relatively humid environment of the genital area. The larvae are then deposited on the penis when the flies feed, particularly in spring and summer (Varner and Schumacher, 1991). The presence of these larvae induces rapid production of eosinophilic granulation tissue, which surrounds the larvae. As a result this whole area becomes friable and tends to haemorrhage at erection and ejaculation. Particular problems arise if significant infestation of the urethral process occurs (Hurtgen, 1987; Varner and Schumacher, 1991; McKinnon and Voss, 1993). Good hygiene and fly control help to control the condition and organophosphates and ivermectins are used to treat it (Bowen, 1986; S. Revell, Wales, 1998, personal communication).

Urethritis, usually due to bacterial infection of the urethra, is a further common cause of haemospermia. The bacteria most commonly associated with the condition are *Pseudomonas aeruginosa*, *Streptococcus* species and *Escherichia coli* (Blanchard *et al.*, 1987; Samper, 1995a). Stallions with a history of infection with *Pseudomonas aeruginosa* have an increased chance of suffering from haemospermia (McKinnon *et al.*, 1988). Viral urethritis causes inflammation and vesiculation which may result in haemorrhage at erection and ejaculation (Voss and Pickett, 1975). Seminal vesiculitis, or infection of other accessory glands, may also be a rare cause of haemorrhaging (Blanchard *et al.*, 1987). Blood contamination of semen associated with migration, through the seminal vesicles, of the parasite *Strongylus endenatus* has been reported (Pickett *et al.*, 1981).

Ulceration of the urethra and rupture of the urethral subepithelial blood vessels have been reported as causes of haemospermia (Schumacher *et al.*, 1995). The latter is reported to be associated with the use of stallion rings to discourage masturbation (Voss and McKinnon, 1993). The use of these rings should be discouraged.

Diagnosis. The extent of the haemorrhage is reflected in the colour of the semen sample, which varies from red to pale pink. However, if no coloration is evident, blood contamination may be shown to be present by microscopic examination for erythrocytes. External haemorrhage is evident from examination of the genitalia for wounds, lacerations, etc. Urethritis, ulceration of the urethra or rupture of the subepithelial vessels around the urethra may be diagnosed by the use of penile and pelvic urethral endoscopy performed on a standing tranquillized stallion. Air or saline may be used to dilate the urethra and aid viewing (Hurtgen, 1987; Voss and McKinnon, 1993). This method may also be used to assist in the diagnosis of accessory gland infection, by examination of the ducts during massage *per rectum*, and examination of the resulting extruded fluid. Biopsies may be taken of suspect tissue. Positive contrast and double contrast radiographic studies, using barium sulphate solution infused into the urethra, may be used to identify occlusions, prolapses, adhesions, strictures, etc. (Walker and Vaughan, 1980; Voss and McKinnon, 1993). The use of ultrasonography allows more detailed examination of accessory gland structure and function, and may be used as an aid to diagnosis (Varner and Schumacher, 1991). Smears of cell debris left on the urethral process immediately *post coitum* can be used, after fixation, for the identification of viruses and neoplastic or other abnormal cells.

The effect of blood contamination on spermatozoa viability. The mechanism by which blood contamination affects spermatozoa quality is unclear. It is thought to be an effect modulated by the erythrocytes, white blood cells or platelets, rather than via the serum (Voss *et al.*, 1976; McKinnon *et al.*, 1988; Varner and Schumacher, 1991). Experimental contamination of semen for AI with serum or whole blood resulted in significantly different pregnancy rates: mares inseminated with semen contaminated only with serum showed significantly higher pregnancy rates than those inseminated with semen contaminated with whole blood (Voss *et al.*, 1976).

Treatment. External lesions can be treated in the normal fashion with topical antibiotics to prevent infection and aid the healing process. Cessation of mating until complete healing is achieved is essential to prevent recurrence. Early treatment of *Habronema* lesions with anti-inflammatory agents (for example, with corticosteroids) may be successful (Hurtgen, 1987).

Subischial urethrostomy has been reported to be successful. The operation involves longitudinal incision of the urethral wall and the enveloping carvenous tissue, in the perineal area. The resulting wound is left to heal naturally (Varner and Schumacher, 1991). This technique was originally developed to investigate the cause of haemospermia prior to the advent of the endoscope. Post urethrostomy and sexual rest stallions who had normal fertility rates prior to the advent of haemospermia are reported to return to their previous breeding ability. However, those with long-term infertility problems before the advent of haemospermia showed no improvement in fertility rates (Sullins *et al.*, 1988). While the reasons for the reported success of this technique are unclear, Voss and McKinnon (1993) have suggested that it may be due to a counter-irritant effect or the temporary diversion of urine.

Bacterial urethritis may be treated, after identification of the bacterium, by using an appropriate specific antibiotic. This is not always successful, due to the relative inaccessibility of the internal genital tract. If persistent genital tract infection is evident, but no pus or haemorrhage is seen, then antibiotic seminal extenders may be used either with AI or by infusion into the mare after natural service (Hurtgen, 1987). Stallions with early habronemiasis lesions or inflammation of the distal urethra may be collected from using an open-ended AV. This allows the blood-free jets to be collected and only these used for insemination (Hurtgen, 1987). Centrifugation of contaminated semen with subsequent resuspension in extenders may also be used (Samper,

1995a). Many conditions leading to urethritis are associated with damp, infected bedding; general improvement in hygiene may have a beneficial effect (Bowen, 1986).

PUS

Accumulation of pus or pyospermia is an important indication of infection. This will be discussed in further detail later in this chapter when the bacteriological evaluation of semen is considered. Regardless of the cause, the presence of a significant amount of pus excludes a sample from use in AI. Natural covering should also be avoided until full recovery is evident (Samper, 1995a).

EJACULATION OF A SMALL VOLUME

Samples with significantly smaller volumes than expected may be obtained consistently from some stallions. These are often the result of incomplete ejaculation (Fayrer-Hosken and Caudle, 1989; Samper, 1995a) and a second ejaculation should be taken to confirm this. It must be remembered that there is considerable variation in the volume of semen produced normally by different stallions. Some stallions consistently produce small volume samples (even as low as 15–20 ml) but these will often have higher than average spermatozoan concentrations.

6.4.3. Osmolarity

The osmolarity of the semen sample should be the next parameter checked. This may be done by means of an osmometer (such as Micro Osmometer Automatic, Type 13 Autocal, Roebling, Germany). The normal osmolarity range should be between 290 and 310 mOsm (Pickett *et al.*, 1976, 1989). Values outside this range cause water to diffuse out of or into the spermatozoon head and tail, down the osmotic pressure gradient across the semi-permeable plasma membrane. Hypotonicity will cause swelling and deformation of the cell, especially the tail, and may result in rupture of the plasma membrane. Hypertonicity causes dehydration, which adversely affects spermatozoon function. An osmolarity greater than 350 mOsm may occur in urospermia and values less than 200 mOsm can cause severe changes in spermatozoon morphology – typically swelling, bending or coiling of the tail (Varner and Schumacher, 1991). The extent of the damage will vary considerably, depending on the hydration staus of the horse (S. Revell, Wales, 1998, personal communication).

6.4.4. Seminal fluid pH

Seminal fluid pH should also be assessed. This may be done by using either a pH meter or, less reliably, short-range pH paper, immediately post collection.

Values should be in the range of 6.9–7.7 (Davies Morel, 1993; Oba et al., 1993), though some authors suggest that a tighter range of 7.35–7.7 is more appropriate (Pickett and Back, 1973; Rossdale and Rickett, 1980; Hurtgen, 1987; Fayrer-Hosken and Caudle, 1989). In general, pH levels tend to be higher in the second ejaculate when two ejaculates are collected in succession. This change in pH is probably due to a reduction in epididymal secretions, after depletion of reserves during the first ejaculate, and a lower concentration of spermatozoa (Pickett et al., 1976, 1988a,b). The reported negative correlation between seminal volume and pH and between the number of spermatozoa and pH (Pickett et al., 1988a) suggests that samples with a high pH may have low spermatozoan concentrations. This relationship may be used to indicate whether ejaculation has been complete. Samples from stallions who appear to have ejaculated fully, due to the large amount of fluid deposited, but which in fact have a low spermatozoan concentration, can be readily identified as the pH of such samples will approach 8 (Pickett et al., 1976; Pickett, 1993a). Elevated pH (above 7.8) may also be indicative of the presence of extraneous material or possible infection. Acidic conditions, as observed naturally within the mare's vagina, are known to be detrimental to spermatozoan viability.

6.5. Microscopic Evaluation

Once gross evaluation of the sample has been completed, microscopic examination is required to enable spermatozoan characteristics to be assessed and evaluated. During microscopic evaluation, as with gross evaluation, it is of paramount importance that the sample is maintained at 38°C, especially as microscopic examination may take some time.

Prior to microscopic evaluation the sample is invariably extended, though an indication of motility is often assessed microscopically using a raw sample. Dilution is primarily used to allow individual spermatozoa to be seen, as spermatozoa within raw semen tend to clump together, making more than a very rough estimate of motility impossible. Extension of the sample also prolongs the life of the spermatozoa, providing them with an additional source of energy and substrates for survival, and the maintenance of viability while evaluation takes place (Yates and Whitacre, 1988). Addition of the extender must be done immediately post collection, ideally within 2 min, to reduce the chance of obtaining erroneous results from a loss of spermatozoa viability due to delay. There are numerous extenders available and used successfully during the evaluation process. In general, these are the same as those used in the preparation of semen for immediate AI or semen storage. Further details of these are given in Chapter 7. The most popular are those based upon either non-fat dried milk solids (NFDSM) or skimmed milk (Tables 6.1 and 6.2). The dilution rate required depends upon the concentration of spermatozoa and also the test being performed.

Table 6.1. Non-fat dried skimmed milk–glucose (NFDSM–glucose) extender 1, for use in semen evaluation (Kenney et al., 1975).

Component	Quantity
NFDSM	2.4 g
Glucose	4.0 g
Penicillin (crystalline)	150,000 units
Streptomycin (crystalline)	150,000 µg
Deionized water (made up to)	100 ml

Table 6.2. Non-fortified skimmed milk extender for use in semen evaluation (Varner and Schumacher, 1991).

Component	Quantity
Non-fortified skimmed milk	100 ml
Polymixin B sulphate	100,000 units
Heat milk to 92–95°C for 10 min in a double boiler, cool and add the polymixin B sulphate.	

6.5.1. Motility

The percentage of motile spermatozoa, and in particular those showing progressive motility, is a good indicator of the number of viable spermatozoa. The correlation between motility and morphologically normal spermatozoa is reported to be 0.63 (Long et al., 1993). Motility may be assessed either visually or using a computerized motility analysis system.

Visual motility analysis

Undiluted raw semen may be placed on a warmed microscope slide and viewed using a light microscope. The wave-like motion seen is assessed and graded. This method allows a quick, rough estimate of motility and, therefore, viability to be made. Dilution of the sample allows a more accurate evaluation as agglutination of spermatozoa tends to occur in the undiluted sample, resulting in erroneously high estimates of motility. Dilution of the sample also counteracts any adverse affects of seminal plasma pH and allows the easier viewing of individual spermatozoa. Assuming a concentration of 200×10^6 spermatozoa, the rate of dilution for this purpose is in the region of 1:20 semen sample:diluent, to give a concentration of approximately 10×10^6 ml^{-1}. Normally 0.25 ml of semen sample, made up to 5 ml with extender, is adequate. The extender and all glassware must be pre-heated to 38°C and maintained at this temperature throughout the procedure. Examination of the sample at a temperature lower than 38°C will result in reduced motility (Amann *et al.*, 1988). This is an important consideration when assessing motility in samples that have been frozen. If the sample is particularly concentrated, then the ratio of semen to extender can be adjusted accordingly. In order to assess the sample for motility, a drop (approximately 0.1 ml) of the extended sample is placed on a clean, dry and pre-warmed microscope slide. Counting chamber slides may be used (Microm Cell VU TM, Thame, UK), ensuring a set volume depth (usually 20 µm) of sample is used. These may be assessed visually but more commonly the assessment is computer aided. The drop of extended sample is then covered with a cover-slip, care being taken not to trap any air bubbles under the cover-slip. The sample should be viewed under 100–400× magnification using a phase contrast microscope, ideally with a heated microscope stage (Jasko, 1992). As mentioned previously, a small pre-warmed glass slab may be used as an alternative to the heated stage. The phase contrast microscope allows clear resolution of cells, such as spermatozoa, which are essentially transparent. Observations from at least five fields of view should be made. The spermatozoa to be noted are those showing progressive motility (PMOT), that is, spermatozoa moving rapidly across the field of vision with each lash of the tail (Yates and Whitacre, 1988). This results in progressive movement and 360° head rotation. Numerous other types of motility may be observed – circular, backwards, intermittent, lashing tails but no accompanying forward movement, etc. These spermatozoa should not be included in the assessment as such abnormal movement is indicative of abnormalities. Motility may be classified on a scale of 0 to 5 (Table 6.3), 0 being very poor and 5 being excellent.

In any evaluation, it must be remembered that not all of the spermatozoa move in a progressively motile fashion all of the time: they often exhibit periods of quiescence or standstill between periods of activity. It cannot be assumed automatically that all stationary spermatozoa are non-viable.

Visual assessment of motility is a quick and easy method to master, but it has the major drawback of subjectivity. Variations are evident due to individual

Table 6.3. Classification of spermatozoa motility assessed microscopically (Davies Morel, 1993).

Grade	Description
0	Immotile
1	Stationary or weak rotatory movements
2	Backward and forward movement or rotatory movement, but fewer than 50% of cells are progressively motile
3	Progressively rapid movement of sperm with slow currents, indicating that about 50–80% of sperm are progressively motile
4	Vigorous, progressive movement with rapid waves, indicating that about 90% of sperm are progressively motile
5	Very vigorous forward motion with strong rapid currents, indicating that nearing 100% of the sperm are progressively motile

judgement, the concentration of spermatozoa, the depth of the sample when viewed, the degree of contamination, the degree of agglutination, the temperature and the diluent. A necessary drawback of visual assessment is the restriction to spermatozoan movement imposed by the relatively shallow sample, required to suit the limited depth of field of the microscope objective lens. The depth of fluid required in order to allow free spermatozoan movement is > 50 µm and probably up to 100 µm is really required. Unfortunately, most microscopes do not possess this depth of field and require ≤ 50 µm (Amann and Hammerstedt, 1980). The depth of solution obtained by placing a small drop of extended semen (7–18 µl) on a microscope slide and covering it with a cover-slip (18 × 18 mm to 22 × 22 mm) is approximately 16 µm. Though not ideal, it seems apparent from the work of Amman et al. (1988) that this depth does not give significantly erroneous results. A number of workers have investigated the different means by which a 15–20 µm depth of solution may be achieved for viewing under the microscope. All of the following are reported to meet this requirement: 7 µl of sample under a 18 mm × 18 mm cover-slip; 10 µl of sample under a 22 mm × 22 mm cover-slip; 12–18 µl of sample under a 22 mm × 22 mm cover-slip (Makler, 1978; Jasko et al., 1990b; Amann and Graham, 1993). Whatever system is used, consistency of approach between samples is of the utmost importance.

In the three examples given, the first two are those used by computerized motility analysers with the use of a special chamber. The last one is that used in visual assessment, as continuous focusing on different levels is possible (Makler, 1978). Visual assessment, therefore, has the advantage of having less of an adverse effect on motility patterns. Unfortunately, the repeatability of visual assessment results between different technicians and laboratories is very poor, limiting its use in accurate comparative work. Conflicting evidence is presented as to whether or not there is a direct relationship between percentage motility and fertility. Some workers have failed to show a correlation (Voss et al., 1981; Dowsett and Pattie, 1982), whereas others have reported a correlation of 0.7–0.8 with visual assessment (Samper et al., 1991; Jasko et al., 1995). This correlation figure apparently declines when the sample has been frozen (Samper et al., 1991).

While motility is being assessed visually, a rough estimate of the ratio of leucocytes to erythrocytes to spermatozoa may be made (Hurtgen, 1987).

Computerized motility analysis

In an attempt to reduce some of the drawbacks of visual assessment of motility, especially the subjectivity, the use of automated assessment has been investigated. Time-lapse photomicrographic assessment was initially attempted and proved to provide more repeatable results than visual analysis, though the results obtained from the two methods of assessment demonstrated a high correlation (0.72–0.81) (Huffel et al., 1985). Multiple-exposure photomicrography and frame-by-frame playback video micrography have also been used in an attempt to reduce the subjectivity of results. However, all these methods are labour intensive. Further work led to the development of computer analysis (Amann, 1988; Jasko *et al.*, 1988, 1991b; Jasko, 1992; Burns and Reasner, 1995). The correlation between the results obtained from assessment via visual analysis compared with computer analysis is very good at 0.92 (Bataille *et al.*, 1990; Malmgren, 1997). Computer analysis is largely restricted to research establishments, due to the expense of the equipment required, but it does provide a convenient and much less subjective assessment. It provides additional information such as a video image along with data on spermatozoan characteristics – for example, the total number of spermatozoa, percentage motility, percentage of progressively motile spermatozoa, linear velocity, linearity, path velocity, lateral head displacement, head size and averages. Such parameters would be impossible to obtain through visual assessment (Palmer, 1991; Burns and Reasner, 1995; Moses *et al.*, 1995). Comparison between computerized systems and frame-by-frame playback video micrography indicates that both systems provide an accurate method of determining spermatozoon motility and velocity, with the computerized system having a slight advantage (Varner *et al.*, 1991a). Two major computerized systems that have been developed are the Hamilton Thorne Motility Analyser (Fig. 6.4) and the CellSoft system (Mortimer *et al.*, 1988; Jasko *et al.*, 1990b; Varner *et al.*, 1991a). The most sophisticated equipment now allows data on individual spermatozoa to be obtained and will provide data on 200 cells within 30 s (HTM-IVOS Semen Analyser, Hamilton Thorne, Beverly, Massachusetts). Computer analysers are now able to provide an automated, rapid and objective result.

For computer analysis the sample to be assessed is prepared by dilution (as for visual assessment) in order to allow individual spermatozoa to be tracked. Low dilution rates lead to erroneous spermatozoon track interpretations, due to overlap of spermatozoa and interlocking between them (Varner *et al.*, 1991a). The dilution rate required depends upon the analyser to be

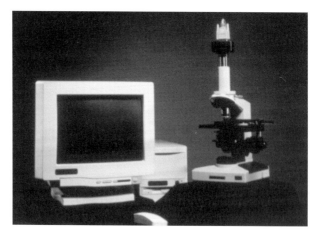

Fig. 6.4. The Hamilton Thorne HTM-IVOS Semen Analyser (by kind permission of Hamilton Thorne, Beverly, Massachusetts).

used but the aim is to obtain a range of 25–50 $\times$ 10^6 spermatozoa ml^{-1}. It is particularly important that the extender is optically clear, and this requirement largely precludes the use of egg-yolk-based extenders with computerized analysis systems (Jasko *et al.*, 1988; Varner *et al.*, 1991a; Ziegler, 1991). A set volume of sample plus extender is put in a small cuvette tube or counting chamber and placed in the machine ready for analysis. Due to the variability in results obtained, both between and within samples, most machines require a sample of 500 cells to be evaluated in order for acceptable results to be obtained. Even so, differences in results obtained between analysers are still a source of inaccuracy when making comparisons (Jasko *et al.*, 1990b).

Many analysers combine motility analysis with an assessment of concentration using double chamber slides. The assessment of samples to gauge spermatozoan concentration will be addressed in a later section of this chapter. One of the major drawbacks with computer-automated spermatozoa-counting machines is their inability to differentiate between spermatozoa and debris; this results in inaccuracies, especially in samples with significant contamination. New systems, using fluorescent differentiation of spermatozoa from debris, have been used for some time in human work (Evenson and Melamed, 1983) but only recently applied to cattle and horses (Garner *et al.*, 1986; Ley *et al.*, 1991). DNA-specific fluorescent dye is added to the sample. Spermatozoan heads, as previously discussed, have significant concentrations of DNA to which the fluorescent dye attaches. Detritus contains little DNA (if any) except for leucocytes or immature spermatocytes, which are diploid in nature and so will have a higher level of fluorescence than the spermatozoa. Appropriate adjustment of the analyser allows discrimination between these cells and an assessment of spermatozoan concentration along with a percentage motility figure assessed from the neighbouring chamber containing non-stained spermatozoa.

Such a system potentially provides a quick, easily repeatable and acceptable range of analysis for spermatozoa for use with AI (HTM-IDENT, Hamilton Thorne Research, Beverly, Massachusetts) but inconsistent and inaccurate results can be easily obtained if the system is not set up correctly (S. Revell, Wales, 1998, personal communication). Similar flow cytometry methods using acrosome-specific fluorescent dyes have been developed and have the advantage of also providing information on either acrosomal or chromatin integrity. Other methods have been used for assessing motility, including a photographic method using dark-field microscopy (Elliott *et al.*, 1973). New methods that are being developed to assess spermatozoan motility include the use of resazurin (7-hydroxy-3H-phenoxazin-3-one 10 oxide) dye, which is reduced to resorufin and then to hydroresorufin by motile spermatozoa. This reaction produces a change in dye colour from purple, through pink to white, which can be compared with a colour chart calibrated to indicate a positive result at a set concentration of motile spermatozoa (Glass *et al.*, 1991). Using this principle, the Stud SCORE Kit is currently being tested for commercial use on horses, calibrated to give a positive result at 140 $\times$ 10^6 spermatozoa ml^{-1} and is reported to be 85% accurate (Stud SCORE Kit™, Microm, Thame, UK). The accuracy of such tests assumes that all spermatozoa respire at the same rate. Evidence in bulls, where differences in ATP content of spermatozoa have been observed, suggest that this assumption is not necessarily correct (S. Revell, Wales, 1998, personal communication).

Separation of spermatozoa on the basis of motility has also been attempted using a bovine serum albumin (BSA) column. Spermatozoa with superior motility are isolated at the lower end of the column and subsequent pregnancy rates using these spermatozoa were reported to be better than controls (Goodeaux and Kreider, 1978). Significant developments are currently being seen in the area of spermatozoan analysis, with whole spermatozoa analysers being developed and marketed at increasingly more affordable prices.

Despite these significant advances and however accurate these analysers become, their ability to predict the fertilizing capacity of a semen sample is restricted by the level of correlation that exists between fertility rates and the parameters they measure. Positive, but low, correlations have been reported between fertility and computer analysis-derived motility (0.34), progressive motility (0.27) and mean velocity (0.3) (Malmgren, 1997). However, the lack of significant correlations remains the major stumbling block to the widespread use of computer analysers.

As a rough guide, semen samples with in excess of 40–50% (ideally not less than 60%) progressively motile spermatozoa may be considered appropriate for use in an AI programme (Pickett *et al.*, 1988a; Fayrer-Hosken and Caudle, 1989; Colenbrander *et al.*, 1992; Davies Morel, 1993; Ricketts, 1993).

6.5.2. Longevity

Assessment of motility over a period of time may be carried out to give an indication of the length of time over which acceptable levels of motility are maintained. The test, using either raw or extended semen, is normally conducted at either 37–38°C, 22°C or 5°C. As the temperature of the test and the dilution rates are known to affect longevity (Samper, 1995b), it is important that they are standardized. Semen:extender dilution rates between 1:1 and 1:4 are recommended. For a standard longevity test, samples of raw semen are kept at either 37°C or 22°C, and an extended sample is kept at 5°C. All are then assessed for progressive motility at 15 min intervals for the first hour, followed by hourly assessments. All samples should be allowed 5 min equilibration time at 37°C before they are viewed. Low temperatures depress spermatozoan velocity, giving erroneous results (Amann, 1988; Amann *et al.*, 1988). It is recommended that raw semen, for use in an AI programme, should show no significant change in percentage progressive motility (PMOT) after 30 min storage at either temperature (22°C or 37°C), greater than 45% PMOT after 3 h, and greater than 10% PMOT after 8 h. The extended semen sample kept at 5°C would be expected to show better motility than the rates given for raw semen; in particular, PMOT of greater than 10% should be evident after storage for 48 h. In general, samples stored at all three temperatures should show no significant change in PMOT over the first few hours.

The purpose of a longevity assessment is to give a guide to the expected survival time for the spermatozoa within the female tract, however it must be remembered that a test carried out *in vitro* may have little relevance to the hurdles, obstacles and support provided *in vivo* by the mare's genital system. Attempts have been made to try to provide a more realistic estimate of longevity by mimicking more closely the conditions within the mare's tract. An example of this is the dilution of a semen sample with 7% glucose solution in a ratio of 1:3, followed by storage at 40°C (Kenney *et al.*, 1975).

6.5.3. Concentration

The concentration of spermatozoa within a sample is one of the most important parameters to assess once viability has been ascertained. Not only will it give an indication as to whether the sample is worth using and an indication of the reproductive capability of the stallion, but it will also determine the number of mares that can be covered from a single ejaculate. This significantly affects the potential value of that sample.

Concentration can be ascertained by one of two methods: a haemocytometer (graduated counting slide), and a spectrophotometer (Jasko, 1992). The former is by far the cheapest option but is more time consuming and is reported to produce less consistent results. However, the calibration of many spectrophotometers relies on haemocytometer counts, and so are themselves dependent upon the accuracy of counting.

Haemocytometer

The haemocytometer (Fig. 6.5), originally developed for counting erythrocytes and leucocytes, can successfully be used for counting other microscopic particles – including spermatozoa. It is able to give a very accurate assessment of spermatozoa concentration, but this is at the expense of time and requires considerable skill and care.

Figure 6.5 illustrates the new improved Neubauer haemocytometer, a graduated slide for assessing erythrocytes and leucocytes using a light microscope. To use this haemocytometer to assess spermatzoan concentrations, a sample of gel-free semen diluted with a formalin saline solution (4% formalin in 0.9% saline) or formal citrate solution (2.9% sodium citrate) is required. The haemocytometer pipettes provided allow for different dilution rates; the rate required depends upon the concentration of the sample (and so may need to be altered after an initial viewing) but a ratio of 1:100 semen:formalin solution is usually appropriate. Once the solution is ready the haemocytometer cover-slip is placed over the grid on the haemocytometer by sliding it with some pressure, so ensuring a good seal. Newton's rings should be visible in both contact areas. When the cover-slip is in place, the etched surface under the cover-slip should be touched with the end of the full pipette and the solution from the pipette allowed to fill up exactly the cavity between the etched surface and the

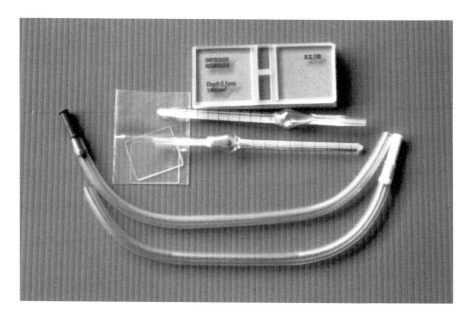

Fig. 6.5. The new improved Neubauer haemocytometer, which may be used to assess spermatozoan concentrations.

cover-slip. The fluid must exactly fill this space: overfilled, underfilled or bubble-containing preparations must be discarded. The slide can then be viewed under a light microscope at 10× magnification. The number of spermatozoa that can be seen within 20 of the smallest counting squares should be counted. The greater the number of replicates of such counts, the greater is the accuracy of the final estimate. Once counting has been completed, the average number (z) of spermatozoa per small square is calculated (Figs 6.6 and 6.7). Then the concentration of spermatozoa ml^{-1} (assuming an initial dilution rate in the pipette of 1:100) is $z \times 4000 \times 100$. (The area of one small haemocytometer square = 1/20 mm × 1/20 mm = 1/400 mm^2; the depth of the chamber = 1/10 mm; therefore, the volume of each square = 1/4000 mm^3 = 1/4000 ml.)

One of the major advantages of such a system is that identification of individual spermatozoa is possible and differentiation between spermatozoa and debris can be made. However, it is evident that the procedure is skilled and time consuming, especially if numerous samples are to be assessed.

Spectrophotometer

A spectrophotometer estimates the optical density of a fluid, placed in a cuvette in the pathway of a beam of light, by measuring the proportion of light absorbed by the fluid. Provided correct calibration has been carried out initially (using haemocytometer counts), then the light absorption recorded can be converted into concentration either manually or automatically.

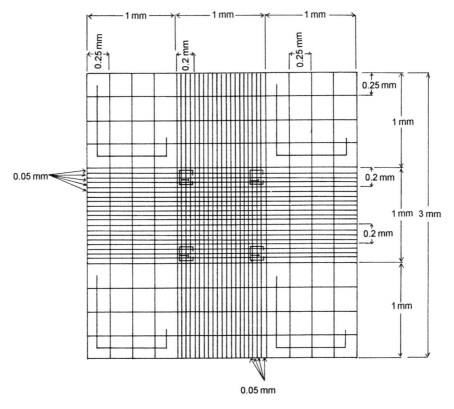

Fig. 6.6. A diagram to illustrate the rules on the base of the counting chamber of the improved Neubauer haemocytometer. The number of spermatozoa within a random sample of 20 of the smallest squares should be counted.

The addition of spermatozoa to a clear solution makes it opaque and the degree of opacity is directly correlated with the number of spermatozoa in the suspension. Since light absorption gives a measure of opacity, a spectrophotometer can be used to assess the concentration of a semen sample (Thomassen, 1988). It is essential that the solution used for diluting the sample is optically clear. Formal saline (10% formalin in 0.9% saline) with a dilution rate of 1:10, 1:20 or 1:30 is most commonly used. Other solutions are advocated, such as lauryl sulphate or borate buffer, which take the shine off the spermatozoa surface, giving less light scattering and hence a more stable reading (S. Revel, Wales, 1998, personal communication). The percentage transmission through the sample at 525 nm is determined. Initial calibration (zeroing of the machine using the diluent alone) is necessary to standardize readings (Pickett and Back, 1973; Hurtgen, 1987).

Equipment has been developed that specifically assesses stallion spermatozoan concentration. The spermiodensimeter is similar to a spectrophotometer and works on the principle of measuring the transparency of a

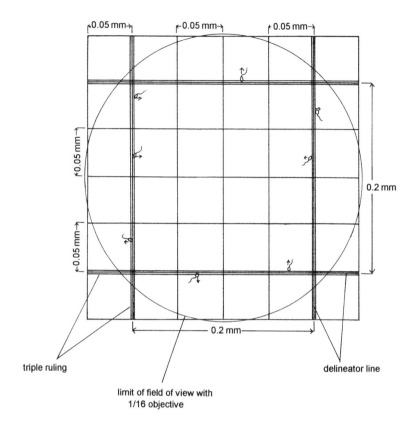

Fig. 6.7. An enlarged view of a group of 16 squares (0.05 mm²) from the centre 1 mm² square on the base of the counting slide of the new improved Neubauer haemocytometer. The large circle indicates the field of view visible under a 0.1666667 objective. The spermatozoa are indicated as small circles. The arrows indicate which spermatozoa should be included (arrows pointing inwards) and which should be excluded (arrows pointing outwards) when counting the spermatozoa present.

semen sample under contrast light, and conversion of that optical density figure to concentration using a table of standards (Bielanski, 1975). This technique is reported to be a simple procedure for use in the field, but with varying accuracy depending upon the experience of the operator (Rossdale and Ricketts, 1980). A further specialization of the spectrophotometer is the sperm counter (Animal Reproduction Services, Chino, California), which works on the same principle as the spectrophotometer but is pre-calibrated to give the concentration of spermatozoa in a given sample. A sample of gel-free semen (0.18 ml) is diluted with formalin saline solution (3.42 ml of 4% formal saline). This solution is made up with care within the cuvette used for measuring the density. The cuvette is place in the spermatozoa counter and a density reading is given within 30 s. Provided an initial standardization of zero has been carried out using the formalin saline solution alone, then the spermatozoa counter will convert the density reading to a spermatozoan concentration reading. When programmed with the appropriate figures (for example, percentage progressive motility observed, and the required number of spermatozoa per inseminate), the machine will also calculate the insemination volume of gel-free semen required (Yates and Whitacre, 1988; Pickett, 1993a). This figure is rather subjective and its accuracy questionable.

Although the use of a spectrophotometer or equivalent is a very quick and efficient method of obtaining spermatozoan concentrations for multiple samples, its results depend upon the measurement of the optical density of the solution presented, and so the presence of other cell debris, which also affect the solution's density, will lead to erroneous results (Yates and Whitacre, 1988). Spectrophotometers are designed for pure colour density measurements, and not for suspensions; therefore, the readings obtained tend to be rather unstable. New systems are now available that claim to be able to differentiate between spermatozoa and debris – for example, the HTM-SPEC-TRA (Hamilton Thorne Research, Beverly, Massachusetts). This differentiation is achieved by means of 'Ident Stain', a DNA-specific fluorescent dye. Spermatozoan heads, with their higher concentrations of DNA, readily absorb the DNA-specific dye and so the spermatozoa are clearly visible under fluorescent lighting. As most debris does not contain DNA, it does not absorb the dye and fluoresce. Some cells that are diploid in nature (for example, leucocytes and early-stage germ cells or spermatogonia) also absorb the dye but result in a higher rate of fluorescence due to the higher DNA content. The fluorimeter analyser is able to differentiate the spermatozoa from the rest, according to brightness of the fluorescence and size. It is reported that such analysers can reduce counting error to ± 2%.

Despite the possible drawbacks, automated spermatozoa counters are becoming increasingly popular and, as discussed previously, complete spermatozoa computer analysis systems are available that automatically give readings for percentage spermatozoan motility and patterns of movement plus concentration.

The values for spermatozoan concentration in semen samples vary widely, ranging from 100 to 200 $\times$ 10^6 ml^{-1}. All values within this range may be considered acceptable and appropriate for use with AI (Kenney *et al.*, 1971; Pickett *et al.*, 1988a; Ricketts, 1993). Factors affecting spermatozoan concentrations include season, ejaculation frequency, testicular size, breed, age, environmental factors, etc. (Chapter 4).

6.5.4. Morphology

Spermatozoan morphology is most commonly assessed by microscopic examination either of an unstained semen sample fixed in buffered formal saline, or of a stained semen sample. The most common stain used is an eosin–nigrosin stain (Dott and Foster, 1972), the components of which are given in Table 6.4.

Table 6.4. Components of eosin–nigrosin stain used to assess spermatozoan morphology (Dott and Foster, 1972).

Component	Quantity
Nigrosin	5.0 g
Eosin	0.84 g
Sodium citrate	1.45 g
Distilled water	50.0 ml

The use of a stain allows the spermatozoa to be highlighted, allowing easier identification of abnormalities. On average 200 spermatozoa are usually examined, allowing a reasonably accurate estimate of the percentage normal and abnormal spermatozoa to be made. Other stains used include indian ink, eosin–aniline blue, bromophenol blue–nigrosin, and Giemsa and Wright stains; and fluorescence may also be used (carboxyfluorescein diacetate and propidium iodide). All of these allow various spermatozoan components or abnormalities to be identified (Yates and Whitacre, 1988; Geisler, 1990; Reifenrath, 1994; Samper, 1995a).

It is important that the stain used has a minimal effect on the osmolarity of the solution and hence avoids causing visible damage to the spermatozoa, which would give an inaccurate impression of spermatozoan abnormalities. The most common causes of erroneously high figures for percentage abnormalities are the adverse affects of the stain itself and failure to ensure that the slide is pre-heated to 37–38°C. These induced abnormalities tend to be evident more as head and mid-piece damage. Results obtained for morphology analysis using such staining techniques must bear in mind these potential areas of error. A further drawback to using these stains is the difficulty of differentiating between early spermatocytes and leucocytes. Staining with hematoxylin–eosin stain, which is specific for leucocytes, may be used (Hurtgen, 1987).

To stain a semen sample, one drop of diluted semen may be placed at one end of a clean, dry and pre-warmed microscope slide alongside of which is placed one or two drops of stain. The stain and semen are mixed using a spreader or cover-slip, and left for 5 min. Alternatively, the stain and the sample may be mixed in a clean, dry, warm test-tube placed in a water bath. After 5 min a drop is transferred to the slide. A thin film of the stained sample is then drawn out along the slide. The slide should be left to dry in a clean, dry and dust-free environment. Provided they are dry, protected slides can be kept for viewing in several days time. After preparation, the slide should be viewed under a light microscope under 10× magnification to check spermatozoan density on the slide. For analysis of spermatozoan morphology, the slide should be viewed under oil immersion at 1000× magnification (objective × eyepiece magnification) (Hurtgen, 1987).

The adverse effects of staining on spermatozoon morphology can be avoided if spermatozoa are viewed unstained. This may be done by fixing a small number of drops of raw semen from a freshly collected sample with warmed, buffered formal saline or 4% glutaraldehyde. An antibiotic may be added if the samples are to be kept for a period of time. These fixed samples may then be viewed as a wet mount using a large cover-slip (23mm × 40mm) under a phase contrast or differential interference contrast microscope (Dott, 1975; Hurtgen, 1987; Malmgren, 1992; Samper, 1995a; S. Revell, Wales, 1998, personal communication).

Staining of spermatozoa to assess the integrity of specific areas may also be carried out. Acrosome stains have been used with some success. Oettle (1986) and Oetjen (1988) reported that Spermac was successful in assessing the structural integrity of the acrosome region. Damage to the acrosome region may also be detected with *Pisum sativum* agglutinin, to which spermatozoa with damaged acrosomal regions become attached, showing a positive correlation of 0.98 (Farlin *et al.*, 1992; Grondahl *et al.*, 1993). Indirect immunofluorescence tests using a monoclonal antibody, which attaches to the acrosome region of the spermatozoa, also allows analysis of the integrity of the spermatozoon acrosome region (Blach *et al.*, 1989; Zhang *et al.*, 1990a). Colloidal gold labelling may be used to identify antibody binding for viewing under transmission electron microscopy (B.A. Ball, California, 1998, personal communication). Alternatively, a fluorescent probe analysis has been used as a rapid assay for acrosome integrity (Ward *et al.*, 1988; Magistrini *et al.*, 1997). The use of direct monoclonal antibodies, specific to an antigen in the acrosome ground substance, has also been reported; the attachment of fluorescent dye to such antibodies facilitates assessment under transmission electron microscopy (Blach *et al.*, 1988). In general, the use of transmission or scanning electron microscopy allows more detailed examination of spermatozoan abnormalities. Work by Veeramachaneni *et al.* (1993) suggested that for the identification of nuclear and acrosomal abnormalities, along with microtubular mass defects, electron microscopic evaluation is required. Such evaluation may be particularly useful for the specific diagnosis of infertility.

The principles of many of these methods have been exploited in automated spermatozoa morphometry analysis systems (Gravance *et al.*, 1996). However, use of such systems, along with electron and scanning microscopes, requires specialist and expensive equipment and so they are not part of a routine evaluation process. For the purposes of research, they are valuable tools.

Some types of abnormalities that may be identified are shown in Fig. 6.9. Chapter 4 discusses the classification of these defects into primary, secondary or tertiary (Dott, 1975; Rossdale and Ricketts, 1980). In summary, primary defects are thought to originate during spermatogenesis and so occur within the testis. They are subdivided into failure at spermatogenesis and failure at epididymal maturation. Failure at spermatogenesis manifests itself as detached heads, head shape abnormalities, double heads, no mid-piece, rudimentary tails, highly coiled tails, and mid-piece abnormalities. These defects may be indicative of a long-term or even permanent problem. Failure during maturation is characterized by cytoplasmic or proximal droplets, indicative of spermatozoan immaturity. Secondary defects originate during the transport of the spermatozoa and

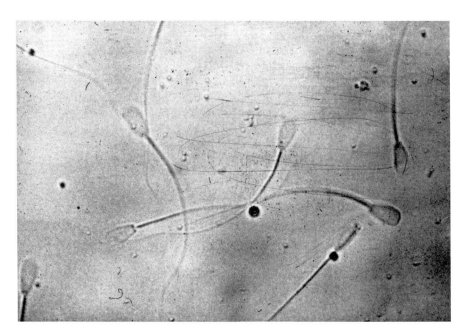

Fig. 6.8. Spermatozoa as viewed under a microscope under high magnification. (Photograph by Julie Baumber and Victor Medina.)

then semen through the epididymis, vas deferens and urethra. These defects are manifest as tail abnormalities such as kinks, bends, coils or swellings, detached heads and tails, and protoplasmic droplets (Rossdale and Ricketts, 1980; Hurtgen, 1987; Varner *et al.*, 1987; Yates and Whitacre, 1988). Finally, tertiary defects result from inappropriate handling post ejaculation (for example, cold shock or exposure to ultraviolet radiation) and are manifest as loss of acrosome, fraying or thickening of the mid-piece, slight bending of the tail, detachment of spermatozoan heads and bursting of spermatozoan heads (Hurtgen, 1987).

The percentage relative occurrence for the more common defects has been reported to be as follow: tail abnormalities 10.1–61%; proximal droplet 9.74–20.9%; distal droplet 7.3–7.55%; head abnormalities 6.7–7.47%; neck abnormalities 4.1%; mid-piece abnormalities 1.07%. All these percentages, apart from the incidence of proximal droplets, are reported to vary with the frequency and time of collection (Dowsett *et al.*, 1984; Szoller *et al.*, 1993). The validity of such classification of spermatozoan defects and the assumptions regarding their origins is disputed. This, plus the differences reported to occur between technicians in the interpretation of apparent defects (Hermenet *et al.*, 1993), has led to the conclusion that it is better to record the actual defects seen, rather than attempt to classify them during the evaluation process (Samper, 1995a).

Reports on the effect that morphological defects have on fertility vary considerably. It is generally accepted that a high concentration of morphologically

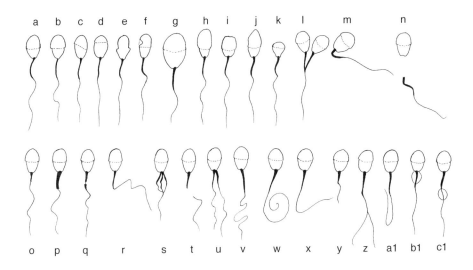

Fig. 6.9. Examples of the types of abnormalities that may be seen when examining spermatozoan morphology. Top row from left to right: (a) a normal spermatozoon is followed by the following abnormalities in order: acrosome defects– (b) swollen, (c) partially lifted, (d) small, (e) lifted, (f) part missing; head defects – (g) big head, (h) elongated, (i) flattened, (j) lanciolated, (k) microhead, (l) double head; neck defects – (m) bent, (n) broken. Second row: mid-piece defects – (o) short, (p) fat, (q) split/constricted, (r) bent annulus, (s) fibrous, (t) broken, (u) double; tail defects – (v) convoluted, (w) corkscrew, (x) bent, (y) small, (z) double, (a1) shoehorn; droplets – (b1) proximal, (c1) distal.

abnormal spermatozoa is associated with a depression in fertility (Adams, 1962; Faulkner and Pineda, 1980). Work has been carried out to try to identify which defects have the most significant effect. Jasko *et al.* (1990a) reported that the number of abnormalities in the head and tail regions, plus the number of spermatozoa with proximal droplets, accounts for as much as 76% of the variation between stallions in fertility to insemination. Those classified as primary abnormalities seem to be the major cause of lowered fertility. Bielanski (1975) reported that samples with in excess of 10% primary abnormalities were associated with depressed fertility rates, and similar results were reported by Szollar *et al.* (1993). Abnormal heads are also reported to have a particularly significant effect on fertility (Jasko *et al.*, 1990a). The correlation between primary abnormalities and fertility is reported to be -0.5. If the effect of different individual types of primary abnormalities are investigated, it seems that mid-piece defects have a greater negative effect on fertility than distal protoplasmic droplets (a secondary defect) (Hurtgen, 1987). However, samples with 30% secondary abnormalities were not consistently associated with lower fertility rates (Bielanski, 1975; Hurtgen, 1987). Work carried out on Dutch Warmbloods indicates a correlation between total number of morphologically abnormal spermatozoa and infertility (Van Duijn and Hendrikse, 1968);

however, the correlation figure obtained was 0.25, which is rather low. It would be difficult, based on these figures, to justify using morphology as an accurate predictor of fertility rates in a stallion.

However, this work did consider both primary and secondary abnormalities together. The predictive value of morphological analysis improves if these abnormalities are categorized. The variability of such predictions is illustrated by Voss *et al.* (1982) who, in a 2-year study, produced varying correlations for morphological abnormalities and fertility. Using 3-year-old stallions, in the first year the correlation between the number of matings per pregnancy and mid-piece morphological abnormalities was 0.53. However, in year two, the correlation had dropped to 0.01. Pickett (1993a) and Pickett *et al.* (1989) reported some correlation between specific morphological abnormalities and the ability of a stallion to pass a fertility assessment, the largest being with head abnormalities: those stallions passing the fertility assessment had on average 9% head abnormalities, compared with 17% for those failing the assessment. The difference in the percentage abnormalities in other regions was much less. Jasko *et al.* (1991a) also attempted to link fertility, as indicated by motility, with spermatozoan parameters including morphology: correlations of 0.42 and 0.32 were reported between normal spermatozoan morphology and total spermatozoan motility, and normal spermatozoan morphology and progressive spermatozoan motility, respectively. A similar relationship has also been reported by Long *et al.* (1993) and this is supported by Jasko *et al.* (1992b). This correlation was slightly enhanced to 0.43 and 0.4, respectively, when computer-aided assessment was used. Their work suggested that the variation in percentage normal morphology between stallion seasons accounted for 15% of variation in fertility rates, and combination of spermatozoan morphology and motility accounted for 37% of the variation. Correlations between different abnormalities have also been reported: for example, neck abnormalities are reported to be correlated with mid-piece and principal piece abnormalities, mid-piece abnormalities with principal piece abnormalities, and finally head abnormalities with mid-piece abnormalities (Long *et al.*, 1993).

In order for a sample to be used for AI, the reported acceptable total percentage abnormalities varies from 40 to 50% (Pickett and Voss, 1972; Kenney, 1975; Hurtgen, 1987; Fayrer-Hosken and Caudle, 1989; Pickett *et al.*, 1989; Davies Morel, 1993; Pickett, 1993a). Fertile stallions will normally show no more than 10–15% of any single morphological defect. In addition, there should normally be less than 5% acrosome or mid-piece abnormalities, less than 1.8% abnormal heads, and less than 6.4 and 11.4% proximal and distal droplets, respectively (Dott, 1975; Rossdale and Ricketts, 1980; Hurtgen, 1987). Similar acceptability parameters were put forward by Bielanski and Kaczmarski (1979), who stated that for full fertilization potential to be achieved a semen sample should show less than 10% cytoplasmic droplets, 30% principle piece loops, 3% separated necks and 1% of other abnormal forms.

To a certain extent, samples with a high percentage of abnormalities can be compensated for if the total semen volume and spermatozoan concentration are high, as abnormal spermatozoa have no apparent detrimental effect on viable spermatozoa in the same sample (Varner and Schumacher, 1991). Volume and concentration of inseminate can be adjusted to ensure that the required number of viable spermatozoa (normally 500×10^6) are deposited at insemination. As far as most AI programmes are concerned, samples with more than 40–50% morphologically abnormal spermatozoa would not normally be considered for use. Stallions passing a broad spectrum fertility test carried out by Pickett et al. (1988a) all demonstrated less than 55% normal spermatozoa in samples ejaculated.

The dispute as to whether a correlation exists between abnormalities and fertility rates, and the significant variation in correlations reported by those demonstrating a link (Voss et al., 1981; Dowsett and Pattie, 1982; Jasko et al., 1988; Chevalier-Clement et al., 1991), ensures that evaluation for morphological abnormalities as a whole is of limited use in providing an absolute prediction of fertility. Evaluation for primary abnormalities has the possibility of being more accurate as a predictor, but again it has limitations. The development of automated morphometric analysis systems may provide possibilities for the future. These machines are now available for use with stallion spermatozoa, but have only recently become anywhere near efficient enough for their use to be considered for domestic animals (Davis et al., 1993; Ball and Mohammed, 1995; Magistrini et al., 1997). Even with the advent of automation, the majority of semen today is not assessed for morphology as its future has already been decided or it has been inseminated before a morphology evaluation has been carried out.

It would seem logical that abnormal morphology should be correlated to motility, and hence assessment for motility would reflect abnormalities and so might suffice as a filter for samples with a high percentage of morphologically abnormal spermatozoa. Correlations of between 0.29 and 0.7 have been reported between morphologically normal spermatozoa and progressive motility (Van Duijn and Hendrickse, 1968; Voss et al., 1982; Pickett et al., 1989; Jasko et al., 1991a; Pickett, 1993a; Long et al., 1993). As would be expected, the correlation between morphologically normal and all motile spermatozoa is at the higher end of the range at 0.61 (Long et al., 1993). Interesting work, reported by Pickett (1993a) and Pickett et al. (1989), indicated that the best correlation was evident in stallions that failed their fertility evaluation assessments. Direct correlations between the number of morphologically normal spermatozoa and conception rates have been reported to be 0.57 (Gastal et al., 1991).

6.5.5. Live:dead ratio

An indication of the percentage of live spermatozoa within a sample may be given initially by a motility assessment. A much more accurate and quantifiable method for assessing the live:dead ratio may be achieved by differential staining with a stain such as eosin–nigrosin, which can also be used in morphology assessments. The eosin is taken up by dead spermatozoa, the plasma membranes of which are permeable to the eosin stain. As a result, the dead spermatozoa appear pink in colour. The spermatozoa are stained and the slide prepared as for morphological examination. A sample of 200 spermatozoa should be evaluated under a light microscope at a magnification of 1000×, and the number of spermatozoa stained and unstained should be noted (Hermenet et al., 1993; M. Boyle, England, 1995, personal communication). An estimate of the ratio of live to dead spermatozoa can then be calculated. Other preparations, such as a mix of ethidium bromide and acridine orange fluorescence or H258 fluorescence staining, have a similar effect and have proved equally successful (Hermenet et al., 1993) (Fig. 6.10). A further development, suggested by Dott and Foster (1972) and reported to give a more accurate evaluation, involved shaking the mounted slides (with cover-slip) in a 5% acetic acid and 0.6 N perchloric acid mix for 10–30 s, after which they were washed off under tap water and briefly immersed in ethanol, followed by ether to remove any water. They could then be viewed microscopically. This system has not gained popularity.

If the results of the live:dead evaluation are not required immediately, then the semen sample can be diluted in an equal volume of 2.9 g of trisodium citrate dihydrate and 0.1 ml of 40% formaldehyde in 100 ml of deionized water. Such dilution is reported to allow samples to be stored for up to 3 months, with no appreciable decline in the percentage of eosinophilic spermatozoa counted (Dott and Foster, 1975).

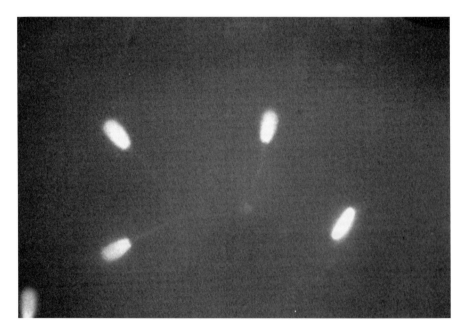

Fig. 6.10. Spermatozoa stained with H258 fluorescence, which is taken up by dead spermatozoa. (Photograph by Julie Baumber and Victor Medina.)

The live:dead ratio gives an accurate indication of spermatozoan viability, and in samples collected from fertile stallions a live:dead ratio of greater than 50% is to be expected (Rossdale and Ricketts, 1980). A sample with a live:dead ratio in excess of 60% is considered appropriate for AI (Davies Morel, 1993; Ricketts, 1993).

6.5.6. Cytology

Cytological evaluation of a semen sample is particularly advantageous if the stallion has a history of, or is currently suffering from, conditions such as haemospermia or genital tract infections. Blood cells, leucocytes and erythrocytes, as well as primary spermatogenic cells, can be identified using either haematoxylin–eosin stain or Wright's stain (Roberts, 1971a; Swerczek, 1975; Rossdale and Ricketts, 1980; Hurtgen, 1987). Prior to staining, the sample needs to be mixed with an equal volume of 50% ethanol or 10% formal saline. After staining, the sample may be viewed under light microscopy. If viewed using a haemocytometer, the blood cells may also be counted and an estimate of total numbers and concentration can be given. The method used for the haemocytometer, and the equation for the conversion of counts to concentration, is the same as that used for the estimation of spermatozoan concentration, discussed previously. The causes of high erythrocyte counts have also been addressed previously. Details of the possible causes of high leucocyte counts are discussed in the following section.

As far as AI is concerned, a sample should not contain more than 1500 leucocytes ml^{-1} and ideally no erythrocytes (though provided the erythrocyte count is below 500 ml^{-1} the sample may be considered for use) (Rossdale and Ricketts, 1980; Davies Morel, 1993, Ricketts, 1993).

6.5.7. Bacteriology

Nearly all semen samples will contain bacteria. These may be non-pathogenic (Burns *et al.*, 1975; Kenney, 1975; Kenney *et al.*, 1975) or may have the potential to cause infection within mares (Bain, 1966; Hughes *et al.*, 1967; Merkt *et al.*, 1975a). The presence of bacteria within a semen sample is not normally evident by gross examination unless the condition has been allowed to develop to such a severe extent that pus is present. Bacteriospermia is the term given to the presence of a high number of potentially pathogenic bacteria. Many stallions with mild cases of bacteriospermia may show neither clinical symptoms nor a depression in fertility rates (Merkt *et al.*, 1975a). However, the area of concern is the potential for such bacteria to cause infection in mares that are covered, which in turn do show clinical signs, including depressed fertility rates. Hughes *et al.* (1967) demonstrated well that despite the absence of clinical signs, such as lesions, 36% of stallions that they tested were positive for *Pseudomonas* spp. It is the potential of such semen samples to cause significant problems in mares that is the main reason for bacteriological evaluation. Of samples collected, apparently only 4–10% are sterile, the majority of the rest (60% of the total) being contaminated with non-pathogenic species, such as *Staphylococcus lentus*, *S. capitis*, *S. haemolyticus*, *S. xylosus*, etc., the remaining 13.5–30% containing potential venereal pathogens (Vaissaire *et al.*, 1987; Danek *et al.*, 1994b; Scherbarth *et al.*, 1994; Madsen and Christensen, 1995).

Bacterial organisms may be broadly classified into three types, according to their effect on the mare:

- Contaminant organisms which are rarely implicated in infection of mares or in the development of endometritis. Examples of such bacteria include diphtheriods, *Neisseria* spp., *Streptococcus faecalis* and *Staphylococcus albus* (Rossdale and Ricketts, 1980).
- Potential causers of acute endometritis. Examples of bacteria capable of causing infection and endometritis in mares with a compromised immune system include *Streptococcus zooepidemicus* (β-haemolytic streptococcus), *Staphylococcus aureus*, haemolytic *Escherichia coli*, *Proteus* spp. and some *Klebsiella pneumoniae* capsule type 7 (Hughes and Loy, 1969a; Peterson *et al.*, 1969; Kenney *et al.*, 1975; LeBlanc *et al.*, 1991).
- Primary causers of acute endometritis. Examples of bacteria which are the prime cause of venereal disease in the mare and appear to be capable of producing acute endometritis in mares with both normal and compromised immunological systems include *Klebsiella pneumoniae* capsule type 1 and 5, some strains of *Pseudomonas aeruginosa* and *Taylorella equigenitalis* (CEM) (Crouch *et al.*, 1972; Crowhurst *et al.*, 1979; Couto and Hughes, 1993; Parlevliet *et al.*, 1997).

The frequency or occurrence for these organisms has been reported to be as follows: β *Streptococcus* spp. (for example, *S. zooepidemicus*) 19%; *Klebsiella pneumoniae* 18.7%; *Bacillus* spp. 11.5%; *Pseudomonas* spp. (for example, *P. aeruginosa*) 9.5–10%; *Micrococcus* spp. 8.5%; *Staphylococcus* spp. (for example, *S. aureus*) 5.5%; haemolytic *E. coli* 2.5–15.5%; followed by *Proteus* spp., *Aerobacter* spp., and α-haemolytic and non-haemolytic *Streptococcus* spp. (Vaissaire *et al.*, 1987; Danek *et al.*, 1993, 1994b; Pickett, 1993a). Chlamydia are also reported to be present in 3.4% of stallion semen sample (Veznik *et al.*, 1996).

The site of infection or colonization in the stallion is normally the accessory glands or the external genitalia, though accessory gland infection is uncommon. However, such infections tend to be persistent. The bacteria present may include *Pseudomonas aeruginosa*, *Klebsiella pneumoniae*, *Streptococcus* spp. and *Staphylcoccus* spp. (Roberts, 1986a; Little and Woods, 1987; Varner *et al.*, 1993). Bacteria colonizing the external genitalia of the stallion (and, therefore, possibly a semen sample) include, *Taylorella equigenitalis*, *Pseudomonas aeruginosa*, *Klebsiella pneumoniae*, *Streptococcus zooepidemicus* and *E. coli* (Couto and Hughes, 1993; Varner and Schumacher, 1993; Danek *et al.*, 1994b). *Taylorella equigenitalis* is of particular importance as the causal agent of contagious equine metritis (CEM) and the only bacterium that is known consistently to cause venereal disease in horses (Parlevliet *et al.*, 1997).

Of all these bacteria, only *Pseudomonas aeruginosa* and *Klebsiella pneumoniae* appear to be transferred by the true venereal route; that is, they are present within the semen ejaculated as opposed to contamination by contact with the external genitalia during mating.

The presence of bacteria within a semen sample may be indicated indirectly by high leucocyte and, more specifically, high neutrophil counts, or alternatively by unusually high erythrocyte counts (Danek *et al.*, 1993). Distinguishing between leucocytes and primordial germ cells may be difficult as they are of a similar size and appearance. Differentiation is made easier by the use of air-dried semen smears stained with Giemsa, Wright or haematoxylin–eosin stains, as discussed previously. Gram staining of semen smears may also be used to aid the identification of bacteria (Varner and Schumacher, 1993). Immunofluoresence has been used to identify contamination within semen samples (Veznik *et al.*, 1996), and bacteria may also be examined by direct plating of semen on to blood agar plates, or the plating of semen swabs (Mackintosh, 1981; Ricketts, 1993). Prior to plating, swabs may be transported in Amies Charcoal Transport Medium and/or stored in Stuart's Transport Medium (Madsen and Christensen, 1995). Differentiation between inherent bacterial contamination and external contamination of a sample can be done by means of an open-ended AV. If the sample still shows high bacterial counts, then inherent bacterial contamination should be suspected and appropriate treatment considered. If external contamination is apparent, then swabbing of the external genitalia and bacterial smear identification or culture can be carried out and appropriate antibiotics dispensed. On many studs it is now standard practice to swab the stallion's urethral fossa, urethral diverticulum and penile sheath before both natural service and AI. In the UK this is in accordance with the Codes of Practice laid down by the Horse Race Betting Levy Board (1997) for use in Thoroughbred studs. Such codes are also used as general guidelines throughout the industry. Swabs of the pre-ejaculatory fluid may be taken, especially if inherent infection is suspected (David *et al.*, 1977). The swabs are submerged in Amies Charcoal Transport Medium for transport to an approved laboratory. The swabs are cultured aerobically and microphilically to screen specifically for *Taylorella equigenitalis*, *Klebsiella pneumoniae* (capsule types 1, 2 and 5) and *Pseudomonas aeruginosa*. Screening of all breeding stallions for *T. equigenitalis* is a statutory requirement under the Infectious Diseases of Horses Order 1987. It is also a requirement of European Community Decisions 95/307/EC and 96/539/EC for the movement of semen within and into EC countries. Similar regulations are in force in many other countries.

Interpretation of bacterial culturing is not simple. Certain capsular strains of *Klebsiella pneumoniae* have been linked to venereal disease outbreaks, but there are many others, about which there is very little information on their ability to cause infertility (Atherton, 1975; Merkt *et al.*, 1975a). It is likely that strains 2, 3 and 5 may be pathogenic, whereas others (for example, strain 7) are non-pathogenic (Kikuchi *et al.*, 1987). Some strains of *Pseudomonas aeruginosa* are known to be associated with venereal disease, whereas others are not; however, the difficulty in differentiating between various strains means that all stallions carrying *P. aeruginosa* are treated as if the strain was pathogenic (Couto and Hughes, 1993). Scherbarth *et al.* (1994) concluded that considerable annual fluctuation did occur, though the incidence of facultative pathogenic organisms was 13.5%. They concluded that the minimal growth in culture throughout the samples collected indicated only a temporary colonization of the genital organs and that, except for *Taylorella equigenitalis* infection, it may be considered to be of little significance unless otherwise indicated by clinical findings.

As far as preventive medicine is concerned there is much debate about the efficacy of routine washing of stallion genitalia prior to covering or collection for AI. Traditionally, the stallion's penis, scrotum and associated areas were washed, often using an antiseptic wash or soaps. The use of antiseptic washes and soaps is now largely frowned upon, due to their spermicidal effects. There is also the possibility that the bacteriocidal nature of such agents will disturb the natural non-pathogenic microfloral balance of the area. Once this natural balance has been disturbed, the area becomes more susceptible to colonization by opportunistic and potentially pathogenic bacteria. This may also occur as a result of regular washing without the use of soaps or antiseptic agents (Bowen *et al.*, 1982). On the other hand, some workers report no significant effect of washing on bacterial colonization (Jones *et al.*, 1984). Alternative antibacterial agents such as chlorhexidine have been used successfully and are reported not to have a spermicidal effect. Nevertheless, they still run the risk of upsetting the natural microfloral balance (Betsch *et al.*, 1991).

The presence of bacteria within a semen sample significantly increases the risk of venereal disease and it has also been suggested that bacterial contamination may have a more direct effect on spermatozoan motility. Such an association has been demonstrated between *Streptococcus zooepidemicus* and motility rates. It has also been linked to the enhanced agglutination of cells and an increase in the number of leucocytes (Danek *et al.*, 1993). However, these findings are not supported by Pickett *et al.* (1988a), who reported that semen samples from stallions that failed a bacterial culture evaluation showed no differences in seminal characteristics compared with those of stallions that had passed the bacteriological examination. The only apparent effect of elevated bacterial counts was an increase in the number of spermatozoa with mid-piece abnormalities ($P < 0.05$), a decrease in spermatozoan concentration in the first of two ejaculates compared ($P < 0.05$) and an increase in total scrotal width ($P < 0.05$). No effect on spermatozoan motility was reported, this being 54% for each group of stallions. However, when *Streptococcus zooepidemicus* organisms were introduced into a semen sample, at a concentration of 500×10^3 bacteria ml^{-1}, and incubated for 60 min, a significant detrimental effect on spermatozoan motility, viability and morphology was reported (Danek *et al.*, 1996b).

The presence of significant numbers of potentially pathogenic bacteria within a semen sample does not necessarily preclude its use in an AI programme. It is evident from the discussion above and from work by Merkt *et al.* (1975a) that the presence of bacteria may not affect spermatozoan quality. Though the inherent fertilizing capacity of semen samples contaminated with bacteria may not be significantly affected, the risk of disease transfer would prevent its untreated use. The addition of antibiotics to extenders may allow the insemination of semen with significant counts of *Pseudomonas* spp. (100% sensitive to streptomycin), *Klebsiella pneumoniae* or *Streptococcus zooepidemicus* (75% sensitive to chloramphenicol, 64% sensitive to penicillin, 50% sensitive to oxytetracycline) without a significant risk to the mare (Simpson *et al.*, 1975; Squires *et al.*, 1981b; Danek *et al.*, 1994b). However, the presence of *Taylorella equigenitalis* would preclude the use of that stallion's semen until the infection had been successfully cleared (Fraser, 1986). Collection of semen using an open-ended AV significantly reduces the contamination of the semen from the external genitalia (Clement *et al.*, 1993) and was reported to reduce the normal contamination rates from nine samples in ten to one sample in three (Tishner and Kosiniak, 1986; Vaissaire *et al.*, 1987; Madsen and Christensen, 1995).

6.5.8. Virology

It is not normal practice to attempt to isolate viruses from a semen sample. Routine good breeding management and veterinary examination for all stallions prior to collection for AI, following the Codes of Practice (Horse Race Betting Levy Board, 1997) and EVA Order, 1995 (EC Decision 95/307/EC; EC Decision 96/539/EC), should ensure that any stallions that may be shedding virus in their semen are picked up. Equine herpes virus (EHV) and equine viral arteritis (EVA) may be shed in the semen of infected or previously infected stallions (Timoney *et al.*, 1988a; Tearle *et al.*, 1996). Of these two, EVA is of particular importance and is a worldwide problem. It has recently become of importance in the UK, where it is now classified as a notifiable disease under certain circumstances, under the EVA Order, 1995 (Horse Race Betting Levy Board, 1997). The virus may be passed in the semen from either symptomatic or long-term asymptomatic carriers (Timoney *et al.*, 1988). Virus isolation from the sperm-rich fraction is normally possible in stallions shedding the virus, due to its persistence in the accessory glands, from which it is shed at ejaculation (Varner and Schumacher, 1993). Evidence suggests that although most seropositive stallions will infect mares via semen, seropositive non-shedding stallions may not (Golnik and Cierpisz, 1994). Serological tests or antibody titres, though again not 100% accurate, do give more consistent results in indicating the presence of the virus and hence the chance of its transfer (Couto and Hughes, 1993). The EVA virus will also survive in chilled and frozen semen and is not affected by any antibiotics added to an extender. For this reason, it is most important, for the control of the disease, that any stallion used in an AI programme should be free of infection (Horse Race Betting Levy Board, 1997) or suitably vaccinated immediately after a clear blood test (Fukunaga et al., 1992), in order to prevent perpetuation of viral transfer (Timoney et al., 1988).

6.5.9. Mycoplasma

Mycoplasma may be isolated from semen samples (Zgorniak-Nowosielska et al., 1984), but no significant work has been carried out on the incidence or implications of mycoplasma contamination. Isolation of mycoplasma is not carried out in routine semen evaluation.

6.6. Functional Tests

Due to the relatively unsuccessful attempts to find a physical characteristic of spermatozoa which shows a consistent and high correlation with fertility, attention has been drawn to the possibility of identifying a suitable test that will assess the functional integrity, rather than the physical integrity, of the spermatozoa. Many such tests have been investigated in isolation or as part of an accumulative model (Wilhelm et al., 1996). However, many are at the experimental stage and are also costly and labour intensive, and so they are not in widespread commercial use at present. They may, however, hold the key to the future of semen evaluation.

6.6.1. Biochemical analysis

Biochemical analysis of semen may give indirect information on the quality of the sample. The total number of spermatozoa present within a sample is reflected in the concentration of enzymes such as aspartate amino transferase, glutamic oxaloacetic transaminase (GOT), lactate dehydrogenase (LDH), hyaluronidase, acid and alkaline phosphatase and acrosin (Castro et al., 1991). Adenosine triphosphate (ATP) may also be used as an indicator. Many of the enzymes such as GOT, LDH, acrosin and hyaluronidase are located in the spermatozoon head, specifically in the acrosome region, or in the mid-piece. As such they can also be used to indicate spermatozoon viability, as elevated free levels of these enzymes are indicative of damage to the spermatozoon (Graham et al., 1978; Samper, 1995b). The techniques require considerable skill if they are to give consistent, accurate results. Other seminal components or characteristics such as proacrosin, fructolysis, GOT, citric acid and pH have also been used as indicators of spermatozoan quality (Cermak, 1975; Gomes, 1977; Fromann et al., 1984). More recent work has demonstrated a significant correlation between acetylcarnitine concentrations and spermatozoan concentration (0.813) (Burns and Casilla, 1990; Stradaioli et al., 1995). Though biochemical analysis can give a good indirect indication of spermatozoan numbers and damage, it is as yet very non-specific and it will remain as such until more detailed work has been carried out to ascertain the normal concentration of these enzymes within the spermatozoa and semen.

Recent research, particularly with human spermatozoa, has concentrated on the assessment of acrosin activity in the spermatozoon head, which is reported to have a significant bearing on the variation in fertility rates reported (Menkveld *et al.*, 1996). Acrosin activity may be assessed via fluoresceinated lectin binding (Meyers *et al.*, 1995a,b). Total acrosin activity reflects an aspect of functional ability of the spermatozoon, as opposed to the ultrastructure, and is reported to correlate positively with fertilization rates (Reichart *et al.*, 1993; Sharma *et al.*, 1993; Francavilla *et al.*, 1994). Protein, fructose and potassium concentrations within seminal plasma have also been used as indicators of likely spermatozoon viability (Castro and Augusto, 1991; Oba *et al.*, 1991). Tests assessing levels of spermatozoan creatine kinase and semen ATP indicate that differentiation between fertile and subfertile spermatozoa may be possible. However, the likely wide variation in ATP within equine spermatozoa, and the involvement of fluorimetric methods and complex chemical reactions, limits their widespread practical use at present (Comhaire *et al.*, 1987; Huszar *et al.*, 1990; S. Revell, Wales, 1998, personal communication).

6.6.2. Membrane integrity tests

The integrity of the spermatozoon membrane is a prerequisite for successful fertilization. It is, therefore, an obvious parameter to be tested to give an indication of spermatozoon viability. Several tests have been investigated to evaluate membrane integrity, involving either the use of antibodies to various components of the spermatozoon membrane, or fluorescent probes, or various stains combined with flow cytometry (Magistrini *et al.*, 1997). Antibodies can be labelled with fluorescent dyes to allow the integrity of the membranes to be assessed visually. Blach *et al.* (1988) developed a monoclonal antibody test for the outer acrosomal membrane. This particular work was investigating the effects of various means of storage on membrane integrity and no fertilization assessment was made. It is possible to use this method to evaluate membrane integrity and they demonstrated an adverse effect of centrifugation, freezing and thawing. The use of monoclonal antibodies in this way was further reported on by Amann and Graham (1993). Using a similar principle, Van Buiten *et al.* (1989) successfully used fluorescein-labelled high molecular weight sugars to indicate the integrity of stallion spermatozoon membranes. Similarly, Casey *et al.* (1993) reported lectins to be a successful tool for assessing the acrosome membrane integrity. Their use in association with a viability stain (for example, Hoechst 33258) would allow differentiation of live and dead spermatozoa and an indication of viability.

Fluorescent probes are increasingly popular and are being developed to assess mitochondria, DNA, and general membrane integrity. For example, calcein AM/ethidium homodimer or carboxyfluorescein diacetate/propidium iodide have been used (Garner *et al.*, 1986; Harrison and Vickers, 1990; Althouse and Hopkins, 1995). Both calcein and carboxyfluorescein are membrane permeable and are de-esterifized to the fluorescent form by non-specific esterases within the spermatozoa. The spermatozoa fluoresce green. The counterstains, ethidium homodimer and propidium iodide, are not membrane permeable and can only enter membrane-damaged spermatozoa, in which they bind to the DNA, resulting in a red fluorescence. These workers concluded that the use of such fluorescent probes provides an accurate evaluation of the integrity of spermatozoan membranes and that they provide a more accurate estimation of the number of viable cells than the more traditional method of motility. These results are supported by Malmgren (1997), who reported a correlation figure of 0.67–0.69, depending upon the duration of semen storage prior to assessment. Since the acrosome fluoresces more brightly than the rest of the head, its integrity can also be evaluated. This fluorescent technique allows easy assessment of the percentage of live spermatozoa and of the percentage of live, morphologically normal spermatozoa even in normally difficult situations. It is now being used in commercial AI laboratories to define the live, normal post-thaw spermatozoa dose (S. Revell, Wales, 1999, personal communication).

6.6.3. Flow cytometry

Flow cytometry tests have been mentioned previously in connection with biochemical analysis. Recent work has investigated the use of flow cytometry, with both single and double staining, to assess the functional capabilities of spermatozoan mitochondria (Wilhelm *et al.*, 1996; Magistrini *et al.*, 1997; Papaioannou *et al.*, 1997). The results are encouraging and suggest a significant positive correlation between spermatozoa with optimally functioning mitochondria and viable spermatozoa (Papaioannou *et al.*, 1997).

6.6.4. Filtration assay

As discussed in Chapter 4, a filter of Sephadex beads over glass wool has been quite successfully used in various animals, including the stallion, to give an assessment of spermatozoan viability. The passing of frozen thawed semen through such a filter causes the less viable spermatozoa to be held within the filter, resulting in a filtrate of highly viable spermatozoa. A positive correlation exists, therefore, between the number of spermatozoa in the filtrate and the fertilizing potential of the sample (Samper and Crabo, 1988; Samper *et al.*, 1988). In human *in vitro* work, such a filter is used to remove less viable spermatozoa and allow the concentration of spermatozoa with high fertilizing potential, so enhancing AI fertilization rates (Jeyendran *et al.*, 1984; Casey *et al.*, 1991; Samper *et al.*, 1988). Cotton, with or without Sephadex, may also be used as a spermatozoan filter, as can a bovine serum albumin (BSA) medium separation column, which is used to filter spermatozoa extended in a BSA–sucrose extender. The percentage motility of spermatozoa in the lower parts of the column was significantly greater than that of both the higher fractions and the sample prior to filtration through the column (Pruitt *et al.*, 1988). Cellulose acetate/nitrate filters have also been used, the suggestion being that the rate of passage of spermatozoa through such filters is an indication of spermatozoan motility and thus of spermatozoan viability (Strzemienski *et al.*, 1987). L4 membrane filters have also been used successfully. In comparative work by Reifenrath *et al.* (1996), L4 filters were compared with centrifugation and it was concluded that spermatozoan motility, viability and morphological integrity were significantly better after filtration than centrifugation, though filtration did result in a greater number of spermatozoa demonstrating membrane defects. Conception rates with filtered spermatozoa were also greater than those obtained with centrifuged spermatozoa.

6.6.5. Hypo-osmotic stress test

The hypo-osmotic test may also be termed a membrane stress test and relies upon the fact that the spermatozoon membrane is semi-permeable, allowing the selective passage of water through it along an osmotic pressure gradient. If spermatozoa are placed in a hypotonic solution (a medium of less than 290 mOsm), water passes into the spermatozoa, resulting in a ballooning of the spermatozoon head and deformation of the tail. The hypo-osmotic pressure test relies upon this deformation of spermatozoa tails, apparent only in spermatozoa that were intact when placed in the hypotonic solution (Jeyendran *et al.*, 1984). As a result, the spermatozoan tails show characteristic deformation in the form of bending and coiling. This effect is only observed in spermatozoa with intact membranes (Jeyendran *et al.*, 1984; England and Plummer, 1993; Correa and Zavos, 1994; Kumi-Diaka and Badtram, 1994; Revell and Mrode, 1994; Correa *et al.*, 1997). These workers showed a significant correlation between the swelling of spermatozoan heads and characteristic tail coiling and spermatozoan motility, and the percentage successfully penetrating an ovum, though other work by England and Plummer (1993) failed to show such correlations.

A modified hypo-osmotic test is widely used in the evaluation of bull semen, where it is reported to give a good range of results (10–60%) which are closely correlated with fertilization rates (Revell and Mrode, 1994). The test has been used in stallions with some success (Zavos and Gregory, 1987; Samper *et al.*, 1991; Samper and Crabo, 1993) and its use as a potential commercial test is being investigated (S. Revell, Wales, 1997, personal communication). Stallion spermatozoa placed in hypo-osmotic solutions (75–129 mOsm) of either NaCl or sodium citrate for 15 min will show characteristic tail coiling (Samper *et al.*, 1991; Samper, 1993). In a modification of the bovine test (Revell and Mrode, 1994), stallion semen post-thaw has been subjected to osmotic stress and then stained with 6-CαD/P1 to assess the proportion of spermatozoa which remain membrane-intact. The range of results obtained was not good (0–35%). Work is currently being carried out on the development of a more suitable form of the test for stallions (S. Revell, Wales, 1999, personal communication). More recent work by Cueva *et al.* (1997b), using hypo-osmotic solution containing citrate (18–96 mOsm) or citrate and fructose (153 mOsm), indicated a significant correlation between the incidence of swollen tails after incubation for 20–30 min, and the incidence of altered acrosomes and total spermatozoan motility. Comparative tests in humans indicate that the results obtained with the hypo-osmotic stress test show a good correlation with results from the zona-free hamster oocyte test (Jeyendran *et al.*, 1984; Rogers and Parker, 1991).

Based on the same principles, hyperosmotic tests have been investigated more recently. However, the correlation between spermatozoon quality parameters of fresh and frozen–thawed semen with hyperosmotic stress resistance was reported to be small (Cueva *et al.*, (1997).

6.6.6. Cervical mucus penetration test

The ability of spermatozoa to penetrate cervical mucus, the first major biological fluid that the spermatozoa naturally come in contact with, has been used to indicate the viability of spermatozoa from several species. A highly significant, positive correlation between penetration and acrosome integrity and also between penetration and total spermatozoan integrity has been reported (Galli *et al.*, 1991). However, no correlation between penetration and fertility was proved, and the test added little information to the standard motility tests. This limits the use of the technique in assessing fertilizing capabilities of spermatozoa (Hafez, 1986). No such investigations, to date, have been reported using stallion spermatozoa.

6.6.7. Oviductal epithelial cell explant test

The reaction of spermatozoa to oviductal epithelial cell monolayers may be tested in the form of a cytofluorescent assay. The effect of spermatozoa on oviductal epithelial cell activity, including protein secretion, is reported to have a high correlation with spermatozoon morphology, and to be a possible prognostic test for *in vitro* fertilization (Thomas *et al.*, 1995b; Thomas and Ball, 1996). The number of spermatozoa bound to explants may also be used as an indicator of spermatozoan viability (Thomas *et al.*, 1995a). However, there is apparent variation in the number of spermatozoa bound, depending on the place from which the explant was extracted and the stage of the mare's oestrous cycle (Thomas *et al.*, 1994a). Little work has been carried out to date to investigate any correlation between spermatozoon–oviduct cell adhesion and subsequent fertilization, though it has been speculated that a correlation may exist (B.A. Ball, California, 1998, personal communication).

6.6.8. Zona-free hamster ova penetration assay

Zona-free hamster oocytes have been used successfully to indicate the viability of human spermatozoa (Yanagimachi *et al.*, 1976; Binor *et al.*, 1980; Overstreet *et al.*, 1980; Hall, 1981) and bovine spermatozoa (Amann, 1984). The ability of a spermatozoon to penetrate a zona-free hamster oocyte depends upon its successful completion of capacitation and the acrosome reaction, resulting in the formation of the male pro-nucleus. The zona-free hamster assay indirectly assesses the spermatozoa's ability to carry out these reactions in a coordinated sequential fashion and thereby indirectly assesses the spermatozoon's viability. Alternatively, spermatozoon capacitation and the acrosome reaction can be induced artificially and full penetration and fertilization of the oocyte assessed, instead of just attachment. Some work on the use of the zona-free hamster oocyte test in stallions has been reported (Samper *et al.*, 1989; Zhang *et al.*, 1990a; Padilla *et al.*, 1991). Pitra *et al.* (1985) failed to demonstrate a correlation between the results of the zona-free hamster oocyte test and equine spermatozoon motility, but a positive correlation with conception rates when semen was inseminated into mares was reported. Incubation of non-capacitated stallion spermatozoa with hamster oocytes resulted in activation of oocyte chromosomes but no decondensation of spermatozoa occurred. It was suggested, therefore, that activation of the oocyte may be used as an indication of stallion spermatozoon viability (Ko and Lee, 1993). Work by Lippe (1986) suggests that there is a wider range (0–59%) in the percentage of spermatozoa that penetrate zona-free hamster oocytes, when compared with a much more reliable range (40–46.7%) achieved in similar tests with hamster and mouse spermatozoa. Time constraints and the intensity of labour required, plus the limited physiological applicability of such tests due to the use of heterologous ova and the absence of the zona pellucida, limit the accuracy of the test in assessing potential fertilization capacity *in vivo*. Binding of equine spermatozoa to equine ova *in vitro* has been successfully achieved, but the relative scarcity of equine ova and the limited research to date precludes its use, at present, as an evaluation test for equine spermatozoa (Hochi *et al.*, 1996).

6.6.9. Heterospermic insemination and competitive fertilization

The pooling of spermatozoa from two males prior to insemination or incubation with oocytes *in vitro* allows direct assessment of the relative fertilizing ability of spermatozoa from two different males. This competitive assessment is termed heterospermic insemination. In order for the procedure to be successful, identification of the source of each spermatozoon is essential. This may be achieved by the use of genetic markers or labelling of spermatozoa (Beatty *et al.*, 1969; Bedford and Overtsreet, 1972). This technique is of limited practical significance.

6.6.10. Hemizona assay

A further development from the zona-free hamster penetration assay and heterospermic insemination is the hemizona assay, in which spermatozoa from different stallions are incubated with hemizona from a single oocyte. A significant relationship has been reported to exist between the number of bound spermatozoa of a stallion and the probability of pregnancy resulting from insemination with that stallion's semen (Fazeli *et al.*, 1995). Further development of this assay to assess the zona pellucida-binding ability of sperm has also been suggested as a means of assessing spermatozoan viability (Pantke *et al.*, 1992, 1995; Meyers *et al.*, 1995a, 1996).

6.6.11. Chromatin analysis

The total amount and structure of chromatin or DNA has been investigated as a possible means of assessing spermatozoon viability (Kenney *et al.*, 1995; Evenson *et al.*, 1995). In the Netherlands the DNA structure and integrity of bull semen is regularly assessed using acrodene orange, which stains undamaged double-helix DNA and split damaged DNA red (S. Revell, Wales, 1999, personal communication).

6.7. Sexing Spermatozoa

Many attempts have been made to pre-select the sex of offspring in humans and other animals. Sexing of spermatozoa has been attempted by various methods with differing success in different animals. Staining with quinacrine mustard, which causes fluorescent staining of the Y chromosome at metaphase (particularly in human and gorilla spermatozoa), is not successful in cattle and rams (Bhattacharya *et al.*, 1977; Ericsson and Glass, 1982; Windsor *et al.*, 1993). Spermatozoon karyotype analysis has also been used (Rudak *et al.*, 1978; Amann, 1989a). No reports on the use of such techniques with stallion spermatozoa have been published.

Most recently the use of flow cytometry has been investigated and appears to have the potential to provide a highly precise method of analysing spermatozoa, and even the differentiation of X and Y chromosome-bearing spermatozoa, by resolving the difference in DNA content. This has been successfully carried out with rabbit and bull spermatozoa (Johnson *et al.*, 1989, 1997) and by practical application has resulted in pre-selected cattle (Cran *et al.*, 1995; Johnson *et al.*, 1998; Seidel *et al.*, 1998). No reports of the use of this procedure with stallions have been published.

6.8. Conclusion

It may be concluded that semen evaluation can be used to provide a mass of information but the correlation of such information with fertility is low, somewhat conflicting and largely unreliable (Jasko, 1992; Malmgren, 1992; Magistrini et al., 1996). More accurate assessment may be gained from the use, and future development, of functional tests rather than the traditional physical evaluation.

At present, on average 50% of stallions undergoing semen evaluation meet the criteria, as discussed in this chapter (Muller, 1987). Much variation exists and semen quality depends upon a whole host of factors, as well as inherent individual variation. It is, therefore, more or less impossible to set down absolute criteria which a stallion must achieve before he can be used for AI. Variation between different laboratories is an additional, and unacceptable, source of variation. Progress has been made towards reducing this variation with the advent of automated computer systems, though these themselves are not without sources of error. The evaluation of physical characteristics, which forms the major part of the traditional semen evaluation process, is time consuming, requires skilled labour and is often a laborious task, even though it requires relatively few facilities. Many of the new advances discussed require significant financial outlay, which is a real problem when considering the commercial application of the techniques. Many of these tests now assess the functional capabilities of the spermatozoa, rather than their physical characteristics, and as such may hold the key to more accurate and reliable semen evaluation in the future.

7 Semen Storage and Transportation

7.1. Introduction

The major advantages of using artificial insemination are the potential to transport semen across the world (and hence eliminate geographical restrictions to the breeding of stock) and the ability to keep semen stored for prosperity. For either of these advantages to be realized, semen must be stored. This chapter addresses the question of semen storage and transportation. For insemination, semen is normally available either as fresh, chilled or frozen; its method of processing determines its potential life span. Many of the principles behind the three methods of preparation are the same and hence some aspects will be considered in general first, followed by details specific to each method.

It is worth mentioning, at the outset, that a considerable amount of research has been carried out into the storage of semen. However, much of this uses only small experimental numbers and lacks specific details of protocol and methods of recording results. This, plus the interaction between the numerous variables in such work, means that many of the results are hard to interpret and seem to produce conflicting evidence. Significant numbers of stallions need to be included in experimental work in order to attempt to compensate for these variables and potentially biasing effects. In addition, most of the work with semen extenders and storage judges success by percentage motility rather than insemination results. Although the motility of spermatozoa is the easiest characteristic to measure by which to judge viability, the correlation between motility and conception rates is low – in the region of 0.63 (Chapter 6). It is essential that work on semen extenders and storage carries spermatozoa assessment through to insemination and pregnancy data, or at least that the work appreciates the relatively low correlation between the two parameters. However, the large numbers of inseminations required to prove significant differences in conception rate makes the cost of equine experiments largely prohibitive.

7.2. Extenders

Regardless of the method of storage, semen should be mixed with an appropriate extender immediately post collection. Some systems allow for the semen to be added directly to the extender within the collection vessel of the artificial vagina. The composition of extenders varies enormously, but they are normally based upon milk or egg yolk products plus antibiotics. The exact components of the extender will vary with the method of storage and possibly between stallions, but the same basic components are common across the board.

In general, the use of an extender enhances the spermatozoon survival rate outside the stallion and, with the addition of antibiotics, reduces bacterial contamination and associated risk of pathogen transfer. An extender may also enhance the viability of spermatozoa from low-fertility stallions and protect them from unfavourable external environmental conditions (Kenney *et al.*, 1975; Douglas-Hamilton *et al.*, 1984, 1987; Varner, 1991; Jasko *et al.*, 1992a). In addition, the use of an extender facilitates the aim of increasing the number of mares covered per ejaculate.

In common with the requirements of seminal plasma, an appropriate extender should:

- provide primarily a medium that is free of pathogenic or infectious agents, with an energy source (often a sugar) and a protein source (often egg yolk or milk) (Samper, 1995b);
- be isotonic with seminal plasma (that is, it should maintain an osmotic pressure compatible with spermatozoa);
- provide a proper balance of minerals;
- contain chemicals capable of neutralizing toxic products produced by spermatozoa;
- provide protection for them from changes in temperature;
- contain elements capable of stabilizing enzyme systems;
- be isothermal (that is, at the same temperature) with the semen at dilution (Varner, 1991).

In general, pasteurized skimmed milk plus antibiotic provides or satisfies all these requirements. Fine tuning of extenders with the addition of various extra components is normally carried out to perfect the final result.

7.2.1. Historical development

The development of extenders for use in equine AI has lagged behind that of AI techniques. Extenders were considered to be of relatively minor importance and so were poorly reported in experimental work, though Dourette (1955) first indicated the importance of the composition of extenders. In small-scale trials involving storage over 48 h, transportation for 150 miles and a number of different extenders, he demonstrated differences in the efficacy of different extenders and claimed good success rates when he extended semen in a ratio of 1:3–1:5 (semen:extender) with a diluent he termed Diluent M (Table 7.1).

Further work by Hejzlar (1957) recorded the best motility with a glucose and egg yolk extender (Table 7.2). However, a higher semen:extender ratio was required (1:2). Higher dilutions did not sustain motility as well.

In these two extenders, egg yolk provides the source of protein. Kamenev (1955) reported on the use of milk as a protein source, and in his experiments the milk used was equine. Kamenev (1955) reported good results using mare's milk alone, having pre-heated it to 30°C and subsequently cooled the milk prior to use. Similar positive results, using mare's milk alone, were reported by Mihailov (1956) and, using mare's milk with egg yolk, by Kuhr (1957).

Nishikawa (1959a,b) further investigated the use of egg yolk extenders which he termed Baken No. 1 and Baken No. 2. Their compositions are given in Table 7.3. Centrifugation was used during the extension process with Baken No. 1 extender. The raw semen was diluted in a ratio between 4:1 and 6:1 (semen:extender) and the sample was centrifuged at 1000–1500 g for 15 min. The supernatant (seminal plasma plus extender) was decanted off and a further 1–2 ml of Baken No. 1 or No. 2 was added prior to storage at 4°C. Of the two extenders, Baken No. 1 proved to be the more successful in maintaining spermatozoon motility at pre-storage levels (70%) after 24 h and showed only a small decline (to 60%) after 48 h storage. Baken No. 2 showed a similar trend but was reported to result in a motilty of 50% after 48 h.

Work in Germany in the mid to late 1950s reported the use of a diluent containing one part sodium bicarbonate to two parts glucose (Vlachos and Paschaleri, 1969). Conception rates of 35% were reported after storage for

Table 7.1. Composition of Diluent M (Dourette, 1955).

Component	Quantity
Glucose	30 g
Lactose	20 g
Sodium potassium tartrate	10 g
Distilled water	1000 g
Fresh egg yolk	200 g
Para-amino benzoic acid	6 g

Table 7.2. Hejzlar's glucose and egg yolk extender (Hejzlar, 1957).

Component	Quantity
Glucose	5.76 g
Sodium potassium tartrate	0.67 g
Distilled water	100.0 ml
Fresh egg yolk	3–5 g

Table 7.3. Composition of Baken No. 1 and Baken No. 2 extenders (Nisikawa, 1959a,b).

(a) Baken No. 1

Component	Quantity
Distilled water	100.0 ml
Potassium chloride	0.08 g
Sodium hydrogen phosphate	0.05 g
To 95 ml of the above mixture add:	
Glucose	5.0 g
Fresh egg yolk	3.0 ml

(b) Baken No. 2

Component	Quantity
Distilled water	100.0 ml
Potassium chloride	0.025 g
Disodium hydrogen phosphate	0.05 g
To 95 ml of the above mixture add:	
Glucose	5.0 g
Fresh egg yolk	2.5 g

96 h and transportation by road. Though fertility rates were relatively low, they compared favourably with general fertility rates from natural service in German studs at the time.

The inclusion of antibiotics in order to satisfy the need for a pathogenically clean substrate was investigated by several researchers in the late 1960s and early 1970s (Fallon, 1967; Hughes et al., 1967; Hughes and Loy, 1969b; Roberts, 1971a,b). The semen from mammals contains a natural microflora and, under normal conditions, this would pose no threat of infection to any females mated (Riley and Masters, 1956; Zemjanis, 1970; Roberts, 1971a,b). Burns et al. (1975) illustrated the significant effect that the addition of 16×10^6 IU of sodium penicillin and 188 mg of gentamicin sulphate per 500 ml of cream gelatin extender had on bacterial populations. A 99% reduction in bacterial counts was reported at 2 h post extension, when compared with a saline control. Since this initial work, several other antibiotics have been used. The use of antibiotics will be considered in detail later (section 7.2.7).

7.2.2. Present use of extenders

The use of extenders allows semen to be stored and the number of mares inseminated per ejaculate to be significantly increased. It also allows the addition of antibiotics and alteration of spermatozoan concentration to enable mares with compromised uterine defence systems to be successfully bred (Kenney et al., 1975; Kotilainen et al., 1994).

Based upon the results of research, most extenders used today include milk products or egg yolk as a source of protein, energy and cryoprotectant, with the addition of an antibiotic. Other common components include tris(hydroxymethyl)-aminomethane (TRIS) as a buffering agent; citric acid, a neutralizing agent, antiperoxidant and buffer; glycerol, a cryoprotectant; and sodium bicarbonate, a buffering solution (Kenney et al., 1975; Pickett and Voss, 1976; Douglas-Hamilton, 1984; Province, 1984; Francl et al., 1987; Pickett, 1993b,c). In broad terms, extenders in use today can be divided into those based on milk or milk products, cream and gelatin, TRIS, or egg yolk.

7.2.3. Milk and milk-product extenders

Table 7.4 gives examples of some of the extenders based on milk and milk products that are in use today. These extenders include non-fat dried skimmed milk–glucose extender (NFDSMG) or Kenney, one of the most popular diluents used in the UK (Kenney et al., 1975), along with E-Z Mixin, which is very similar in composition but is more popular in the USA. Both are classified as NFDSMG extenders and are commonly used for extending semen prior to storage, but also for evaluation, as they are optically clear, maintain spermatozoon motility and fertility well and are relatively straightforward and cheap to prepare. Their clarity, which is important as far as microscopic evaluation is concerned, is due to the lack of fat globules. These extenders, though slightly more expensive than the straightforward skimmed milk preparations, are normally the preferred option. The straightforward skimmed milk extender requires the milk to be heated to 92–95°C for 10 min in order to inactivate lactenin, an antistreptoccal agent naturally found in milk which is toxic to bovine and equine spermatozoa (Flipse et al., 1954; Householder et al., 1981). Heating milk to and maintaining it at this constant temperature for 10 min is a potential source of error in preparation. In addition, heating milk to 92–95°C at altitudes greater than 5000 feet (1524 m) is difficult. Skimmed milk–gelatin preparations tend to be less popular than skimmed milk alone, as they are not as simple to prepare and do not provide such an optically clear medium for microscopic examination.

Pregnancy rates with all-milk extenders are reported to be quite acceptable. In experiments conducted by Kenney et al. (1975), semen extended with Kenney extender resulted in pregnancy rates of 58.3% (7/12). Slightly higher pregnancy rates of around 62.5% have been reported by other workers for skimmed milk extenders (Householder et al., 1981).

Province (1984) demonstrated the superior nature of E-Z Mixin over skimmed milk, as far as motility is concerned after 4 h storage. This small trial was developed to investigate fertilization rates post insemination: using 100×10^6 spermatozoa in total, diluted in E-Z Mixin or skimmed milk, Province demonstrated that E-Z Mixin was again superior to the skimmed milk extender, pregnancy rates at 6-day embryo recovery being 52% (13/25) and 40% (10/25), respectively. However, a second experiment failed to show a

Table 7.4. Milk and milk-product-based extenders in use today.

(a) E-Z Mixin (Province et al., 1984, 1985)

Component	Quantity
Non-fat dry milk	2.4 g
Glucose monohydrate	4.9 g
Sodium bicarbonate (7.5% sol)	2.0 ml
Polymixin B sulphate (50 mg ml^{-1})	2.0 ml
Distilled water	92.0 ml
Osmolarity (mOsm kg^{-1})	375.00 ± 2
pH	6.99 ± 0.02

Mix the liquids first and then add the powders.

(b) Non-fat dried skimmed milk–glucose extender II (NFDSMG) or Kenney extender (Kenney et al., 1975)

Component	Quantity
Non-fat dry skimmed milk	2.4 g
Glucose	4.9 g
Gentamicin sulphate (reagent grade)	100.0 mg
8.4% NaHCO$_3$	2.0 ml
Deionized water	92.0 ml

To avoid the antibiotic curdling the milk, mix the liquids first before adding the NFDSM.

(c) Heated skim milk extender (Voss and Pickett, 1976)

Component	Quantity
Skimmed milk	100.0 ml

Heat to 92–93°C for 10 min in a double boiler.
Cool to 37°C before use

(d) NFDSM–glucose extender I (Kenney et al., 1975)

Component	Quantity
NFDSM	2.4 g
Glucose	4.0 g
Penicillin (crystalline)	150,000 units
Streptomycin (crystalline)	150,000 µg
Deionized water (made up to)	100 ml

(e) Kenney's modified Tyrodes extender (Ijaz and Ducharme, 1995)

Component	Quantity
Glucose	4.9 g
NFDSM	2.4 g
Sodium penicillin	150,000 units
Dihydrostreptomycin	150.0 mg
Distilled water (made up to)	100 ml

Add 65% of the above Kenney extender to 35% (v/v) Tyrode's medium

Table 7.4. Milk and milk-product-based extenders in use today (*continued*).

Tyrode's medium:	
NaCl	420.0 mg
KCl	187.0 mg
Na_2HCO_3	210.0 mg
Sodium lactate syrup (98%)	190.0 µl
$CaCl_2$	29.0 mg
MgCl	8.0 mg
HEPES	238.0 mg
Sodium pyruvate	11.0 mg
Bovine serum albumin fatty acid free (BSAAFF)	600.0 mg
Water	100.0 ml

(f) Non-fortified skimmed milk extender (Varner and Schumacher, 1991)

Component	Quantity
Non-fortified skimmed milk	100 ml
Polymixin B sulphate	100,000 IU

Heat the milk to 92–95°C for 10 min in a double boiler; cool, and add the polymixin B sulphate.

(g) CGH-27 (Nishikawa, 1975)

Component	Quantity
Distilled water	100.0 ml
Gelatin	0.1 g
Sodium phosphate (dibasic)	0.05 g
Potassium chloride	0.025 g
Potassium sodium tartrate	0.25 g
Glycine	0.7 g
Glucose	4.5 g
Casein	0.5 g
Tranquillizing drug	0.01 g
Egg yolk	7% v/v

pH adjusted to 6.7–6.9 by 2% citric acid

(h) Skimmed milk extender (Samper, 1995b)

Component	Quantity
Powdered skimmed milk	2.4 g
Glucose	4.9 g
Ticarcillin (*r* Timentin)	100.0 mg
Distilled water	100.0 ml

(i) Skimmed milk gel extender (Voss and Pickett, 1976; Pickett, 1993; Householder *et al.*, 1981)

Component	Quantity
Skimmed milk	100.0 ml
Gelatin	1.3 g

Add the gelatin to the skimmed milk and agitate for 1 min.
Heat the mixture in a boiler for 10 min at 92°C, swirling the mixture periodically

Table 7.4. Milk and milk-product-based extenders in use today (*continued*).

(j) INRA 82 (Magistrini *et al.*, 1992a,b; Ijaz and Ducharme, 1995)

Component	Quantity
Glucose	5.0 g
Lactose	300.0 mg
Raffinose	300.0 mg
Trisodium citrate dihydrate	60.0 mg
Potassium citrate	82.0 mg
HEPES	952.0 mg
Penicillin	10.0 IU ml^{-1}
Gentamycin	10.0 µg ml^{-1}
Water	100.0 ml
UHT skimmed milk	100.0 ml
Osmolarity (mOsm kg^{-1})	326.0
pH	7.1

(k) Mare's milk extender (Lawson and Davies Morel, 1996)

Component	Quantity
Mare's milk	100 ml
Osmolarity (mOsm kg^{-1})	303.00
pH	7.07
Heat the mare's milk at 62.8°C for 30 min	

(l) NFS mare's milk (Lawson, 1996)

Component	Quantity
Mare's milk	100.0 ml
Glucose	4.9 g
Streptomycin sulphate (crystalline)	0.1 g
Penicillin (crystalline)	0.1 g
Sterile deionized water	88.0 ml
Osmolarity (mOsm kg^{-1})	307.00
pH	7.14
Heat the mare's milk at 62.8°C for 30 min	

difference in pregnancy rates between the groups, both being 40%. These apparently conflicting results may have been due to the compromised nature of the reproductive tracts of the mares selected for insemination in the second experiment (Province *et al.*, 1985). Similar results were achieved by Francl *et al.* (1987) using a greater number of spermatozoa per insemination (250×10^6). At embryo recovery 6.5 days post insemination, pregnancy rates of 62% and 56% were recorded for E-Z Mixin and skimmed milk, respectively. This work indicates that the superior ability of extenders such as E-Z Mixin and Kenney in maintaining motility post storage is also reflected in better conception rates post insemination. The use of glucose, an extra source of energy, and sodium bicarbonate, a buffer, in Kenney and E-Z Mixin may account for their superior ability to maintain progressive motility and achieve better conception rates. A more recent addition to the range of milk-based extenders is INRA 82, developed and widely used in France (Magistrini *et al.*, 1992a). This extender is reported to maintain motility better than Kenney or E-Z Mixin, with a motility of 56.1% after 96 h at 5°C (Ijaz and Ducharme, 1995).

The previous discussion in this section relates to the use of cow's milk and cow's milk products. Work has also been carried out on the use of mare's milk. Kamenev (1955) compared mare's milk as a semen extender with those based on glucose or on egg yolk with glucose. The mare's milk extender proved to be superior, giving pregnancy rates of 74% (109/147). The mare's milk was only heated to 30°C, which, from more recent research, would seem not to be high enough to inactivate the lactenin component of milk (Thacker and Almquist, 1953; Filpse *et al.*, 1954; Householder *et al.*, 1981). Mihailov (1956) and Lawson (1996) indicated that heating mare's milk similarly to cow's milk was required to inactivate the lactenin component. Mihailov (1956) boiled mare's milk prior to cooling, and reported pregnancy rates of up to 93.7% using dilutions of between 1:4 and 1:5 (semen:milk). Dilutions of less than 1:4 proved not to be successful. Lawson (1996) demonstrated that after skimming and pasteurizing mare's milk according to the Vat method (heating to 62.8°C for 30 min) (Zall, 1990), 40% motility was maintained up to 24 h post collection. This compared favourably with the control, which was Kenney. The addition of glucose, antibiotic and 2% egg yolk to the mare's milk extender further improved it and resulted in significantly better motility rates at 48 h post collection than those evident with Kenney. This would indicate a possible advantage of mare's milk over cow's milk as a major component of milk-based extenders for use with equine semen (Lawson, 1996).

Work has been carried out in an attempt to discover which element of milk affords protection to spermatozoa. It is apparent that at least two fractions, native phosphocaseinate and β lactoglobulin, are responsible individually, but their effect was not synergistic (Batellier *et al.*, 1997).

7.2.4. Cream–gel extenders

Table 7.5 gives some examples of cream and gel extenders in use today. The addition of gelatin to an extender is thought to act as a membrane stabilizer. Work by Householder *et al.* (1981), comparing skimmed milk with skimmed milk plus gelatin, demonstrated an advantage with gelatin: pregnancy rates of 43.8% and 62.5%, respectively, were obtained. Previous work by Pickett *et al.* (1975a) had indicated a similar advantage for cream–gel extenders over raw semen. The pregnancy rates reported were significantly higher than Householder's, at 95.8% (cream–gel extender) and 91.7% (raw semen). The major disadvantages in using cream–gel extenders, and the reason why

Table 7.5. Cream- and gel-based extenders in use today.

(a) Cream–gel extender (Voss and Pickett, 1976; Lawson, 1996)

Component	Quantity
Gelatin	1.3 g
Distilled water	10.0 ml
Half-and-half cream	90.0 ml
Penicillin	100,000 IU
Streptomycin	100,000 IU
Polymixin B sulphate	20,000 IU
Osmolarity (mOsm kg^{-1})	280.00
pH	6.52

Add the gelatin to distilled water and autoclave for 20 min. Heat the cream in a boiler at 92–95°C for 10 min and add the cream to the gelatin solution after removing scum from heated cream, to make a total of 100 ml.

(b) CGH-27 extender (Lawson, 1996)

Component	Quantity
Glucose	4.5 g
Glycine	0.7 g
Casein	0.5 g
Egg yolk	7.0 ml
Gelatin	0.1 g
Sodium phosphate	0.05 g
Potassium chloride	0.025 g
Sodium potassium tartrate	0.25 g
Distilled water	100.0 ml
Osmolarity (mOsm kg^{-1})	305.00
pH	6.72

(c) Gelatin–cream extender (Hughes and Loy, 1969b)

Component	Quantity
Gelatin	1.3 g
Sterile distilled water	10.0 ml
Half-and-half cream	90.0 ml
Streptomycin	100,000 µg
Penicillin	100,000 IU
Polymixin B sulphate	20,000 IU

Dissolve the gelatin in water and sterilize. Heat the cream to 92–95° for 4 min; remove scum from the surface and add to the gelatin solution. Cool; add antibiotics.

(d) Gelatin–cream–glucose extender (Burns *et al.*, 1975)

Component	Quantity
Half-and-half cream	450.0 ml
Gelatin	16.6 g
Sodium penicillin (buffered with calcium carbonate)	10^6 IU
Physiological saline	20.0 ml
Gentamycin sulphate	188 mg

Warm the cream to 43°C in a double boiler; add the gelatin warmed to 65°C; cool to 43°C; then add antibiotics.

(despite their apparent success) they are not widely used as commercial semen extenders, are their difficulty in preparation and the presence of fat globules, which makes microscopic examination difficult.

7.2.5. TRIS extenders

One of the major problems encountered in storing semen is the accumulation of metabolic by-products, or waste, which subsequently have an adverse or toxic effect on spermatozoa. Buffers have been used in an attempt to counteract this. The most popular buffer used in equine semen extenders is tris(hydroxymethyl)-aminomethane (TRIS). Table 7.6 gives an example of an extender based on TRIS.

In common with TES [N-tris(hydroxymethyl) methyl-2-aminomethane-sulphonic acid], MES [2-(N-morpholino)ethanesulphonic acid], HEPES (N-2-hydroxyethylpiperazine-N-2-ethanesulphonic acid), sodium phosphate, sodium citrate, MOPS [2-(N-morpholino)propane sulphonic acid] and BES [N,N-bis(2-hydroxyethyl)-2-aminoethane sulphonic acid], TRIS has been used quite successfully as the buffer component of mammalian semen extenders (Filpse *et al.*, 1954; Salisbury *et al.*, 1978). The addition of these electrolytes to a milk or egg yolk extender confers a buffering capacity to the solution, in addition to being relatively harmless to the various cells. In the case of semen extenders, they provide an efficient means of buffering the accumulation of spermatozoan metabolites over a range of temperatures and pH (Davis *et al.*, 1963; Magistrini and Vidament, 1992).

Unfortunately TRIS does not appear to be as successful in the case of equine semen extenders as it is with other species (Pickett *et al.*, 1975a). Buffalo semen has been extended quite successfully using egg yolk and TRIS extenders (Mohan and Sahni, 1991). It can be used successfully in bulls (Salisbury *et al.*, 1978) and in boars (Pursel and Johnson, 1975a,b; Revell and Glossop, 1989), but similar success has not yet been achieved with stallion semen: the use of TRIS and egg yolk extenders is reported to be less successful at maintaining motility than with other species. In comparative work, it proved to be less successful than skimmed milk, caprogen (CAP), Cornell University,

Table 7.6. TRIS-based extender (Magistrini and Vidament, 1992).

Component	Quantity
TRIS (hydroxymethyl) aminomethane	2.44 g
Citric acid monohydrate	1.36 g
Glucose	0.82 g
Egg yolk	20.0 ml
Polymixin B sulphate	100,000 IU
Distilled water	100.0 ml
Osmolarity (mOsm kg^{-1})	301.0
pH	6.55

or egg yolk and bicarbonate extenders (Pickett *et al.*, 1975a). This lack of success in maintaining motility is borne out in the lower fertility rates reported by Pickett *et al.* (1975a). In comparisons with cream–gel extenders and raw semen, fertility rates of 37.5% were reported for a 2.4% TRIS extender compared with 75% for the other treatments. The adverse effect of TRIS is apparent within 1 h of extension. Pace and Sullivan (1975) showed a similar adverse effect even after short-term storage of stallion semen in TRIS (or other hydrogen ion) extenders.

7.2.6. Egg yolk extenders

Table 7.7 gives examples of egg yolk-based extenders. These are used primarily for storage rather than semen evaluation, due to their optical opacity. Their success in preserving semen was first identified in work with bull semen by Phillips (1939), from which it was evident that egg yolk was effective at preserving bull semen at low temperatures but above freezing. Subsequent work has indicated its successful use as a cryoprotectant in long-term storage of semen by freezing.

The specific use of egg yolk as a cryoprotectant will be discussed in section 7.5. In summary, its success is attributed to its low-density lipoprotein fraction, more specifically the phospholipids, which become firmly bound to the spermatozoan plasma membrane (Butler and Roberts, 1975; Watson, 1976; Foulkes, 1977). Many extenders are based on egg yolk and many have found popularity for commercial use in Europe (De Vries, 1987; Klug, 1992). The Dimitropoulos extender in Table 7.7 is reported to result in progressive motility rates of 56% at 24 h and 40% after 48 h of storage at 5°C (Ijaz and Ducharme, 1995). The same extender was used by DeVries (1987), who achieved pregnancy rates of 76% after 3–4 days of storage followed by insemination of 500 × 10^6 motile spermatozoa every 48 h during oestrus. In work by Braun *et al.* (1993) this egg yolk extender, used with 5% seminal plasma, outperformed the standard skimmed milk extenders plus seminal plasma in maintaining progressive motility up to 50% after 24 h and 28% at 72 h. This contradicted previous work that had indicated no difference between the two types of extender in their ability to maintain motility (Wockener *et al.*, 1990). An egg yolk extender prepared similarly to the one above, but used without centrifugation, was reported as successful with stallion semen (Samper, 1995b). Jasko *et al.* (1992a) suggested that the use of NFDSMG extender prepared similarly plus 4% clarified egg yolk, added to seminal plasma-free spermatozoa, maintained high progressive motility in equine semen stored beyond 12 h at 5°C. This advantageous effect of egg yolk was disputed by Bedford *et al.* (1995b), but in this case seminal plasma was not removed prior to extension. Bedford *et al.* (1995a) demonstrated that this apparent negative effect of egg yolk was due to an adverse interaction between egg yolk and seminal plasma, but insemination results using seminal plasma-free spermatozoa extended in NFDSMG plus 4% egg yolk resulted in disappointing conception rates. When Amann and Pickett (1987)

Table 7.7. Egg yolk based extenders in use today.

(a) Dimitropoulos extender (De Vires, 1987; Braun et al., 1993; Ijaz and Ducharme, 1995)

Component	Quantity
Solution A	
Anhydrous glucose	2.0 g
Fructose	2.0 g
Distilled water	100.0 ml
Solution B	
Sodium citrate dihydrate	2.0 g
Glycine	0.94 g
Sulphonilamide	0.35 g
Distilled water	100.0 ml
Egg yolk	20.0 ml
Osmolarity (mOsm kg^{-1})	280.0
pH	6.9

Make up solution A and B separately, then mix 30 ml of A and 50 ml of B. To the combined solution, add the egg yolk; centrifuge for 20 min at 1200 g. Use the resulting supernatant as the extender.

(b) Glucose–lactose extender (Martin et al., 1979)

Component	Quantity
Glucose	30.0 g
Sodium citrate	1.85 g
Sodium EDTA	1.85 g
Sodium bicarbonate	0.6 g
Distilled water	100.0 ml

Add 50 ml of this solution to 50 ml of 11% lactose solution; supplement with 20% egg yolk and 4% glycerol.

(c) Glucose–sucrose extender (Boyle, personal communication, Wales, 1996)

Component	Quantity
Skimmed milk	2.4 g
Glucose	2.65 g
Sucrose	4.0 g
Distilled water	100.0 ml

Supplement 92.5 ml of this solution with 4% egg yolk and 3.5% glycerol.

reduced the amount of egg yolk to 2% and used this extender with spermatozoa plus seminal plasma, no adverse affects of egg yolk on motility were reported. It seems, therefore, that a fine balance has to be struck in the concentration of egg yolk. This must be such as to allow seminal plasma to be present and benefit to be gained from the cryoprotectant properties of egg yolk, but without the adverse reaction that seems to occur between egg yolk and seminal plasma.

7.2.7. Other major components within extenders

The four main types of extender for use with stallion semen have been considered. However, much work has been carried out on the addition of other components to improve what are still relatively poor and very variable conception rates. This section will consider some of the other types of extenders used and the components included.

Citric acid

The inclusion of citric acid in an extender for stallion semen is relatively common. It acts like a buffer, in a similar way to TRIS. However, it also acts as an antiperoxidant, reducing the peroxidation of lipids in the spermatozoan plasma membranes and thereby reducing the detrimental effect of storage on membrane integrity. It also forms complexes with calcium ions, which is of particular importance during the cooling of semen, as calcium leakage, due to increased membrane permeability when cooled, can be mopped up by the citric acid component of an extender (Jones *et al.*, 1979).

Bovine serum albumin

Bovine serum albumin (BSA) has also been used quite effectively as a component of stallion (Kreider *et al.*, 1985), boar (Dixon *et al.*, 1980) and ram (Harrison *et al.*, 1982; Klem *et al.*, 1986) semen extenders. It is reported that BSA, like citric acid, may act as an antiperoxidation agent, reducing the deleterious effects of membrane lipid peroxidation. Inclusion of 3% BSA in a skimmed milk extender, or even the addition of 3% BSA to a raw semen sample, is reported to help to maintain motility (Pickett and Voss, 1975; Padilla and Foote, 1991). Addition to skimmed milk seems particularly successful in terms of motility over the first 6 h of cooled storage (Kreider *et al.*, 1985). This supported previous encouraging work by Goodeaux and Kreider (1978) when the addition of 3% BSA to seminal plasma was compared with spermatozoa in Tyrode's solution (a common buffer). The results indicated an 80% pregnancy rate for both extenders at 45 days but a 60% foaling rate for BSA extender compared with a 40% foaling rate from the Tyrode's solution. As far as pregnancy rates are concerned, this advantage was not supported in work by Klem *et al.* (1986).

Equine serum has also been included in extenders. Work by Palacios Angola and Zarco Quintero (1996) replaced egg yolk in freezing extenders with BSA, equine serum or bovine serum. Unfortunately no improvement on post-thaw motility was observed and it was concluded that egg yolk diluent provided the best cryopreservation. The use of serum is frowned upon by many health authorities, as it can carry pathogens (S. Revell, Wales, 1998, personal communication).

Goodeaux and Kreider (1978) also investigated the effect of separation of spermatozoa, using a BSA column. The spermatozoa isolated in the lower fractions exhibited superior motility and resulted in 45-day pregnancy rates of 70%, which was carried through to a 70% foaling rate.

Energy sources

Most extenders use glucose as the major source of energy for metabolic activity and movement of spermatozoa (Katila, 1997), but the use of other non-glycolysable sugars has been investigated. Nishikawa (1959a) included sucrose and reported relative success in maintaining a higher percentage of motile spermatozoa for stallion semen stored at 4°C for 24 h in a sucrose extender when compared with lactose. A cryoprotectant role for sugars has been indicated in work on ram semen: diluents containing arabinose, galactose, ribose, xylose or lactose proved superior for spermatozoa storage at 5°C compared with diluents containing glucose, mannose, fructose or sucrose, which were in turn superior for storage at 37°C (Jones and Martin, 1965; Lapwood and Martin, 1966). These findings led to similar work in horses (Arns *et al.*, 1987), the results of which indicated that there was no difference in the ability of BSA diluents, containing lactose, raffinose or sucrose, to maintain stallion spermatozoan motility when stored at either 37°C or 5°C. However, they all outperformed similar diluents containing arabinose or galactose, which appeared to be detrimental to spermatozoa. No differences were detected in comparing spermatozoan viability post freezing and thawing in BSA diluents containing lactose, raffinose or sucrose.

It is apparent that the metabolizable substrate for the production of ATP to supply energy for equine spermatozoa is most commonly provided by glucose, and it is evident from the work by Arns *et al.* (1987) and Katila (1997) that fructose, lactose, raffinose, sucrose and pyruvate may all function in a similar manner.

Antibiotics

Many of the aforementioned extenders include antibiotics. Semen from mammals invariably contains a natural microflora which, under normal conditions, would pose no threat of infection to females mated (Riley and Masters, 1956; Hughes *et al.*, 1967; Zemjanis, 1970; Roberts, 1971 a,b; Clement *et al.*, 1995). The inclusion of antimicrobial agents, such as antibiotics, is necessitated not only by the risk of infection but also by the ideal nature of most extenders, both for spermatozoan survival and microbial growth. This microbial growth presents a risk of infection in immunologically compromised mares and acts as competition for substrates within the extender (Gunsalus *et al.*, 1941; Almquist, 1949a,b; Bennet, 1986; Evans *et al.*, 1986; Watson, 1988). In addition, mares with compromised uterine defence systems may succumb to uterine infections following insemination with heavily contaminated samples, reducing the potential to conceive. Kenney *et al.* (1975) reported the beneficial use of extenders for such mares. There is also some evidence, though it is disputed, that an increase in bacterial counts *per se* may cause a decline in motility (Pickett *et al.*, 1988).

Despite all the precautions taken at collection, semen samples are invariably contaminated with both pathogenic and non-pathogenic bacteria (Bain, 1966; Crouch *et al.*, 1972; Burns *et al.*, 1975; Merkt *et al.*, 1975a; Clement *et al.*, 1995). Thorough hygiene procedures during preparation, at collection and during handling can reduce microbial counts but cannot eliminate them, and so antibiotics are required to deal with this contamination, whatever the storage method used. Microbial contamination and growth are evident in raw, fresh, chilled or frozen semen (Lorton *et al.*, 1988a,b). As discussed in Chapter 6, numerous microorganisms are reported to be present in stallion semen, including *Klebsiella pneumoniae*, *Streptococci*, *Pseudomonas*, *Proteus*, *Staphylococcus*, *E. coli* and *Aerobacter* as well as yeasts, mycoplasma and viruses (Roberts, 1971b; Crouch *et al.*, 1972; Kenney *et al.*, 1975; Foote, 1978). The use of antibiotics will control bacteria, especially *Pseudomonas aeruginosa*, but may not eradicate them. The success of antibiotics is enhanced by cooling the sample to 5°C rather than to just 20°C (Vaillancourt *et al.*, 1993). The inclusion of antibiotics in extenders is, therefore, largely a safety precaution.

The most popular antibiotics, either in isolation or as a combination, were traditionally penicillin, streptomycin and polymixin B. These still find favour in many countries today. More recently ticarcillin, amikacin and gentamicin have been used (Samper, 1995b; Hurtgen, 1997). The presence of both Gram-positive and Gram-negative organisms in stallion semen and the recent suggestion that resistance to antibiotics is becoming evident have led to the use of several different combinations of antibiotics, rather than the inclusion of a single one (Back *et al.*, 1975; Burns *et al.*, 1975; Squires *et al.*, 1981b). For example, in bull semen the use of a new combination of gentamicin, tylosin and linco-spectin has been recommended to overcome the development of bacterial resistance (Lorton *et al.*, 1988a,b; Shin *et al.*, 1988). For stallion semen, combinations of some of the following are currently used: amikacin sulphate, gentamicin sulphate, streptomycin sulphate, sodium or potassium penicillin, ticarcillin disodium and polymixin B sulphate (Back *et al.*, 1975; Varner, 1991; Clement *et al.*, 1995).

Antibiotics undoubtedly reduce the proliferation of microorganisms within stallion semen (Burns *et al.*, 1975; Clement *et al.*, 1995), but their use has also been reported to be associated with detrimental effects on spermatozoa, in particular motility (Back *et al.*, 1975; Squires *et al.*, 1981b; Varner, 1991; Varner *et al.*, 1992; Jasko *et al.*, 1993a). The extent of those adverse effects depends upon the antibiotics used – some have only a negligible effect (Arriola and Foote, 1982; Back *et al.*, 1975; Varner *et al.*, 1992).

Antibiotics such as gentamicin and amikacin, being acidic in nature, are known to affect the pH of the extender. This requires the addition of a buffer, such as sodium hydrogencarbonate (sodium bicarbonate), to try to counteract this acidic effect. Acid conditions are not conducive to spermatozoan survival (Jasko *et al.*, 1993a).

This effect on the pH may also curdle the milk in milk-based extenders, and so special precautions have to be taken during preparation (as indicated with some of the extenders given in Table 7.4). Antibiotics such as penicillin, polymixin B sulphate and streptomycin do not appear to have the same detrimental effect.

Some antibiotics are known to have adverse side effects on general body systems. For example, gentamicin is associated with damage to neural transmission and polymixin is associated with renal toxicity. These two antibiotics are also reported to be associated with significant depression in motility. Comparing the effects of various antibiotics on motility, Jasko *et al.* (1993a) demonstrated that gentamacin had the most significant effect, followed by polymixin B. Amikacin sulphate and ticarcillin disodium had no apparent effect. The exact mechanism by which the toxicity of antibiotics affects cellular mechanisms is unclear, but it is evident that the significance of the effect increases with the duration of storage (Varner, 1991). Most research has concentrated on the use of gentamicin and polymixin B, which were two of the most popular antibiotics used in initial work.

Using computer image analysis, Jasko *et al.* (1993a) investigated the effect of two inclusion rates of gentamicin sulphate: 1000 and 2000 μg ml^{-1} of extender. After storage for only 1 h at 23°C, an adverse effect of both inclusion rates was evident compared with controls. Longer-term storage at 5°C resulted in a 7% reduction in progressive motility at 24 h and a reduction of 8% was observed at 48 h for both inclusion rates when compared with controls. No such changes were reported in similar work by Back *et al.* (1975) and Squires *et al.* (1981b). These researchers did not have the benefit of computer-aided analysis to assess motility, which would suggest that the adverse effects of gentamicin are quite subtle but may still affect spermatozoan viability. Clement *et al.* (1993) suggested that gentamicin may be used at inclusion rates below 500 μg l^{-1}.

As mentioned, the inclusion of polymixin B sulphate in stallion semen extenders has also been demonstrated to have an adverse effect. Inclusion rates of 1000 IU ml^{-1} or greater were reported to reduce progressive motility by 7% after 24 h of storage at 5°C. This was similar to the effect demonstrated with 1000 μg ml^{-1} of gentamicin (Jasko *et al.*, 1993a).

It may be concluded that the adverse effects of gentamicin sulphate and polymixin B sulphate limit the applicability of their use as antibiotic in stallion semen extenders, at least at concentrations equal to, or in excess of, 1000 μg ml^{-1} and 1000 IU ml^{-1}, respectively.

Inclusion of 2000 μg ml^{-1} of amikacin sulphate or ticarcillin disodium has been reported to have no adverse effects on motility after short-term storage (24 h) at 23°C or longer-term storage at 5°C (Jasko *et al.*, 1993a). Inclusion of less than 50 mg l^{-1} of penicillin is also appropriate (Clement *et al.*, 1993),

Other less commonly used components

Various other additions have been made to extenders in an attempt to improve survival of spermatozoa. These include vitamin E, which is reported to protect spermatozoa successfully against cold shock and lipid peroxidation via an increase in endogenous respiration (Augero et al., 1994). Similarly the inclusion of ascorbic acid (vitamin C) in a skimmed milk or glycine extender is reported to act as an antioxidant and to have a beneficial effect on the percentage of membrane-intact spermatozoa at 5°C. Its effect on motility is more variable, depending upon its inclusion rate (Aurich et al., 1997). Inclusions of methylxanthanines such as theophylline, caffeine and HEPES solution have been associated with improvement in motility at 5–7°C, probably by inhibition of phosphodiesterase. However, inclusion of theophylline and HEPES in extenders for storage over 24 h failed to produce any improvement (Heiskanen et al., 1987).

Work with hamster spermatozoa indicated that the inclusion of taurine (a free β amino acid) enhances motility and fertilization capacity. Based upon this evidence, Ijaz and Ducharme (1995) successfully improved equine motility after 24 h of storage at 5°C in Kenney and INRA 82 extenders by including 100 mM of taurine and ensuring that the resultant extender was isotonic with the spermatozoa. It was postulated that this effect may be due to an inhibition of lipid peroxidation (Alvarez and Storey, 1984a) and by control of the sodium–potassium pump across the spermatozoon plasma membrane, maintaining higher intracellular potassium levels by maintaining membrane stability (Ijaz and Ducharme, 1995). Melatonin has also been reported to protect membrane ultrastructure when included at high concentrations (10–20 µg per 10×10^6 spermatozoa) (Anwar et al., 1996).

The inclusion of Tyrode's solution, a common buffer, has been investigated and showed some success in maintaining motility, especially Tyrode's solution high in potassium (Padilla and Foote, 1991). The use of Tyrode's solution on its own as a extender is reported to give pregnancy rates of 80% at 45 days post insemination (Goodeaux and Kreider, 1978).

Minimum essential medium (MEM) has been suggested as a suitable diluent for semen to be inseminated into mares that habitually demonstrate an allergic reaction to the more conventional diluents. With the addition of antibiotics such as penicillin or streptomycin, MEM has been shown to be more successful in controlling bacterial growth than a milk–glucose diluent (Reis, 1991).

The use of membrane protectants, such as lecithin and phenylmethane sulphonylfluoride, in diluents has been tried but with no associated improvement in motility (Schrop, 1992; Katila, 1997). The use of Merck extender has also been attempted, but to no advantage in terms of conception rates after cool storage (Witte, 1989). Polyvinyl alcohol, which is thought to have cryoprotectant properties, has been used as a component of semen extenders and was shown to have some advantageous effect on maintaining motility (Clay et al., 1984). The inclusion of metabolic inhibitors such as caprogen and proteinase inhibitors has been investigated, with the aim of reducing the metabolic rate without the need to cool spermatozoa beyond ambient temperature. The use of both of these metabolic inhibitors was associated with only limited success (Pickett et al., 1975a; Province et al., 1985; Katila, 1997).

Lastly, glycerol, or glycine, is a common component of extenders, providing protection from cold shock. Numerous authors (Polge *et al.*, 1949; Pickett and Amann, 1987; Tekin *et al.*, 1989; Witte, 1989; Arnold, 1992; Katila, 1997) have demonstrated the successful use of glycerol and other cryoprotectants in counteracting the adverse effects of cold shock and freezing. The inclusion of such components for cryopreservation will be discussed at length in section 7.5.

7.2.8. Removal of seminal plasma

The components and function of seminal plasma have been discussed at some length in Chapter 4. In summary, seminal plasma has several functions: it modifies spermatozoa to allow progressive motility during ejaculation and within the female tract; it acts as a transport medium providing metabolic substrates to maintain the viability of spermatozoa up to the time of fertilization; and finally, due to its volume, it increases the chance of even distribution of spermatozoa within the uterine body and horns (Metz *et al.*, 1990).

It is also apparent that some components of seminal plasma may have a detrimental effect on spermatozoa, and even hasten their death during preservation. This effect is evident as a reduction in motility (Pickett *et al.*, 1975a; Varner *et al.*, 1987; Webb *et al.*, 1990; Jasko *et al.*, 1991c; Pruitt *et al.*, 1993), and not necessarily as a detrimental effect on spermatozoon morphology (Sanchez *et al.*, 1995). The fraction of seminal plasma responsible for this effect appears to originate from the vesicular glands (Varner *et al.*, 1987; Webb *et al.*, 1990). It has been further suggested that the effect may be due to elevated concentrations of sodium chloride, which affect osmotic pressure changes, especially during cooling and freezing (Nishikawa, 1975). Differing levels of sodium chloride between stallions may account for the apparent differences in the ability of different stallions' semen to survive cool and frozen storage.

It can be concluded that the removal of seminal plasma and its replacement with an appropriate extender would potentially allow the requirements of spermatozoa to be met and may enhance spermatozoan survival during storage. Work by Palmer *et al.* (1984) demonstrated spermatozoan survival rates of 27.3%, 44.2%, 50.8% and 55.9%, respectively, for extended semen that included 50%, 20%, 10% and 0% seminal plasma. Varner *et al.* (1987) and Webb *et al.* (1990) demonstrated that spermatozoa extended after the removal of seminal plasma and then stored at 25°C, 5°C or −196°C showed better longevity than spermatozoa extended in the presence of seminal plasma. Finally, work carried out in Argentina by Sanchez *et al.* (1995) demonstrated an advantage, as regards motility, in removing seminal plasma prior to storage at 4°C for 72 h.

These findings have not been supported by all work. Jasko *et al.* (1992a) failed to show an advantageous effect of removing seminal plasma prior to extension with NFDSMG extender. There was even a suggestion from their results that such treatment of semen resulted in an adverse effect on progressive motility immediately upon extension, though removal of seminal plasma prior to extension with an egg yolk extender did prove beneficial. This further supports the evidence, presented earlier, suggesting that there is an adverse reaction between seminal plasma and egg yolk to the detriment of spermatozoan survival. The work by Jasko *et al.* (1992a) even suggested that there may be advantages in the inclusion of some seminal plasma (0–20%). Similarly Pool *et al.* (1993) found no detrimental effect on motility from the presence of seminal plasma when an inclusion rate of 1:4 (semen:extender) was used. The extender was based on skimmed milk and glucose, and the semen was stored at 5°C for 24 h. The motility characteristics reported were similar for semen extended and stored whole and a spermatozoa-rich fraction diluted with the same extender.

These results are rather conflicting and confusing, and the reasons for this include the fact that many of these extenders may have included egg yolk. The possibility of an adverse reaction between egg yolk and seminal plasma has been discussed previously and has again been suggested in work by Bedford *et al.* (1995a,b). It is also apparent that the composition of seminal plasma varies considerably between stallions (Jasko *et al.*, 1992a; Charneco *et al.*, 1993; Bedford *et al.*, 1995a,b) and so the adverse effect of seminal plasma, in combination with various extenders, will vary between stallions (Katila, 1997). It may be advocated that different extender regimes should be developed for different stallions in order to maximize success.

Lastly, it may not be beneficial to remove seminal plasma completely. Several workers have suggested that the inclusion of some seminal plasma may be an advantage. Pruitt *et al.* (1993) and Braun *et al.* (1994a) suggested that the inclusion of up to 20% seminal plasma is associated with improved motility, but levels greater than 20% cause motility to be depressed. Palmer (1984) suggested that inclusion of 0–10% seminal plasma was better than 20–50%, and Ahlemeyer (1991) suggested that an inclusion rate of 5% seminal plasma was beneficial in the freezing of semen.

Centrifugation

One of the major methods for removal of seminal plasma is centrifugation. This allows the concentration of the sample into a spermatozoa-rich and a spermatozoa-poor fraction, allowing the use of the former. Centrifugation also provides a more accurate alternative to aspiration, the method that used to be used to remove seminal plasma. Centrifugation at 400 g for 9–15 min successfully fractionates a semen sample, allowing the supernatant (mainly seminal plasma) to be aspirated off, leaving a soft plug of spermatozoa-rich semen in the bottom (Jasko *et al.*, 1992a; Bedford *et al.*, 1995a,b). Such treatment is reported to result in an 80% spermatozoa recovery rate (Heitland *et al.*, 1996). Centrifugation also allows dilution for cool storage and freezing to be more strictly controlled, giving a standard concentration of spermatozoa per volume of diluent and small concentrated volumes for efficient storage. These are then readily available at standard dose rates for direct insemination. This is particularly important when considering the freezing of semen, where small volumes of 0.5–4 ml are regularly used.

Centrifugation may be successfully used to remove seminal plasma but it is not without detrimental effects. The process may damage spermatozoa, with reductions in motility and adverse changes to spermatozoon morphology (Pickett *et al.*, 1975d; Baemgartl *et al.*, 1980; Gather, 1994). There are four ways in which the detrimental effects of centrifugation can be reduced: by the addition of a primary extender; by minimizing the centrifugal force; by minimizing the time of centrifugation; and by underlayering the semen with a dense, isotonic liquid cushion (Equi Prep, Genus PLC) on which the centrifuged spermatozoa float as a dense band (Oshida *et al.*, 1967; Nishikawa *et al.*, 1972a,b, 1976; Pickett *et al.*, 1975d; Martin *et al.*, 1979; Revell *et al.*, 1997). It is apparent that there is much variation in the extent to which spermatozoa from different stallions are affected by centrifugation (Nishikawa, 1975; Cochran *et al.*, 1983; Pickett and Amann, 1993).

Initially, centrifugation was carried out on a raw sample in the absence of an additional extender. This successfully allowed the removal of seminal plasma (Buell, 1963) but resulted in a significant depression in motility post centrifugation. The advantage of adding an extender to the raw semen sample was demonstrated by Cochran *et al.* (1984) where extension 1:1 with a glucose EDTA solution, followed by centrifugation, resulted in a higher percentage of progressive motility than in a similarly treated sample of raw semen. The adverse effect has been postulated to affect fertility as well as motility, possibly by an effect on the function of the acrosome. Baemgartl *et al.* (1980) supported this hypothesis of irreversible damage to the ultrastructure of the plasma membrane and acrosome region, but Vieira *et al.* (1980) failed to find damage to the acrosomal region as a result of centrifugation, though they did demonstrate a 25% reduction in maximal acrosin activity in samples frozen and thawed after centrifugation.

Much of the work concerning centrifugation and the removal of seminal plasma has been done in association with cryopreservation of spermatozoa. In such cases a primary extender is used during centrifugation, and a secondary extender is used after centrifugation for storage and cryoprotection. Details on the composition of primary and secondary extenders are given in section 7.5.2. The absolute and relative compositions of primary and secondary extenders are particularly important when considering cryopreservation.

If centrifugation is carried out with fresh and cooled semen, the primary and secondary extenders are invariably the same as there is no need for cryoprotectants. This general discussion will confine itself to the use of extenders for centrifugation, of which there are numerous examples used at various dilution rates. In general, the majority of the extenders considered in the previous sections are appropriate for use with centrifugation. All have disadvantages and advantages that need to be balanced in order to achieve a satisfactory result.

Various methods for the addition of primary extenders for centrifugation have been investigated. The extender may be fully mixed with the sample or may be used as an initial layer or lower cushion in the centrifuge tube, on top of which the semen sample is placed. Direct mixing of the semen sample with the extender has proved successful in maintaining motility post centrifugation (Martin *et al.*, 1979). However, the results obtained for motility post centrifugation vary considerably. This is likely to be due not only to the nature of the primary extenders used but also to variation in experimental protocol. Cochran *et al.* (1984) demonstrated that time of motility evaluation post centrifugation had a significant effect on the results obtained, with an improvement in motility of up to 10% being evident between time 0 and time 90 min post cessation of centrifugation. This presumably reflects the time taken for the spermatozoa to recover from the trauma of the centrifugation process. The timing of evaluation is invariably not indicated in research results presented, especially in the older work.

Initial investigations into protocols, other than direct mixing, used a layering process. This was first done by Buell (1963), who layered raw semen on top of the primary extender, enhancing the removal of seminal plasma as the spermatozoa collected in the lower layer of extender. Cochran *et al.* (1984) developed the technique further by using 0.25 ml of a dense glucose solution placed at the bottom of the centrifuge tube as a cushion. On top of this was added the semen sample, in a citrate–EDTA extender. They reported a further refinement of this technique: using a glucose–EDTA extender as a cushion, they demonstrated improved motility rates (46%) post centrifugation than was the case without the use of the cushion (35%). This work also used different dilution rates and centrifugal forces for the two different treatments, which could themselves account for the differences reported. Volkmann and Van Zyl (1987) and Volkmann (1987) concluded that, based upon post-thaw motility, there was no advantage in using such a cushion.

Pickett and Amann (1993) suggested that the use of lactose–EDTA–egg yolk extender as a cushion worked just as well as a dense glucose solution. This extender was reported to be as successful as a cushion, both with or without glycerol. Pickett and Amann (1993) also supported the previous conclusions that the use of lactose (4.5–6.5%) and/or glycerol (0–4%) had no detrimental effect on motility post centrifugation, nor was there any apparent interaction between these two components. A subsequent study reported by the same authors suggested that 4% glycerol in the cushion medium was advantageous (Christanelli *et al.*, 1985). Revell *et al.* (1997) reported 30% more live, morphologically normal post-thaw spermatozoa in fractions of ejaculates centrifuged at 1000 g for 25 min over a dense liquid cushion, compared with fractions from the same ejaculates pelleted at 600 g for 15 min. Assessment of spermatozoa was by fluorescent staining with carboxyfluorescein diacetate/propidium iodide. The use of a cushion is now widespread, in the belief that it reduces damage from spermatozoa being forced against the bottom of the tube during centrifugation. Genus, the UK's leading equine AI company, reports considerable success using a specifically formulated centrifuge cushion, and plans to market a system of three standard proprietary preparations, to include primary semen extender (EquiPrep A), a cushion (EquiPrep B) and a secondary or freezing extender (EquiFreeze) (S. Revell, Wales, 1997, personal communication).

The adverse effects of centrifugation are also dependent upon the centrifugal force used and the duration of centrifugation. It can be safely surmised that the lower the centrifugal force and the shorter the duration of centrifugation, the less damage to spermatozoa is likely to occur. However, or dense concentration of spermatozoa in the bottom of the tube, so that the supernatant can be decanted easily. Several gravitational forces or centrifuge speeds have been used. Nishikawa (1975) used 2000–2200 r.p.m. for 10 min, but with the occasional need to increase this to 2500–3000 r.p.m. for 10 min if the viscosity of the semen was particularly high, often due to incomplete removal of the gel fraction. Other workers have used lower gravitational forces: 400–600 g, but for longer (15 min) (Cochran et al., 1984); 350–450 g for 12 min at 20°C (Piao and Wang, 1988); 400 g for 15 min (Amann and Pickett, 1984); 303 g for 10 min at 20°C (Piao and Wang, 1988); or 800 g for 10 min or 1000 g for 15 min are reported to be successful (S. Revell, Wales, 1998, personal communication). Hurtgen (1997) reported the successful use of a high centrifugal force (660 g) for a shorter period (80–90 s) but carried out twice per sample.

When considering these results it should be noted that live, membrane-intact spermatozoa have a lower buoyant density than those that are dead. That is, the dead spermatozoa will pellet first and any reduction in the number of spermatozoa recovered due to low-speed spins will be disproportionately due to the failure of live cells to pellet. This is further demonstrated by the successful use of the supernatant (after centrifugation at 800 g for 10 min) for fresh inseminations by some establishments (Celle, Hannover, Germany: S. Revell, Wales, 1998, personal communication).

The volume of the centrifuge tubes has also been reported to affect post-centrifugal motility. Normally standard 15–50 ml tubes are used, though evidence has been presented which suggests that smaller (10 ml) tubes may reduce centrifugal damage (Pickett and Amann, 1993). A double centrifuge system has been used and was reported to be successful, resulting in a greater spermatozoan recovery rate (90%). A high rate of centrifugation (660 g) for a short period (80–90 s) is followed by decanting of the supernatant followed by resuspension in a freezing extender containing 16% egg yolk and ticarcillin (200 mg ml^{-1}). This process is repeated, followed by freezing in straws (Hurtgen, 1997). Ahlemeyer (1991) suggested that such double centrifugation resulted in an elevation in spermatozoan abnormalities and a decrease in motility.

Once centrifugation has been completed, the pelleted sample is resuspended in the secondary extender by swirling. Standardization of sample concentrations appropriate for insemination is made at this stage. Appropriate dilution rates are calculated in order to obtain the desired number of spermatozoa per insemination dose: 100–800 × 10^6 (Hurtgen, 1997).

It can be concluded that although centrifugation provides a successful means of removing seminal plasma and concentrating the semen sample, allowing standardization of storage or insemination doses, it is not without cost – the most significant being a possible decline in the motility and potential fertilizing capacity of the spermatozoa.

Alternative methods for removing seminal plasma

Removal of seminal plasma may also be achieved by semen collection using an open-ended artificial vagina (AV). Such an AV allows isolation of the sperm-rich fraction. This technique was reported by Bader and Huttenrauch (1966) and has since been used by several other workers (Tischner *et al.*, 1974; Klug *et al.*, 1975; Merkt *et al.*, 1975b; Tischner, 1979; Love *et al.*, 1989; Heiskanen *et al.*, 1994b). Langkammer (1994) demonstrated a significant increase in motility post storage for 72 h at 5°C, compared with centrifuged unfractionated samples. Rather disappointing conception rates have been reported, with figures as low as 24–33% (Heiskanen *et al.*, 1994b). Normally the sperm-rich fraction collected has a concentration of $300–500 \times 10^6$ spermatozoa ml^{-1}, which is appropriate for use without additional concentration. Such samples can be diluted in a ratio of between 1:1 and 1:10 (semen:extender) with a secondary extender prior to use, cooling or freezing. One of the major disadvantages of this system is the relatively large volumes that result, making them inappropriate for freezing unless they can be split. Even so, larger storage space per inseminate is required. For this reason the use of an open-ended AV is not popular in commercial work.

Another alternative to centrifugation is filtration after dilution. Rauterberg (1994) demonstrated the advantage over centrifugation of filtering fresh semen diluted in skimmed milk extender through a glass wool Sephadex filter, in terms of motility, acrosome intactness and lack of abnormal spermatozoa, after 48 h storage at 5°C. Conception rate to first service for centrifuged and filtered semen (38 and 41%, respectively) showed a similar advantage in favour of filtering semen. Much of the recent research into filtering semen has been carried out in association with freezing. Semen filtered prior to freezing showed better motility and conception rates (69.2 and 80.6%) than centrifuged semen (47.6 and 66.6%). Work by Spreckels (1994), again with semen destined for freezing, supported the reports of advantageous effects over centrifugation, on motility pre freezing, of filtration through a glass wool Sephadex filter after dilution in a glycine or skimmed milk extender. However, this advantage was not carried through to post-thaw motility.

Column separation has been suggested as a means of improving the quality of a semen sample prepared for cooling or freezing (Casey *et al.*, 1991). Significant correlations between pregnancy rates per cycle and glass wool Sephadex filtration (0.93) and Sephadex filtration (0.84) were reported, but no correlation with motility was observed (Samper *et al.*, 1991). Finally, the use of a 'swim up' method of separating highly viable spermatozoa from seminal plasma and poor quality spermatozoa has been reported (Henri-Petri, 1993).

7.2.9. Osmolarity

The osmolarity of any extender is important, primarily to prevent the movement of water across the semi-permeable spermatozoon plasma membrane. Influx of water into the spermatozoa due to low osmolarity of an extender (hypotonic solution) will lead to rupture of the plasma membrane. The opposite will occur if the osmolarity of the extender is too high (hypertonic solution) (Cueva *et al.*, 1997a). The result will be a passage of water out of the spermatozoan head and into the surrounding medium, resulting in cytoplasmic hypertonicity and death. Osmolarity of seminal plasma in the stallion is 300–310 mOsm kg^{-1} (Pickett *et al.*, 1976, 1989; Katila, 1997) and it is advisable that all extenders should be adjusted to ensure an osmolarity near this range. Special care should be taken with egg yolk extenders, which tend to raise osmolarity.

The fact that osmolarity is not consistently reported for extenders indicates that its importance is not always appreciated. The osmolarity of most extenders ranges from 250 to 400 mOsm (Katila, 1997) but 350 mOsm appears to be the most appropriate, based upon 12 h storage at 5°C in a milk-based extender (Varner, 1991). In general, hypertonic extenders tend to be less toxic than hypotonic extenders (S. Revell, Wales, 1998, personal communication).

7.2.10. pH

The pH of an extender is also of importance. Some antibiotics, such as gentamicin, are known to reduce the pH of the medium and require the addition of a buffer, such as calcium bicarbonate (Burns *et al.*, 1975). The pH of an extender should be similar to that of seminal plasma; that is, 6.9–7.8. A tighter range of 7.3–7.7 is advocated by some (Pickett and Back, 1973; Pickett *et al.*, 1976). Most milk-based extenders naturally have a pH in the range 6.5–7.0 (Pickett and Amann, 1987). The mechanism by which pH adversely affects spermatozoan function is unclear, but it is possible that it is mediated via creatine. In common with other active cells, spermatozoa have relatively high concentrations of arginine and creatine, both of which act as an energy transport medium for contractile energy to drive the movement of the spermatozoon tail. Of the two, creatine is reported to be significantly affected by pH. Acidic conditions inhibit its function as an energy transport medium. This effect of pH could explain not only the adverse effect of inappropriate pH in extenders but also the detrimental effect that vaginal pH has on motility and viablility (M. Boyle, Wales, 1996, personal communication).

7.3. Fresh or Raw Semen

Insemination of mares with raw, untreated semen is not usually carried out, or advised, unless the indication for insemination is due to physical complications in either the mare or the stallion which preclude natural covering. In order to obtain acceptable results, insemination with raw semen should be carried out between 30 s and 1 h after collection.

Insemination with raw semen provides none of the advantages of AI. Even the division of raw semen to allow the insemination of a number of mares with a single ejaculate is less successful than if the division of the sample occurs after the addition of an extender. The use of an extender is, therefore, recommended (Varner, 1986).

Many of the extenders appropriate for use with fresh semen have been reviewed in the previous section. Pregnancy rates of 75% have been reported with the use of fresh semen, comparing very favourably with fertilization rates to natural service (Pickett, 1993b). Similar conception rates for fresh semen extended with Kenney extender (73.7%) were reported by Silva Filho *et al.* (1994). These were better than the conception rates reported for raw semen (64.8%) in the same experiment. Extended semen can be stored quite successfully at room temperature (15–20°C), ideally in the dark, for 12–24 h without a significant drop in spermatozoan viability (Coulter and Foote, 1977; Francl *et al.*, 1987; Varner and Schumacher, 1991).

7.3.1. Extenders for use with fresh semen

Details on the types of extenders that are used have already been given. As regards their success rate when specifically used with fresh semen, the following conclusions may be drawn. When semen is extended in two of the more common extenders and inseminated immediately, cream–gel extender is reported to yield the highest 50-day pregnancy rates (95.8%), compared with raw semen (91.7%) and semen extended in TRIS diluent (75%) (Pace and Sullivan, 1975). This generally poor performance of TRIS extenders has been attributed to the fact that TRIS is a hydrogen ion extender and to the common inclusion of glycerol in TRIS extenders (Pickett *et al.*, 1975a; Pace and Sullivan, 1975; Demick *et al.*, 1976). With short-term storage these figures remain excellent throughout but TRIS extenders still remain relatively poor (Pace and Sullivan, 1975). Comparative work with cream–gel, skimmed milk and skimmed milk–gel yielded lower conception rates than those reported by Pace and Sullivan (1975), but they did demonstrate that both skimmed milk and skimmed milk–gel performed better as extenders for fresh semen than cream–gel (Householder *et al.*, 1981).

With storage over time, comparative work between NFDSMG and skimmed milk indicated the superior nature of NFDSMG in maintaining motility (Province, 1984). Motility rates after 12 h of cooled storage were 15% for semen extended with skimmed milk and 31% for that extended with NFDSMG. This difference persisted until 36 h, at which time spermatozoa stored in skimmed milk showed only minimal motility, compared with 14% for spermatozoa stored in NFDSMG.

It may be generally concluded that NFDSMG provides the best extender for fresh semen, followed by skimmed milk–gel and cream–gel. TRIS extenders, though reasonably successful in maintaining motility, do not compare favourably with other extenders as far as pregnancy rates are concerned.

7.3.2. Removal of seminal plasma

Removal of seminal plasma is not normally practised with raw or fresh semen as the immediacy with which the collection and insemination should ideally take place means that the detrimental effects of seminal plasma are minimized. If removal of seminal plasma is required, then the protocols involved are the same as previously discussed.

7.3.3. Dilution rates

Dilution rate with all insemination is an important consideration. Inadequate dilution will not provide adequate support for spermatozoa. On the other hand, excessive dilution has been reported to be associated with depressed fertilization rates (Katila, 1997). A minimum semen:extender ratio of 1:1 has been recommended (Brinsko and Varner, 1992). Other workers have also demonstrated an effect of dilution rate on spermatozoan survival. Semen dilution rates of 1:2, 1:5, 1:10 and 1:20 were compared and resulted in survival rates of 41.2%, 54.9%, 57.2% and 56.4%, respectively, after 24 h of storage (Palmer *et al.*, 1984). However, it is impossible to draw absolute conclusions from this work as the original spermatozoa concentration (and, therefore, the final concentration post dilution) is not always apparent. In general, semen for use fresh is diluted in a ratio of between 1:2 and 1:4, depending upon the concentration of the raw semen sample.

7.3.4. Conception rates

Work by several authors indicates that conception rates for mares inseminated with extended fresh semen, immediately post extension, vary from 50 to 75% depending upon the extender used (Yi *et al.*, 1983; Silva Filho *et al.*, 1991a; Jasko *et al.*, 1993b). This is reported to be similar to that obtained with undiluted raw semen (65%) (Silva Filho *et al.*, 1994). Extension with E-Z Mixin followed by insemination within 1 h resulted in 6-day conception rates of between 56.2% and 68.1%. This compared well with 62.5% reported after extension with skimmed milk extender (Francl *et al.*, 1987). The mares in this experiment were inseminated on every other day, commencing at day 2 or 3 of oestrus or when a 35 mm diameter follicle was detected at rectal palpation. Insemination with freshly collected and extended semen continued until oestrus ended. Conception rates were recorded as day 6 embryo recovery rates. Using a similar protocol, the use of treated skimmed milk extender gave slightly lower conception rates at 40–62.5% (Province *et al.*,

1985; Francl *et al.*, 1987). Similar figures for conception rates of 65%, after extension with E-Z Mixin and insemination within 1 h, were reported by Squires *et al.* (1988), and a figure of 75% (7-day recovery rate) was reported by Pickett *et al.* (1987) after insemination with semen stored in E-Z Mixin for 1 h at 37°C. Singhvi (1990) reported disappointing conception rates of 39.1% for fresh insemination using skimmed milk–glucose extender. Other workers have reported conception rates of 43–73% (Julienne, 1987; Silva Filho *et al.*, 1994).

7.3.5. Conclusion

The period of viability of fresh semen stored at body temperature or non-refrigerated is limited and the use of fresh semen is largely confined to veterinary work (Kenney *et al.*, 1975; Kotilainen *et al.*, 1994). Very occasionally it may be used for splitting ejaculates for immediate insemination into several mares on a single stud. This is often in an attempt to reduce the covering load of a stallion, or as an aid to management, if the stallion is required to work or compete on the day he is needed for covering. This is still practised to some degree in the USA.

Stallion semen is, therefore, invariably chilled or frozen to allow for 3–4 days or more long-term storage.

7.4. Chilled Semen

In order to extend the time span of the viability of spermatozoa, their metabolic rate has to be slowed down, so reducing the rate at which the substrates within the surrounding extender are used and the rate at which toxins are produced. As a general rule the metabolic rate of cells (and, therefore, spermatozoa) can be reduced by cooling, though carbon dioxide and other metabolic inhibitors, such as proteinase inhibitors, have been used to produce a similar but less successful effect on spermatozoa. Likewise, caprogen may be used, its nitrogen saturation acting to reduce spermatozoan metabolism (Pickett *et al.*, 1975a; Province *et al.*, 1985; Heiskanen *et al.*, 1987; Katila, 1997). Cooling of stallion spermatozoa successfully depresses their metabolic rate and prolongs survival, a fact appreciated ever since Spallanzani's original work in 1776 (Perry, 1945; Varner, 1986); therefore, the simplest method to prolong the life of spermatozoa is chilling. No particularly sophisticated equipment is required and storage by this method allows adequate time for the transportation of semen over reasonably long distances, but it does not allow the storage of semen for future use.

The semen sample to be chilled should be filtered prior to extension, to remove the gel fraction, and extended immediately at 37°C. Filtering may be done at collection, via an in-line filter in the AV liner or the neck of the collecting vessel, or at the laboratory prior to extension. Numerous extenders have been experimented with over the years. Most of them have been considered in the previous section and include the usual milk, milk-product and egg yolk extenders, often with the addition of extra components to regulate osmotic pressure and pH.

Semen stored after chilling to 5–8°C will survive for 24–48 h without a significant decline in motility (Malmgren *et al.*, 1994). Samples have been reported to survive for up to 96 h without a significant drop in fertilization rates (Hughes and Loy, 1970). Chilling semen provides an efficient and successful means of short-term storage, but chilling itself has some adverse effects on spermatozoa, manifested as a depression in survival rate, motility and conception rates (Braun *et al.*, 1994b; Malmgren *et al.*, 1994). These adverse effects of cooling need to be understood in order to devise protocols that minimize them.

7.4.1. Cold shock

Cold shock is the term that describes the stress response shown by spermatozoa as a reaction to a drop in environmental temperature. The severity of the cold shock depends upon the final temperature and the rate of temperature drop. The resulting cellular damage affects both the structure and function of the cell and can be categorized as direct or indirect (Amann and Pickett, 1987; Watson, 1990). Direct damage is more definable and is the type normally associated with the term cold shock: it is evident shortly after the drop in temperature has occurred and is affected by the rate of cooling. Indirect or latent damage is more difficult to quantify and may not be initially apparent; it tends not to be dependent upon the rate of cooling. Direct cellular damage is normally irreversible and is evident as abnormal motility in samples cooled rapidly to 5°C. The resulting alteration in membranes is accompanied by a loss of intracellular components and a reduction in cellular metabolism. It is this damage to the cellular membranes that is of most significance and has a carry-over effect on other cellular structures and functions.

The membranes of spermatozoa have been considered in some detail in Chapter 4. In summary, these consist in the main of a phospholipid bilayer with the occasional area of hexagonal phase configuration. Within this matrix are found protein complexes. The composition is unique to each membrane and relatively stable. The relative concentrations of proteins and lipids within equine spermatozoon membranes is similar but some variation is seen (Eddy, 1988; Bedford and Hoskins, 1990). This unique structure enables them to carry out the specialized functions of each specific area of membrane. In general, protein and lipid concentrations are equal within spermatozoon membranes. The lipid composition is the most important when considering the effects of cooling. In stallion spermatozoa the major lipids are phospholipids and cholesterol, present in a ratio of 0.64:0.36, respectively (Parks and Lynch, 1992; Amann and Graham, 1993).

Membranes are naturally fluid, a prerequisite for their efficient function. Two main factors are known to affect fluidity: firstly, the relative concentrations of phospholipids and cholesterol (the greater the relative concentration of phospholipids, the more fluid are the membrane); and secondly, the temperature of the membrane. As membranes are cooled, lipids undergo transition from their normal fluid state to a liquid crystalline state, in which the fatty acyl chains become disordered. With further cooling this liquid crystalline state is transformed to a gel state, in which the fatty acyl chains have become re-ordered in a parallel fashion, producing a rigid structure. The phase transition temperature for these changes varies with different lipids and depends upon their structures. In general, the longer the fatty acyl chains, the higher is the phase transition temperature. As each lipid class within a membrane reaches its phase transition temperature it conforms to the gel configuration and tends to aggregate together with other similarly conformed lipids within the membrane. The remainder of the lipids within the membrane may still be fluid, so areas of gel membrane can be identified within a mainly fluid structure.

The peak phase transition temperature for phospholipids (the main lipids) within horse membranes is 20.7°C, which is quite low compared with 24.0, 25.4 and 24.5°C for the same lipid within the spermatozoa of boars, bulls and fowl, respectively. Similarly, the peak transition temperature for glycolipids in stallion spermatozoon membranes is 33.4°C, compared with 36.2 and 42.8°C for boars and bulls, respectively (no transition temperature was detected for fowls) (Parks and Lynch, 1992). These differences in peak transition temperatures account for the variable tolerance to cold shock exhibited by spermatozoa from different species and the high tolerance to cold shock demonstrated by fowl spermatozoa.

The isolated areas of gel lipid within an otherwise liquid membrane, which occur at cooling, result in what were once randomly distributed lipids becoming aggregated and unable to interact normally with their associated proteins and other lipids within the normal structure. In addition, the junction areas between the gel and the other lipid and protein fractions become areas of weakness, subject to fusion and rupture as well as being permeable to ions (Hammerstedt *et al.*, 1990). In addition to an effect on the fluidity of the membrane, cold shock causes changes to the distribution of the phospholipids across the bilayer. The normal fluid membrane configuration has, for stability, a roughly even distribution of total phospholipids in both the outer and inner (cytosolic) layers of the bilayer. This distribution is not necessarily the preferred distribution for all phospholipids. For example, the major phospholipids within the spermatozoon plasma membrane are phosphatildylcholine, sphingomyelin and phosphatidylethanolamine (Amann and Pickett, 1987; Hammerstedt *et al.*, 1990). These have differing positions within the membrane bilayer. Phosphatidylcholine and sphingomyelin are associated with the outer layer of the bilayer, whereas phosphatidylethanolamine has an affinity for the inner, cytosolic layer. These affinities are not normally evident except when the membrane is under stress. Cooling is a major stressor, as a result of which these phospholipids (among others) reorientate themselves into their preferred layer of the bilayer, or even into a hexagonal phase II state as illustrated in Fig. 7.1 (Quinn, 1989; Amman and Graham, 1993). These changes, especially the formation of a hexagonal phase II configuration, disrupt membrane function and permeability (Hammerstedt *et al.*, 1990).

Another major component of the spermatozoon membrane is protein. The protein–lipid interactions are critical in the efficient functioning of the membrane. It is important to ensure even distribution and moulding of proteins into the bilayer, thus eliminating pores and membrane faults. These interactions may also be required for the efficient functioning of these proteins as enzymes, receptors or channels for the movement of ions. Over time these configurations of, and interactions between, membrane components have evolved to be those ideal for membrane functioning at a normal body temperature of 39°C. As discussed previously, cooling the membranes past the transition phase temperature results in a change to the gel state and a gradual aggregation of specific lipids within certain areas of the membrane. As a result, these protein–lipid interactions are disrupted and so the proteins no longer act efficiently as enzymes, receptors or ionic channels. The membrane as a whole loses some of its structural and functional integrity.

Outside the bilayer configuration of the membrane lies the glycocalyx, a polysaccharide outer coat probably involved in antigenicity, cell adhesiveness, specific permeability, triphosphate activity and attachment of peripheral

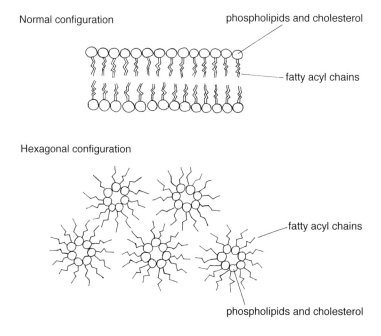

Fig. 7.1. The hexagonal phase II state, seen in spermatozoon membranes post cold shock (Hammerstedt *et al.*, 1990).

proteins (Winzler, 1970; Hammerstedt *et al.*, 1990). These peripheral proteins, provided by the seminal plasma, confer stability to the spermatozoon in its passage through the female system and may be involved in capacitation (Hernandez-Jauregui *et al.*, 1975). Disruption of the membrane configuration also interferes with the function of the glycocalyx components and hence peripheral protein attachment, causing them to aggregate largely in the areas of membrane still in the fluid state once gel formation has begun to occur (Housley and Stanley, 1982).

Unfortunately, many of these changes to membrane configuration involving lipids are irreversible. Subsequent warming of the cooled spermatozoa does not restore the original membrane configuration.

Cold shock damage manifests itself as a decline in cell metabolism, altered membrane permeability, loss of lipids, ions and other elements, irreversible loss of spermatozoan motility and an increase in the number of dead spermatozoa. Many of these changes seen in spermatozoa are allied with those occurring during senescence. However, with the exception of motility, they occur at a slower rate. Though motility rates are adversely affected by cooling, the correlation between motility and fertility is reduced post thaw compared with that observed in fresh and chilled samples (Amann and Pickett, 1987; Watson, 1990). Motion characteristics also apparently change: spermatozoa show an increased incidence of backward motion because of an overbending of the tail area. This is due to irreversible changes to the mid-piece and coiling of the tail (Watson and Morris, 1987; Watson, 1990).

The decrease in correlation between motility and fertility has been suggested to be due to a difference in the change of permeability of the acrosome and mitochondrial membranes to calcium ions. The acrosome membrane suffers most from cold shock. As seen previously, cooling spermatozoan membranes results in areas of gel state, with permeable borders. Under normal conditions the leakage of sodium and calcium ions into the cell is counteracted by sodium and calcium ion transporters. However, at a low temperature (for example, 5°C) the permeability of the membranes to calcium significantly increases and the leakage of calcium into the cell can no longer be counteracted by the ion pumps, so calcium accumulates within the spermatozoon (Ellery and Hall, 1987; Watson and Morris, 1987). It is possible that the acrosome membrane suffers more from this change in permeability than the mitochondrial membranes, and that these differences account for the greater loss in fertility than would be expected from the corresponding reduction in motility (Amann and Graham, 1993).

Microscopic examination of stallion spermatozoa indicates that the function of the mitochondrial cristae is also affected by cold shock. This damage to mitochondrial function within the spermatozoa is likely to account for the decrease in motility observed (Watson *et al.*, 1987). In addition, expansion or swelling of the acrosomal area was observed to be a consequence of cold shock, which would indicate a loss of integrity as membranes are normally unable to stretch. However, Watson *et al.* (1987) failed to demonstrate any lack of continuity in these membranes, indicating a possible addition of membrane substance, the origin of which is unclear. The apparent disruption to acrosomal membranes was reported to be less than that observed in similarly treated spermatozoa from other ungulates (Watson *et al.*, 1987). This supports other evidence which suggests that stallion spermatozoa may be more resistant to cold shock than those of rams, boars and bulls (Bogart, and Mayer, 1950).

The effects of cold shock are more evident when freezing semen, with spermatozoa undergoing more changes similar to those of capacitation and senescence (Ball *et al.*, 1997). Hence when cryopreservation is considered later in this chapter, cryoprotectants are discussed at length.

Some of the effects of cold shock are evident in cooled semen. Inclusion of components such as egg yolk, milk, glycerol, BSA, polyvinyl alcohol and liposomes in extenders affords some protection to spermatozoa (Pickett and Amann, 1987; Katila, 1997). It is increasingly apparent that the lipoproteins and phospholipids in milk have a significantly beneficial effect in protecting spermatozoa in fresh and cooled storage (Pickett and Amann, 1987), hence the continued popularity and success of milk-based extenders such as Kenney and E-Z Mixin.

7.4.2. Cooling rates

The extent of damage to a spermatozoon as a result of cold shock depends upon not only the drop in temperature but also the speed with which this drop is attained. In general, the faster the rate of cooling, the more severe is the damage (White and Wales, 1960; Kayser, 1990). There is further evidence which suggests that the rate of temperature drop also determines the subsequent active life of the spermatozoon (Douglas-Hamilton *et al.*, 1984; Varner *et al.*, 1988). Work by Chang and Walton (1940) initially indicated that the lower the final temperature of storage, the slower was the rate of cooling required to minimize the effects of cold shock. It is also apparent that the type of extender has a bearing on the susceptibility of spermatozoa to cold shock and the rate of temperature decline required for successful storage.

Several regimes for the rate of cooling have been investigated. Douglas-Hamilton *et al.* (1984) successfully inseminated mares with semen cooled from 37°C to 4–6°C, at a rate of 0.3°C min^{-1}. Further work supported this, indicating that a cooling rate of 0.5°C min^{-1} was less successful than a slower cooling rate of 0.05°C min^{-1} when spermatozoa were subsequently stored at 5°C for 96 h (Kayser, 1990). Slowing the rate of cooling to 0.012°C min^{-1} had no beneficial effect over 0.05°C min^{-1}, again in spermatozoa stored at 5°C for 96 h (Kayser, 1990). In fact, it has been suggested that such slow cooling rates may, in themselves, be detrimental to the survival of spermatozoa (Douglas-Hamilton *et al.*, 1984). However, it is apparent from motility assessment that the rate of temperature drop is most critical over the specific temperature range of 20–5°C (White and Wales, 1960; Kayser, 1990; Kayser *et al.*, 1992; Katila, 1997). Therefore, a cooling regime consisting of rapid cooling from

37°C down to 20°C, followed by a slower cooling rate of 0.1–0.05°C min^{-1} as the sample temperature drops from 19–22°C down to 5°C for storage, has been suggested (Kayser, 1990). Further work around this basic protocol has been carried out. Bedford *et al.* (1995a) varied the rate of cooling from 20°C to 5°C, and reported that no significant differences in terms of motility was evident at rates in the range of 0.5°C to 0.05°C min^{-1}. Further work in this area suggests that the critical range can be defined more specifically as 18°C to 8°C; outside this range rapid cooling rates could be resumed (Moran *et al.*, 1992; Katila, 1997). However, due to inter- and intra-stallion differences no hard and fast rule as regards cooling rates can be proposed. It is highly likely that the most beneficial method is the design of specific cooling regimes for individual stallions. This would pose serious difficulties for commercial work and so most practices use the protocol suggested by Kayser (1990).

Traditionally, semen was cooled by being placed in a refrigerator or cooling unit. Today more accurate control over cooling rates may be achieved by computerized cooling systems, but refrigerators are still commonplace in most commercial practices.

7.4.3. Storage temperature

The final storage temperature is known to have an effect on subsequent fertility. Semen is commonly cooled to between 4 and 5°C, though for short-term storage it may be cooled to room temperature (20°C). Good results have been achieved with semen stored for 12–24 h at 20°C (Province *et al.*, 1985; Francl *et al.*, 1987). As might be expected, the lower the temperature, the lower is the depression in metabolic rate and, therefore, the longer the preservation of spermatozoan viability. However, cooling spermatozoa below 20°C results in exposure to cold shock and its potentially detrimental effects. Despite the risk of cold shock, the best conception rates are generally obtained with storage at the lower temperature over 24–48 h (Pickett *et al.*, 1987; Squires *et al.*, 1988; Varner *et al.*, 1988, 1989; Kayser *et al.*, 1992). In the work by Pickett *et al.* (1987) and Squires *et al.* (1988), 500 × 10^6 spermatozoa were inseminated into mares after extension in E-Z Mixin, following storage at either 5°C or 20°C. Pregnancy rates were assessed at embryo recovery post insemination. A similar experimental protocol was used by Varner *et al.* (1989) but with a skimmed milk–glucose extender. They failed to note a difference in 15-day pregnancy rates between cooling to 5°C or 20°C but observed a difference in motility between the two temperatures: spermatozoa stored at 20°C showed lower motility than those stored at 5°C. Moran *et al.* (1992) investigated the potential benefit of reducing the temperature of storage still further, to below 5°C, and demonstrated an adverse effect on motility when semen was stored at temperatures of 0°C or 2°C, compared with 4–6°C.

An accumulation of data presented by Pickett (1993b) further illustrated the advantage of storing semen at 5°C over 20°C for both 24 and 48 h. The difference between the two temperatures at 24 h was relatively small (59% compared with 51% embryo recovery). Insemination of semen after storage at 20°C for 48 h failed to result in any conceptions, but the number of mares used was relatively low at 16. Other storage temperatures have been investigated. Province *et al.* (1984) and Magistrini *et al.* (1992a) suggested that storage temperatures of 10–15°C may be superior to 0, 4 and 5°C.

It is apparent that, for short-term storage (24 h), cooling to 20°C provides a viable method. For longer-term storage, further cooling to 4–6°C is most appropriate (Zidane *et al.*, 1991). There are obvious advantages to storage at room temperature in places where refrigeration is difficult, but fluctuations in ambient temperatures do occur and are not advantageous to the semen sample. Storage of semen at ambient temperature does not in itself significantly reduce the spermatozoan metabolic rate, thus limiting the length of storage. If storage at ambient temperature is to be extended, an alternative to dropping the temperature further needs to be found. In order to achieve this, various agents have been used to reduce the metabolic rate. The first effective ambient temperature storage medium was used in cattle, based on the known ability of carbon dioxide to reduce spermatozoan metabolism. Spermatozoa in an egg yolk–citrate extender were exposed to carbon dioxide by bubbling the pure gas through the medium to saturation. This successfully reduced the metabolic rate and increased spermatozoan survival over a period of time. This medium was termed illini variable temperature (IVT) extender (Salisbury *et al.*, 1978). Other extenders containing metabolic inhibitors have been used in bull, boar and ram semen. These include caprogen extender (Shannon, 1972) and Cornell University extender (Foote *et al.*, 1960). The use of such metabolic inhibitors allows the storage of bull semen for up to 3 days at ambient temperature. There are a few reports of using such metabolic inhibitors with stallion semen, but with only limited success. The use of caprogen is reported by Province *et al.* (1985) and of proteinase inhibitors by Katila (1997) but with only limited success. More recently Genus AI, UK have reported no significant decline in spermatozoan viability over 10 h using a specific equine ambient temperature extender (S. Revell, Wales, 1999 personal communication).

7.4.4. Length of storage time

In general, the longer the storage time, the greater are the potential detrimental effects on the spermatozoa. Prolonged exposure to a cooled environment increases the effect of cold shock. In addition, though the metabolism of the spermatozoa is depressed by cooling, it does not cease altogether, and so eventually the build-up of toxins and the exhaustion of the nutrient supply result in death. Most studies indicate that insemination of mares with spermatozoa that have been stored at either 20°C or 5°C for up to 24 h will result in acceptable fertilization rates (50–60%) when compared with fresh semen (Douglas-Hamilton *et al.*, 1984; Francl *et al.*, 1987; Squires *et al.*, 1988,

Varner *et al.*, 1989). Conception rates with spermatozoa stored at 20°C tend to decline after 12 h. It is generally accepted that storage of semen (from the vast majority of stallions) at 20°C for up to 12 h has no detrimental effect on fertilization rates, and that storage at 20°C for 12–24 h is acceptable for semen from the majority of stallions but not neccessarily all. Storage beyond 24 h at either 5°C or 20°C results in a depression in fertility rates (Francl *et al.*, 1987; Pickett *et al.*, 1987; Squires *et al.*, 1988) but the depression is slower at 5°C, so allowing a significantly longer storage time. Heiskanen *et al.* (1994a) reported conception rates of 77% for semen stored at 5–7°C for as long as 80 h. This ability of semen to survive at 5°C for prolonged periods is used in the development of specialized insulated semen transport containers, the first of which was the Equitainer developed by Douglas-Hamilton *et al.* (1984).

It is interesting to note that a more significant decline in motility has been reported with prolonged storage than might be expected from conception rates (Squires *et al.*, 1988). Motility rates as low as 10% have been reported for spermatozoa stored for 24 h at 20°C (Province *et al.*, 1984, 1985), but other workers have reported much higher motility rates post cool storage (Douglas-Hamilton, 1984). These discrepancies may be due to the length of time that elapsed between removal of spermatozoa from the cooling unit and evaluation. Delaying evaluation so that a 10 min equilibration time is allowed for the spermatozoa to reach 37°C is reported to result in a 20 percentage point improvement in motility rates (Francl *et al.*, 1987). This, yet again, underlines the importance of standardization of methods for the evaluation of spermatozoa so that more valid comparisons may be made between the work of different researchers.

Although the most consistent results are obtained with cool storage up to 24 h, there are several reports of successful insemination of mares with semen stored in excess of 48 h. For example, Hughes and Loy (1969b) reported a conception rate of 73% after semen storage for 96 h, and Van der Holst (1984) reported a conception rate of 87% in mares inseminated with semen stored at 5°C for 4–5 days. Heiskanen *et al.* (1994a) reported conception rates of 77% for semen stored at 5–7°C for 80 h.

7.4.5. Extenders for use with chilled semen

Details of the types of extenders that are used have been given earlier in this chapter. Some work has been carried out to investigate the relative efficacy of different extenders for use specifically with stored chilled semen. A similar advantage in the use of NFDSMG extender over other common extenders is evident with cooled semen as with fresh semen (Francl *et al.*, 1987). Comparisons between skimmed milk and NFDSMG extenders supported this apparent superiority of NFDSMG, with motility rates of 24.8% and 37.3%, respectively, being evident after 72 h cooled storage (Sanchez *et al.*, 1995). As far as motility rates are concerned, it is apparent that, after 48 h storage, the addition of 4% glycerol to NFDSMG extender has a beneficial effect. As an extender for cooled semen, such a mix performs better than NFDSMG alone or with the addition of 4% egg yolk or the addition of 4% egg yolk and 4% glycerol. Unfortunately, the beneficial effect of adding 4% glycerol is not reflected in fertility rates, where NFDSMG alone performed best (Bedford et al., 1995b). The superiority of NFDSM for cool semen storage is further reported by Katila (1997).

Storage in E-Z Mixin (identical to NFDSMG extender but with the substitution of gentamicin with polymixin B sulphate as the antibiotic) at 20°C for 12 h resulted in a conception rate of 50%, which is comparable with fresh insemination results using E-Z Mixin or skimmed milk (Francl et al., 1987). In further work reported by the same authors a conception rate of 61.7% was obtained for semen stored in E-Z Mixin for 24 h but this time at 20°C. Wockener and Collenbrander (1993) demonstrated the superiority (as far as motility was concerned) of skimmed milk extender, the most common extender used in the USA, over a glycine–egg yolk extender, widely used in European countries, for storage at 5°C. This apparent advantage was not evident when spermatozoon morphology and acrosome integrity were assessed: using these criteria, the glycine–egg yolk extender proved superior.

Investigations into the possibility of using heated mare's milk as a semen extender for cooled storage have yielded quite encouraging results. Though the experiment was not taken far enough to assess conception rates, it was apparent that motility of spermatozoa stored at 5°C in heated mare's milk and egg yolk extender was 24.4% after 48 h storage. In this work a comparison was also made with heated mare's milk alone, and with mare's milk extender without egg yolk and NFDSM. The mare's milk and egg yolk extender outperformed all the other extenders at 48 h, though the performance of NFDSM was equivalent up to 24 h (Lawson, 1996; Lawson and Davies Morel, 1996). Recently the use of INRA 82 extender has become more widespread. Ijaz and Ducharme (1995) demonstrated the advantage of INRA 82 plus taurine over Kenney (NFDSM), Dimitripoulos, egg yolk–citrate–taurine extender and E-Z Mixin for storage at 5°C, with motility rates of 40% being reported for INRA 82 plus taurine after 144 h storage. EDTA–glucose extenders have also been used for cool storage, but with limited success (Sanchez et al., 1995). The inclusion of other components such as thymus α 1 or oxytocin and thymus extract in semen diluents for cooling have been reported to be beneficial as far as motility is concerned (Pesic et al., 1993).

Ijaz and Ducharme (1995) suggested a new system for the extension of semen for cooled storage. They stored semen in various extenders for 24 h at 5°C, followed by washing of the spermatozoa. A small volume (3.5 ml) of the semen extender solution was washed twice, by centrifugation at 350 g for 5 min at room temperature, in 6.5 ml of SP-TALP medium supplemented with 0.6% BSA fatty acid free (BSAAFF). The resultant pellet was resuspended in SP-TALP to obtain a concentration of 25×10^6 spermatozoa ml^{-1} and incubated for 24 h at 39°C (5% CO_2) in a humidifier incubator. Though successful in maintaining good motility rates, this system of washing did not result in consistent improvement over the same extenders but without the washing process.

It has been reported that at low temperatures (below 8°C) the removal of oxygen-rich air is beneficial for the survival of spermatozoan (Katila, 1997).

7.4.6. Removal of seminal plasma

Removal of seminal plasma is often carried out with semen destined for cool storage. This is invariably done in commercial practice by centrifugation, though the use of an open-ended AV for collection can yield suitable sperm-rich samples for extension and cool storage. The protocol and reasoning behind seminal plasma removal has already been discussed earlier in section 7.2.8.

7.4.7. Packaging for chilled semen

Spermatozoa may be packaged for cool storage in a number of ways. Recommendations have included sterilized polyethylene bags (Whirl-Pak) (Douglas-Hamilton *et al.*, 1984), plastic bottles and baby-bottle liners, but care should be taken as some types of plastic can be spermiotoxic (Katila, 1997). Syringes may also be used, but those with rubber components should be avoided, as an adverse effect has been reported on spermatozoa in contact with rubber for as little as 30 min (Broussard *et al.*, 1990), though more recent work by the same workers casts doubt on this association (Broussard *et al.*, 1993).

The presence or absence of air (oxygen) during storage is reported to have an effect. Storage at 4°C in the absence of oxygen is reported to result in better motility, though at 37°C oxygen is required for spermatozoan survival (Katila, 1997). For storage at 4°C, therefore, it seems that air should be excluded from plastic bags or syringes and tubes should be well filled prior to storage (Douglas-Hamilton *et al.*, 1984; Magistrini *et al.*, 1992a). Spermatozoa can use both anaerobic and aerobic pathways in metabolism, but long-term access to oxygen may result in lipid peroxidation (Magistrini *et al.*, 1992b). Motility was enhanced in spermatozoa stored in 2 ml semen tubes placed horizontally (that is, with a greater surface area and less distance between the spermatozoa and the environmental oxygen) than in 2 ml tubes stored vertically (Magistrini *et al.*, 1992b). Katila (1997) suggested that the storage of semen in tubes on a roller bench (5 turns min^{-1}) may prolong survival.

7.4.8. Methods of transportation

If the advantages of chilling semen are to be fully realized, then an effective means of storage and transportation while still maintaining a cool environment needs to be devised. Several containers have been designed in an attempt to satisfy this requirement. The one in most widespread use is the Equitainer (Hamilton-Thorn Research, Danvers, Massachusetts) (Figs 7.2 and 7.3).

Fig. 7.2. The Equitainer, a common method of cooling and transporting cooled semen, fully assembled and ready for use. (Hamilton-Thorn Research, Danvers, Massachusetts.)

The Equitainer is designed to permit controlled cooling of the stored semen sample at 0.3°C min^{-1} and subsequent maintenance of the sample at 4°C for up to 36 h. It comprises a sturdy outer case, the lumen of which is used to hold the coolant and semen sample (Figs 7.2 and 7.3). It is advised that the sample for storage should contain 100–500 × 10^6 motile spermatozoa per insemination dose, but in the USA larger doses of up to 1000 × 10^6 are typical (B.A. Ball, California, 1998, personal communication). The semen, which should be diluted in NFDSMG extender (or equivalent) in a minimum ratio of 1:2 semen:extender, is placed in the isothermalizer (a conductive container). Prior to packaging, the coolant cans or block are cooled to below 0°C; they are then placed at the bottom of the lumen of the container and covered by a thermal impedance pad. The isothermalizer and ballast bags are

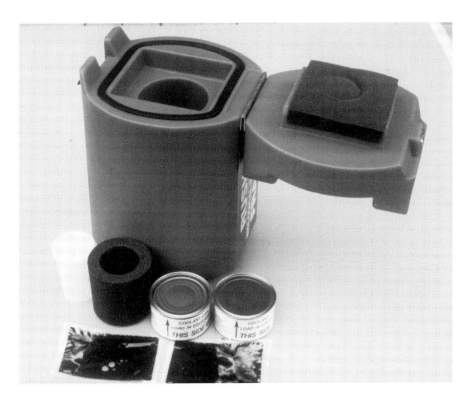

Fig. 7.3. The component parts of the Equitainer shown in Fig. 7.2. The ballast bags can be seen in the foreground; behind them are the two coolant cans or blocks; and to their left is the black foam insulating jacket (which acts as a thermal impedance pad and foam rubber protective padding) into which the plastic semen container (isothermalizer), far left, is placed. (Hamilton-Thorn Research, Danvers, Massachusetts.)

placed on top of the thermal impedance pad, followed by a foam-rubber protective padding, and the lid is closed. The use of the ballast bags enables a total volume of 120–170 ml to be achieved, to ensure the proper cooling rate.

Once the container is closed, the contents of the isothermalizer are gradually cooled by the coolant block at an initial rate of $-0.3°C\ min^{-1}$. With time, and as the isothermalizer gets cooler, the rate of temperature decline falls. The temperature of the contents of the Equitainer lumen, and hence the semen sample, equilibrates at 4–5°C after about 10 h. The insulating nature of the Equitainer ensures that the sample remains at this temperature for 24–48 h (Douglas-Hamilton *et al.*, 1984). This container is now a standard form of transporting semen within and between countries and is designed to allow carriage by courier companies, postal services, etc.

Though widespread in use, the Equitainer does have some disadvantages. Its relative durability and robustness means that it is relatively expensive to produce and, as it is not disposable, incurs the extra cost of return transportation to the stud or AI centre. Bearing these drawbacks in mind, experiments have taken place with other disposable containers. These include the Expecta Foal (Expecta, Parker, Colorado), developed in the USA and now becoming more widespread in its use. The Expecta Foal is a disposable container consisting of an insulated semen receptacle, into which the semen sample is placed in a plastic bag, normally after cooling to 5°C. This whole receptacle is then placed in a styrofoam box, along with three pre-cooled (to −20°C) coolant bags, and the styrofoam lid is fitted. The styrofoam container is then placed in a cardboard box ready for transportation. Another disposable alternative is the Equine Express (M.P. and J. Associates, Des Moines, Iowa), which is based on what was previously known as the Salisbro container, developed in Sweden for the transportation of human spermatozoa. It is now gaining popularity in Europe and the USA. It consists of two 20 ml all-plastic syringes which are filled with the semen. The syringes are fitted with rubber caps and loaded into compartments in a styrofoam box. A thin styrofoam plate is placed over the syringes, and a rectangular coolant can (previously cooled to −20°C) is placed into the box, with the styrofoam plate separating the coolant can from the samples. The styrofoam box lid is then closed. Other models of containers are also available – for example, the Celle, which is popular in Germany, and the Starstedt, which is popular in The Netherlands – but they are not as widespread in their use (Huek, 1990; Katila et al., 1997).

Katil et al. (1997) have compared the performance of the three most popular containers: Equitainer, the Expecta Foal and the Equine Express. The Equitainer proved to maintain a cool temperature (5.5°C) longer than either of the other two containers (48 h compared with 24 h). The cooling rate of the sample contained within the Expecta Foal was more erratic than the other two and the lowest temperature recorded was −0.1°C. The Equine Express performed in a similar manner to the Equitainer, but failed to maintain a low temperature for longer than 24 h. Progressive motility rates for all three containers after 24 h were similar at 61–65%. It was concluded from this work that all three containers provide an adequate means of cooling and transporting semen for 24 h. If storage and transportation is required for longer than 24 h, the Equitainer is the only really suitable container (Katila et al., 1997).

Comparisons between the Celle, the Starstedt and the Equitainer were reported by Katila (1997). The Equitainer outperformed the other two, maintaining semen temperature at 5°C for 48 h. The Celle container maintained 5°C for 26 h and both models were, therefore, considered appropriate for storage up to 24 h, but only the Equitainer was appropriate for storage in excess of 24 h. The Starstedt was poor in its performance, with a temperature below 10°C being maintained for only 12 h.

Once in an appropriate container, semen can be transported via air, sea and land. The only restriction to transportation distance is the time of storage and this can depend on many factors, including the packaging. In the UK, the normal postal or parcel service and specialized courier services will deliver Equitainers. A specialized stallion semen parcel service is available in The Netherlands, using a car with a refrigerated container (Katila, 1997). Similar vehicles are routinely used to transport cooled semen between collection site and insemination centres in France (Davies Morel, personal observation). Sweden has a specialized mail service that will deliver semen the following day, provided it arrives at the post office by a specified time (Katila, 1997).

7.4.9. Dilution rates

Semen extension normally occurs within a range of 1:1 to 1:10 concentrated semen:extender, though a tighter range of 1:2 to 1:4 is normally advised. Dilution rates depend largely upon the initial concentration of the sample and mean very little as a simple extension ratio. The aim of extension is to store the sample at an appropriate concentration and volume for immediate insemination, without the need for further treatment. When considering dilution, adequate extender is required to ensure maximum survival rates and so, in general, the greater the dilution, the greater is the survival potential (Samper et al., 1988). However, excessive dilution results in low concentrations may require large volumes to be inseminated to ensure that adequate numbers of spermatozoa are inseminated for acceptable fertilization rates. It is generally agreed that between 100×10^6 and 500×10^6 spermatozoa (normally 250×10^6 are used) are required in order to obtain optimal fertilization rates (Pace and Sullivan, 1975; Householder et al., 1981). Insemination of volumes in excess of 100 ml have been reported to be associated with depressed fertility rates (Rowley et al., 1990). A happy medium, therefore, has to be struck between insemination volume and spermatozoan concentration. The optimal concentration for cooled storage has been reported to be 25×10^6 spermatozoa ml^{-1} of extender (or extender plus seminal plasma) (Varner et al., 1987; Webb et al., 1993). Varner et al. (1987) reported better motility rates with spermatozoan concentrations of 25×10^6 when compared with 50×10^6 and 10×10^6. Similar work by Webb et al. (1993) compared different rates of dilution of semen in skimmed milk–glucose extender and showed there to be an effect on progressive motility and the rate of forward movement. Dilution rates were adjusted to give concentrations of 25, 50, 100, 150 or 250×10^6 spermatozoa ml^{-1}. These were then stored at 5°C for 24 or 48 h. The most successful concentration was 25×10^6. In practice, the need to minimize insemination volumes and the desire to pack large numbers of doses in transport containers have led to the suggestion that 100×10^6 spermatozoa ml^{-1} might be the most appropriate concentration (Webb et al., 1993). Dilution rates to achieve concentrations of 25×10^6, 50×10^6 or 100×10^6 have been reported to be equally appropriate for storage at 5°C in NFDSMG extender for 24 h (Jasko et al., 1991c; Katila, 1997).

Careful consideration needs to be given to dilution rates to ensure that insemination volumes used are not in excess of 100 ml (Rowley et al., 1990) when concentrations greater than $2.5–5 \times 10^6$ spermatozoa ml^{-1} are used (Jasko et al., 1992c). As discussed previously, centrifugation may need to be considered in order to increase the concentration and control dilution rates (Cochran et al., 1984; Jasko et al., 1991c, 1992a).

In addition to the actual dilution rates, timing of dilution is important. The sooner the extender can be added, the better will be the results. Braun *et al.* (1994b) demonstrated that there was a significant advantage (in terms of progressive motility, at 0 and 48 h storage at 5°C) in adding the extender immediately on collection – that is, in the AV – compared with a 10 min delay between ejaculation and extension.

7.4.10. Conception rates

Conception rates depend upon so many different factors that it is impossible to arrive at an overall average conception rate for cooled semen. In general, conception rates for cooled semen stored for any period in excess of 24 h (Francl *et al.*, 1987; Pickett *et al.*, 1987; Squires *et al.*, 1988) are lower than those attained for insemination with fresh semen. Conception rates ranging from 50 to 80% are regularly reported for cooled semen, in comparison with rates of 70–76% for fresh semen insemination (Hughes and Loy, 1970; Palmer *et al.*, 1984; Van der Holst, 1984; Francl *et al.*, 1987; Silva Filho *et al.*, 1991b; Jasko *et al.*, 1993b; Heiskanen *et al.*, 1994a). However, wide variations exist. Lower conception rates, nearer to 30–40%, have been reported for single oestrus inseminations (Domerg, 1984), whereas conception rates of 72–77% have been reported for mares inseminated twice using semen from four stallions that has been kept at 4°C for 72 h (Yurdaydin *et al.*, 1993). Heiskanen *et al.* (1993, 1994a) demonstrated conception rates per cycle of 60–100%, depending upon the diluent used and the timing of the inseminations relative to ovulation.

7.4.11. Conclusion

As with other types of semen storage, results are so variable and depend upon so many interacting factors that it is hard to draw any absolute conclusions. However, transportation of cooled semen is now commonplace and has gone a long way to reducing the geographical barriers to breeding horses. The major development that has allowed the practical application of so much investigative research was the Equitainer, which is now widely used throughout the world and has formed the basis of many of the alternative insulated transport/storage containers that have been developed subsequently.

7.5. Cryopreservation of Semen

The use of cooled semen is very successful for short-term storage but long-term storage by this method is not possible, viability in general being maintained for only 48 h. In order to realize many of the potential advantages of AI, long-term storage is necessary. This is only really possible by freezing, a system which halts the metabolic processes of the spermatozoa, allowing (in theory) indefinite storage without loss of fertility. The discovery of the cryoprotectant properties of glycerol by Polge *et al.* (1949) made cryopreservation possible. As a result, the spermatozoa of many species can now be stored at −196°C in liquid nitrogen for indefinite periods, while still retaining acceptable fertilization rates post thaw (Salisbury *et al.*, 1978; Foote, 1982, 1988; Fomina and Miroshnikova, 1989; Braun *et al.*, 1990).

7.5.1. Principles

Even under ideal conditions, it is inevitable that some damage will occur to spermatozoa during the freezing process. The two main reasons for damage are internal ice crystal formation, resulting in alterations to spermatozoan structure, and increases in solute concentrations, possibly to toxic levels, as pure water is withdrawn to form ice both extra- and intra-cellularly (Salisbury *et al.*, 1978; Mazur, 1984; Graham, 1996). Changes in plasma membrane permeability to calcium have also been demonstrated: semen post thaw has a significantly higher calcium flux than fresh semen (Leopold, 1994). This damage is largely manifested as a depression in motility, and possibly acrosome morphology (Palacios Angola *et al.*, 1992; Wockener and Colenbrander, 1993; Graham, 1996). The use of cryoprotectants such as glycerol, first identified for use with bovine spermatozoa, prevents some of this damage. However, the success of cryopreservation depends upon many other factors, including interactions between cryoprotectant, extender, cooling rate, warming rate and packaging, as well as individual variation, which appears to be particularly significant in the stallion (Christanelli *et al.*, 1984; Loomis *et al.*, 1984; Graham, 1996). Some loss in spermatozoan viability must be expected, due to the processing procedures prior to freezing as well as the actual freezing process (Loomis, 1993). The success of insemination with spermatozoa post cryopreservation varies significantly. However, these figures are affected by the method of experimentation and recording, which is notoriously unstandardized in equine reproductive research. Information such as pregnancy rate to a single insemination, timing of insemination, number of spermatozoa inseminated, volume of inseminate used or type of extender used are often not reported. In addition, motility of spermatozoa has proven to be an even poorer indicator of fertility in freeze–thaw samples (Samper *et al.*, 1991).

Regardless of all these considerations, for cryopreservation to be considered a success the process should enable a spematozoon to retain its fertilizing capacity post thaw. To achieve this it must retain its ability to produce energy via metaboloism; to show progressive motility; to maintain plasma membrane configuration and integrity (in order that it will survive in the female tract and attach to the oocyte plasma membrane); and to retain enzymes, such as acrosin, within the acrosome (to allow penetration of the ova). Disruption of any of these functions or abilities will significantly affect the spermatozoon's ability to achieve fertilization. The greatest risk to the maintenance of these attributes is presented by the formation of ice crystals and the resultant movement of water up osmotic gradients.

During the process of freezing, several biophysical changes are evident within the semen sample. As the temperature drops to below freezing the sample undergoes supercooling, whereby the temperature of the surrounding medium drops below freezing, to about −5°C. Below this, as the temperature drops from −6°C to −15°C, extracellular ice crystals begin to form from water within the surrounding medium. This ice formation increases the concentration of solutes, such as sugars, salts and proteins. In response to this newly developed osmotic pressure (OP) gradient and the fact that water within the spermatozoa is slower to form ice crystals than the water in the surrounding medium, water passes out of the spermatozoa, particularly the spermatozoon head, across the semi-permeable plasma membrane. As a result the spermatozoon becomes increasingly dehydrated. The rate of efflux of water from the spermatozoa also depends upon the speed of temperature drop: the slower the drop, the greater is the time allowed for the efflux of water, and hence the greater is the dehydration. This does reduce the chance of ice crystal formation within the spermatozoon, which could cause considerable physical damage (Mazur, 1985; Amann and Pickett, 1987; Hammerstedt *et al.*, 1990), but it has to be weighed against the greater damage due to increased intracellular dehydration and solute concentration. On the other hand, if the cooling rate is rapid, water has little time to move out of the spermatozoon, across the plasma membrane, and hence large intracellular ice crystals form, causing physical damage to cell membranes and components. However, the problems of dehydration and solute concentration are less evident with rapid cooling. The aim is to arrive at an optimum cooling rate which will provide a compromise between all these factors, but this optimum rate is likely to change with the composition of the medium surrounding the spermatozoa – that is, with the seminal plasma, and hence the stallion, and with the extender used.

There are two main temperature ranges of concern regarding damage to spermatozoa during freezing: the period of supercooling (0 to −5°C) and the formation of ice crystals (−6°C to −15°C). Excessive supercooling results in rapid ice formation, with the possibility of physical damage. This problem can be overcome by a technique termed seeding, which is designed to induce ice formation more gradually over a greater temperature range. There is no evidence that seeding a semen sample during the freezing process has any advantageous effects (Fiser *et al.*, 1991; S. Revell, Wales, 1998, personal communication) and so it is not practised. The second area of concern – the formation of ice crystals and accompanying change in OP and solute concentrations – is known to have a significant effect on spermatozoan function post thaw. In an attempt to overcome some of these problems, the use of cryoprotectants (or antifreeze agents) has been investigated.

Cryoprotectants may be divided into two types, depending upon their action: they are either penetrating or non-penetrating. Penetrating cryoprotectants are able to penetrate the plasma membrane of the spermatozoa and, therefore, act intracellularly as well as extracellularly. The second type of cryoprotectant is non-penetrating and can only act extracellularly. The first cryoprotectant identified was glycerol (Polge et al., 1949). Glycerol remains one of the most favoured cryoprotectants, especially with bovine semen. It is a penetrating cryoprotectant, acting as a solvent and readily taken up by spermatozoa, entering the cell within 1 min of addition to the surrounding medium (Pickett and Amann, 1993). Its presence, both intra- and extracellularly, acts to lower the freezing point of the medium to a temperature much lower than that of water. This in turn reduces the proportion of the medium which is frozen at any one time, reducing the effect of low temperature on solute concentrations and hence on osmotic pressure differences (Mazur, 1984; Amann and Pickett, 1987; Watson and Duncan, 1988). It is also reported to provide channels of unfrozen medium, between ice crystals, in which spermatozoa may exist while at low temperatures (Mazur, 1984; Mazur and Cole, 1989; Pegg and Diaper, 1989). A further effect of glycerol may be a salt buffering action (Mazur, 1985). Other penetrating cryoprotectants include dimethyl sulphoxide (DMSO) and propylene glycol (Salisbury et al., 1978).

Non-penetrating cryoprotectants include many of the sugars – lactose, mannose, raffinose, trehalose – also polyvinylpyrrolidone and some proteins, such as egg yolk lipoprotein. These cryoprotectants are thought to act by increasing the osmotic pressure of the extracellular fluid and hence drawing water out from the spermatozoa, decreasing the risk of ice crystals forming and causing physical damage. However, they do not alleviate, and may even exacerbate the problem of dehydration and increases in solute concentration (Steinmann, 1996).

Other alternatives have been used as cryoprotectants, including Orvus ES paste, a mix of anionic detergents (S. Revell, Wales, 1998, personal communication), and the synthetic detergent OEP, an amino-sodium lauryl sulphate. OEP apparently alters the composition of egg yolk, improving its cryprotectant properties. Its inclusion was initially tried in extenders for use with boar semen, with some success (Pursel and Johnson, 1975b; Garner, 1991). It was subsequently used by Christanelli et al. (1984) in an extender containing 5% egg yolk, 2.5% glycerol and 0.4% OEP, along with lactose, fructose, glucose, ethylenediamine tetraacetic acid (EDTA), sodium citrate and sodium bicarbonate. This extender was used with equine semen plus 25% seminal plasma and it was reported that 90% of samples, post thaw, demonstrated a minimum of 50% motility. However, due to the large number of different components (many of which are known to have cryoprotectant properties), it is unclear whether such encouraging results were due to a direct effect of the inclusion of OEP.

An alternative investigated more recently is the use of liposomes composed of phosphatidylserine and cholesterol, but no advantage over the addition of HEPES-buffered saline was observed (Denniston et al., 1997).

It has been evident for sometime that cryoprotectants, both penetrating and non-penetrating, do themselves damage spermatozoa (Demick *et al.*, 1976; Fahy, 1986; Fiser *et al.*, 1991). This may be due to physical damage, as a result of the changes in osmotic pressure gradients, or to biochemical disruption of subcellular components. The adverse effect of cryoprotectants is evident as a reduction in motility rather than a reduction in fertility. In fact, the correlation between motility and fertility rates is lower in post-thaw samples than is evident in fresh or cooled semen samples. The addition of cryoprotectants such as glycerol has a more severe effect on motility than on mortality (Blanc *et al.*, 1989). The use of motility as an indication of viability is not, therefore, a very accurate assessment in freeze–thaw samples. This may be due to a greater detrimental effect of the cryopreservation agent on the mitochondria than on the acrosome region of the spermatozoan head. The detrimental effects of glycerol on such functions is apparently more evident in stallions than in other species such as cattle. A reduction in this effect may be achieved by altering the freezing protocol and timing of the addition of glycerol.

A further detrimental effect of the use of glycerol is the rapid efflux of glycerol from the spermatozoa into the glycerol-free secretions of the mare's tract at insemination. This rapid exit across the plasma membrane is likely to cause damage and, therefore, loss of function (Pickett and Amann, 1993), and may account in part for the apparent reduction in the positive correlation between motility and fertilization rates in equine spermatozoa post thaw.

The protocol for the use of cryoprotectants is ultimately a compromise between the advantageous and detrimental effects of their inclusion. Ideally, the exact protocol may well vary with individual stallions in order to obtain optimal results. However, such individual tailoring is not practical in a commercial situation and hence further compromise is normally required.

7.5.2. Extenders for use with frozen semen

Many of the extenders used for the cryopreservation of stallion semen are based upon those used for cool storage, as detailed earlier, but with the addition of cryoprotectants. Initially, heated whole cow's milk plus 10% glucose was used as an extender for freezing stallion semen. In fact this extender was used to store the semen that resulted in the first pregnancy, in a mare, as a result of AI with frozen semen (Barker and Gandier, 1957). Out of the seven mares inseminated, one pregnancy resulted. In this initial work semen was collected from the cauda epididymis and diluted in a single diluent prior to freezing.

Once the deleterious effects of seminal plasma had been noted (Pickett *et al.*, 1975a; Varner *et al.*, 1987; Webb *et al.*, 1990) a double diluent procedure was proposed. It is common practice today that the preparation of semen for cryopreservation involves the use of two extenders: a primary extender for initial dilution, which is aspirated off after centrifugation, prior to the addition of a secondary extender and freezing. Numerous extenders have been used as primary or secondary extenders.

Primary extenders

The prime function of a primary extender is to maintain motility, but it also acts to protect spermatozoa during the process of centrifugation. There is no real requirement for a cryoprotectant in such extenders. Examples of primary extenders are given in Table 7.8. Investigations into the addition of other elements, including lecithin and phenylmethanesulphonyl fluoride, to skimmed milk primary extenders have been carried out, but with little advantage being reported (Heffe, 1993).

Secondary extenders

Many of the secondary extenders are also based upon those given for fresh and cooled semen storage and may be similar to the primary extenders used, but in order to protect spermatozoa from cold shock, the addition of a cryoprotectant is needed. These secondary extenders may be used in isolation without prior centrifugation, or used after centrifugation. After centrifugation the secondary extender is normally (though not always) added after the primary extender has been aspirated off. Several cryoprotectants have been investigated for use with stallion semen.

Glycerol and egg yolk extenders were amongst the first to be used for freezing semen (Table 7.9) (Polge *et al.*, 1949; Nishikawa, 1975; M. Boyle, Cambridge, 1996, personal communication). More recent work has demonstrated the cryoprotectant nature of many other components, including sugars and liposomes (Heitland *et al.*, 1995). Many extenders used for freezing contain a mixture of many components in varying ratios. Examples of extenders for freezing include those based on skimmed milk with egg yolk, and egg yolk (Table 7.10), and are commonly used as secondary extenders. The addition of a detergent (or surfactant) to the extender was investigated, with some success, under the impression that it had a beneficial effect on the lipoproteins in egg yolk and within the plasma membrane of the spermatozoa, affording the spermatozoa additional cryoprotection (Jimenez, 1987; Pickett and Amann, 1993) (Table 7.11).

Other secondary extenders (Table 7.12) use sugars as a major component. Use of sugar-based secondary extenders has been reported both with and without initial dilution with a primary extender followed by centrifugation. If they are to be used without centrifugation then, as would be expected, the best results are obtained by fractionating semen and using the sperm-rich fraction (Tischner, 1979; Naumenkov and Romankova, 1981, 1983). The inclusion of mannitol and glucose in addition to lactose in the second extender given in Table 7.12 was reported to increase the percentage of motile spermatozoa post thaw. These last two extenders are popular for use in Eastern Europe. In comparative work carried out by Barsel (1994), an

Table 7.8. Examples of primary extenders for use at semen centrifugation prior to cryopreservation.

(a) BSA primary extender (Samper, 1995b)

Component	Quantity
Sucrose	5.0 g
Glucose	3.0 g
BSA	1.5 g
Distilled water	100.0 ml

(b) Glucose-EDTA primary extender I (Cochran et al., 1984)

Component	Quantity
Glucose	1.5 g
Sodium citrate dihydrate	25.95 g
Disodium EDTA	3.699 g
Sodium bicarbonate	1.2 g
Deionized water	1000.0 ml
pH	6.89
Osmotic pressure (mOsm kg^{-1})	290.0

(c) HF-20 primary extender (Nishikawa, 1975)

Component	Quantity
Glucose	5.0 g
Lactose	3.0 g
Raffinose	3.0 g
Sodium citrate	0.15 g
Sodium phosphate	0.05 g
Potassium sodium tartrate	0.05 g
Egg yolk	0.5–2.0 g
Penicillin	25,000 IU
Streptomycin	25,000 µg
Deionized water (made up to)	100.0 ml

(d) EDTA Glucose extender II (Cochran et al., 1984)

Component	Quantity
Glucose	59.985 g
Sodium citrate dihydrate	3.7 g
Disodium EDTA	3.699 g
Sodium bicarbonate	1.2 g
Polymixin B sulphate	10×10^6 IU
Deionized water	1000.0 ml
pH	6.59
Osmotic pressure (mOsm kg^{-1})	409.0

(e) Skimmed milk and sugar primary extender (Pickett and Amann, 1993)

Component	Quantity
Sterile skimmed milk	50.0 ml
Salt-and-sugar solution	50.0 ml
Gentamicin	5.0 mg
Penicillin	1000 IU
Salt-and-sugar solution:	
Glucose	2.5 g
Lactose	0.15 g
Raffinose	0.15 g
Sodium citrate dihydrate	0.03 g
Potassium citrate	0.04 g
Water (made up to)	50.0 ml

Table 7.9. Glycine egg yolk extender (M. Boyle, Cambridge, 1996, personal communication).

Component	Quantity
Extender 1	
Glucose monohydrate	12.0 g
Fructose	12.0 g
Distilled water	600.0 ml
Extender 2	
Sodium citrate	20.0 g
Glycine	9.4 g
Distilled water	1000.0 ml
Mix extenders 1 and 2; supplement with 20% egg yolk and clarify.	

Table 7.10. Skimmed milk egg yolk secondary extender (Samper, 1995b).

Component	Quantity
Skimmed milk	2.4 g
Egg yolk	8.0 ml
Sucrose	9.3 g
Glycerol	3.5%
Distilled water	100.0 ml

Table 7.11. Secondary extenders, incorporating detergent (Equex STM, Nova Chemicals Sales, Scituate, Massachusetts), used for the cryopreservation of stallion semen (Cochran *et al.*, 1984).

Component	Quantity
Lactose solution (11% w/v)	50.0 ml
Glucose EDTA solution (EDTA Glucose extender II primary extender)	25.0 ml
Egg yolk	20.0 ml
Glycerol	5.0 ml
Equex STM	0.5 ml

egg yolk, EDTA and lactose extender, similar to those given in Table 7.12, outperformed NFDSMG extender as far as motility and acrosome integrity were concerned.

A sugar-based extender with the addition of raffinose has been used with success in Japan (Table 7.13). This extender may be used without the 10% glycerol as a primary extender for centrifugation, or with the addition of the 10% glycerol as a freezing extender. Motility rates of between 49 and 73% were reported post thaw in 80% of the samples frozen, resulting in up to 100% conception rates in small trials (Nishikawa, 1975). Good success has been reported with the use of trehalose as a cryoprotectant within a skimmed milk–egg yolk extender. It is suggested that trehalose has a stabilizing effect on the spermatozoon plasma membrane (Steinmann, 1996). In France a

Table 7.12. Secondary extenders, with the addition of sugars, for use in the cryopreservation of equine semen.

(a) Lactose-based secondary extender (Tischner, 1979)

Component	Quantity
Lactose	11.0 g
Disodium EDTA	0.1 g
Sodium citrate dihydrate	0.089 g
Sodium bicarbonate	0.008 g
Deionized water	100.0 ml
Egg yolk	1.6 g
Glycerol	3.5 ml
Penicillin	50.0 IU
Streptomycin	50.0 mg

(b) Lactose-, mannitol- and glucose-based secondary extender (Naumenkov and Romankova, 1981, 1983)

Component	Quantity
Lactose	6.6 g
Mannitol	2.1 g
Glucose	0.7 g
Disodium EDTA	0.15 g
Sodium citrate dihydrate	0.16 g
Sodium bicarbonate	0.015 g
Deionized water	100.0 ml
Egg yolk	2.5 g
Glycerol	3.5 ml

Table 7.13. HF-20 extender which may be used as a primary extender (without the 10% glycerol) or as a secondary extender (with the 10% glycerol) (Nishikawa, 1975).

Component	Quantity
Glucose	5.0 g
Lactose	0.3 g
Raffinose	0.3 g
Sodium citrate	0.15 g
Sodium phosphate	0.05 g
Potassium sodium tartrate	0.05 g
Egg yolk	0.5–2.0 g
Penicillin	25,000 IU
Streptomycin	25,000 µg
Deionized water (made up to)	100.0 ml
Glycerol	10%

skimmed milk, sugar and salt solution extender is popular for use as a secondary extender (Table 7.14). A very simple, sugar-based secondary extender (Table 7.15) has been used successfully, in a ratio of 1:1, after primary extension with a simple 11% sucrose solution followed by centrifugation and aspiration (Piao and Wang, 1988).

As mentioned previously in the context of fresh and cooled semen storage, hydrogen ion extenders may also be used. Examples are given in Table 7.16. They may be used as primary extenders as they are, or as secondary extenders with the addition of 2–7% glycerol. Of these extenders, TCA 325 was reported to provide the best conception rates after storage in ampoules (Pace and Sullivan, 1975).

It is evident that many of these extenders use glycerol as the major cryoprotectant. Because the use of glycerol may itself be detrimental to spermatozoan viability (Pace and Sullivan, 1975; Demick et al., 1976; Fahy, 1986), a compromise has to be reached as regards the concentration of glycerol and the length of time that the glycerol is in contact with the spermatozoa prior to freezing, in order to maximize the beneficial effects of glycerol as a cryoprotectant but minimize its toxic effects. Many variations in contact time and concentration of glycerol in diluents have been investigated. In addition, the efficiency of glycerol may be affected by the diluent to which it is added, as well as the method of storage. Nagase (1967) reported that the

Table 7.14. A skimmed milk, sugar and salt solution secondary extender used in France.

Component	Quantity
Sterile skimmed milk	50.0 ml
Salt-and-sugar solution	50.0 ml
Egg yolk	2.0 ml
Glycerol	2.5%
Gentamicin	5.0 mg
Penicillin	5000 IU
Salt-and-sugar solution:	
Glucose	2.5 g
Lactose	0.15 g
Raffinose	0.15 g
Sodium citrate dihydrate	0.03 g
Potassium citrate	0.04 g
Water (made up to)	50.0 ml

Table 7.15. A simple sugar-based secondary extender, used after primary extension with an 11% sucrose solution (Piao and Wang, 1988).

Component	Quantity
11% Sucrose solution	100.0 ml
Skimmed milk	45.0 ml
Egg yolk	16.0 ml
Glycerol	6.0 ml

Table 7.16. Hydrogen ion primary extenders (without glycerol) and secondary extenders (plus 2–7% glycerol) (Pace and Sullivan, 1975).

(a) Basic components. For use as primary extenders all diluents contain the following four basic components, to which are added the specified extra ingredients, and the volume is made up to 100 ml with distilled water. For use as secondary extenders 2–7% glycerol is added to each one.

Component	Quantity
Fresh egg yolk	20.0%
Penicillin	500.0 IU
Dihydrostreptomycin sulphate	1000.0 µg
Polymixin B sulphate	500.0 IU

(b) BIS-TRIS [BIS(2 hydroxyyethyl)imino-TRIS(hydroxymethyl)methane] extender

Component	Quantity
BIS-TRIS	2.532 g
Citric acid	0.292 g
Sodium citrate	0.656 g
Dextrose	0.630 g
pH	7.0
Osmotic pressure (mOsm kg^{-1})	325.0

(c) BES [N,N-BIS(2 hydroxyethyl)-2-aminoethanesulphonic acid] extender

Component	Quantity
BES	1.96 g
TRIS[a]	4.79 g
Sodium citrate	0.656 g
Dextrose	0.630 g
pH	7.0
Osmotic pressure (mOsm kg^{-1})	325.0

(d) GCT extender

Component	Quantity
Glycyl-glycine	2.16 g
TRIS	0.25 g
Sodium citrate	0.328 g
Dextrose	0.630 g
pH	7.0
Osmotic pressure (mOsm kg^{-1})	325.0

(e) GGBT extender

Component	Quantity
BIS-TRIS	0.970 g
Glycyl-glycine	1.70 g
Dextrose	0.630 g
pH	7.0
Osmotic pressure (mOsm kg^{-1})	325.0

Table 7.16. *continued.*

(f) TCA 325 extender

Component	Quantity
TRIS	2.487 g
Citric acid	1.39 g
Dextrose	0.838 g
pH	6.85
Osmotic pressure (mOsm kg^{-1})	325.0

(g) TCA 390 extender

Component	Quantity
TRIS	2.9 g
Dextrose	1.0 g
pH	6.85
Osmotic pressure (mOsm kg^{-1})	390.0

[a]TRIS = TRIS(hydroxymethyl)-aminomethane

optimum effect of glycerol was obtained with an equilibration time of 3–5 h. However, it has also been reported that exposure of spermatozoa to glycerol for only a few seconds still resulted in cryopreservation when glycerol was included in a glucose–skimmed milk extender or a glucose–egg yolk extender and the sample was stored in ampoules (Rajamannan *et al.*, 1968) – hence the popularity and success of many of the extenders listed. Further work supported this finding that exposure of spermatozoa to glycerol under the correct circumstances for just a few seconds could be as beneficial (as regards cryopreservation) as exposure over 2–3 min or over 4 h (Berndtson and Foote, 1969; Nishikawa *et al.*, 1972a,b). Nishikawa (1975) found that 5–10% final glycerol concentration in an HF-20 extender with an equilibration time of 30 s ensured good fertility. Conception rates of up to 60% were reported after storage in 1 ml straws. Again this work supported the significant variability found in the freezing capacity of spermatozoa from different stallions.

As indicated, Nishikawa (1975) obtained good results with 10% glycerol diluent. Other work suggested that glycerol levels could just as successfully be lowered to 5% but inclusion rates of 0, 1 and 3% were less effective in maintaining motility (Nishikawa *et al.*, 1968; Nishikawa, 1975). Other workers have come up with similar results. Piao and Wang (1988) suggested that 6% glycerol was the most beneficial. Pace and Sullivan (1975) demonstrated that reducing glycerol from 7% to 2% was advantageous, obtaining conception rates of 12% and 46%, respectively. Barsel (1994) suggested that 5% glycerol was superior to both 3% and 7% in a lactose–EDTA–egg yolk extender. Cochran *et al.* (1984), investigating the effect of extender, freezing rate and thawing temperature on motility, demonstrated that an inclusion rate of 4% in an extender containing 20% egg yolk was superior to 2% or 6% glycerol as regards progressive motility. This superiority of 4% glycerol was further confirmed by Christanelli *et al.* (1985). Burns and Reasner (1995) concluded that a sucrose–glucose–dried skimmed milk freezing extender with 2% glycerol was most beneficial as regards motility and progressive motility, plus lateral head displacement, but an inclusion rate of 1% glycerol was best for velocity and progressive velocity. Unfortunately, no inseminations were carried out. It has also been suggested that glycerol accelerates the rate at which maximal post-thaw motility is reached (Nishikawa, 1975). In addition to altering the inclusion rates, it has been suggested that the slow addition of glycerol, or addition at 4°C, may alleviate some of the adverse effects (Pickett and Amann, 1993), but this has not been fully investigated.

Clearly it is impossible to recommend either an absolute inclusion rate for glycerol or an optimum equilibration time. Both are highly dependent upon other factors – for example, other components of the extender, individual stallions, or freezing and thawing rate. It was once thought that an adverse reaction may occur between glycerol and egg yolk, though some more recent research casts doubt on this assumption (Loomis *et al.*, 1983; Cochran *et al.*, 1984).

In an attempt to provide additional protection for acrosome membranes and help to preserve motility post freezing and thawing, concanavalin A has been added to the freezing diluent. Concanavalin A has the ability to coat and thus protect membranes. Betaine, included at 0–3%, has also been used as a possible cryoprotectant, with some limited success (Koskinen *et al.*, 1989). More recently the inclusion of liposomes, which have proved successful in bulls, has been tried with some success with equine semen (Heitland *et al.*, 1995). Phenolic antioxidants have proved successful in rams (Erokhin *et al.*, 1996).

7.5.3. Removal of seminal plasma

Several workers have demonstrated a direct adverse effect of seminal plasma on samples destined for freezing (Buell, 1963; Rajamannan *et al.*, 1968; Nishikawa, 1975) and many of the protocols for freezing equine semen involve the use of seminal plasma-free samples. This may be achieved most successfully by centrifugation, but centrifugation itself may cause damage; therefore, some protocols call for the use of an open-ended AV to allow fractionation of the semen sample and use of spermatozoa-rich fractions only. However, the damage due directly to centrifugation is reported to be minor compared with the damage caused by the freeze–thaw process itself (Blach *et al.*, 1989). Revell, comparing cushioned with conventional centrifugation using 11 split ejaculates, reported that the number of live normal spermatozoa recovered after centrifugation represented 81% and 67% of the respective pre-centrifugation populations. The respective post-thaw live, normal numbers were 77% and 63% of the pre-freezing live, normal populations, indicating that the loss or damage to live, normal spermatozoa undergoing centrifugation and freezing–thawing was only slightly greater than the loss or damage resulting from freeze–thaw alone (S. Revell, Wales, 1998, personal communication).

The protocol for and reasoning behind the removal of seminal plasma in samples to be frozen has been discussed at length in section 7.3.2.

7.5.4. Packaging for frozen semen

Several methods are available for the packaging of spermatozoa for freezing. They include: glass ampoules or vials (Fig. 7.4); polypropylene, polyvinyl or plastic round or flat straws (usually 0.5–1.0 ml in volume) (Fig. 7.5), flat aluminium packets (10–15 ml); pellets (0.1–0.2 ml); and macrotubes (Merkt *et al.*, 1975b; Loomis *et al.*, 1984; Amann *et al.*, 1987; Blach *et al.*, 1989; Haard and Haard, 1991; Piao and Wang, 1988).

In initial work, semen was frozen in glass ampoules or vials with a volume of 1–10 ml. Later the use of straws became more widespread. These were based upon those in use in cattle work (Nagase and Niwa, 1964), having the advantage of being smaller in volume and, therefore, taking up less storage space (Merkt *et al.*, 1975b). Both ampoules and straws were traditionally frozen by suspension over liquid nitrogen, followed by plunging into liquid nitrogen at −196°C. Subsequent work investigating the effect of the rate of freezing (discussed in the following sections) led to the present use of computer-controlled freezing units.

Pellets are efficient to store, due to their size, which also enables a more rapid decline in temperature to be achieved. Pellets are frozen by the placing

Fig. 7.4. Glass vials are one of the means of storing frozen semen. (Photograph by Julie Baumber and Victor Medina.)

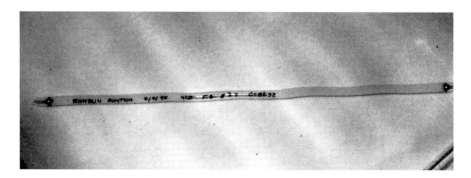

Fig. 7.5. A plastic round straw is one of the most common methods of storing frozen semen. (Photograph by Julie Baumber and Victor Medina.)

of small drops (approximately 0.1 ml) into small indentations in a block of solid carbon dioxide (Pace and Sullivan, 1975). Alternatively, a metal plate cooled to −75°C, with appropriate indentations, may be used. The temperature is then dropped to the required −196°C for long-term storage. Although pellets have the advantage of allowing a rapid drop in temperature to be achieved, they are not suited to easy identification once frozen. In addition, the re-use of the carbon dioxide block or metal plate runs the risk of cross-contamination with spermatozoa from the previous freezing batch. The use of vials or straws (the method adopted by most countries these days) more readily allows the accurate identification of samples and considerably reduces the risk of cross-contamination.

As several different methods of storage have been used, the question of whether the means of storage has an effect on the success rate has been raised. Pace and Sullivan (1975) directly compared spermatozoa stored in ampoules and in pellets. Their results indicated a slight advantage in terms of pregnancy rates in the use of ampoules (44%) over pellets (36%). However, they used six different extenders, and a significant interaction between extender and means of packaging was demonstrated. Some extenders froze significantly better in pellet form than in ampoules – such as BIS-TRIS and TCA_{390} ($P < 0.05$); but vice versa for TCA_{325} ($P < 0.05$). On average, pelleted semen performed the best. Merkt *et al.* (1975b) also evaluated the use of pelleted semen, collected as a fractionated sample using an open-ended AV. They reported that stallion semen may be stored as such for several years with no significant loss of motility or fertility.

The use of vials or pellets is the preferred method in China but is less popular in other parts of the world, though considerable research into their use has been conducted in Germany. Semen may also be stored in straws. Straws are normally made of plastic, polypropylene, polyvinyl or polyvinyl chloride. They may be sealed with an aluminium plug or ball bearing, or (more popularly today) they are sealed at one end by two cotton plugs with PVA (polyvinyl alcohol) powder between them, and are ultrasonically sealed or plugged with PVA powder at the other end. They vary in size quite considerably, with volumes ranging from 0.25 ml straws to 10 ml macrotubes, and diameters from 1.5 mm to 6 mm (Martin *et al.*, 1979). A comparison between pellets and 10 ml polypropylene straws was reported by Aliev (1981a,b). Pregnancy rates of 55% and 53%, respectively, were reported for pellets and straws in one study, and 60% and 73%, respectively, in another. At 10 ml, the straws were larger than those commonly used, which are normally 0.25–4 ml in volume. Differences in post-thaw motility between semen frozen in ampoules or in straws have been reported, with semen stored in straws surviving better (Yi *et al.*, 1983). Polyvinyl 1 ml straws are quite commonly in use in Europe and America; with such straws, dilution rates need to be adjusted to give 150–300 × 10^6 spermatozoa ml^{-1}, and three to five of these straws are pooled on thawing prior to insemination. The use of smaller (0.5 ml) straws is also widespread; their use was first reported by Veselinovic *et al.* (1980) and since then they have been adopted by many workers, especially in the USA (Christanelli *et al.*, 1984; Loomis *et al.*, 1984; Amann *et al.*, 1987; Blach *et al.*, 1989; Haard and Haard, 1991).

The effect of different straw volumes is disputed. Cochran *et al.* (1983) and Delius (1985) demonstrated that increasing the volume of the straw from 0.5 ml to 1 ml by doubling the length but retaining the same diameter had no effect on motility or fertility rates post thaw. Work by Wockener and Schuberth (1993) showed no such differences between round 4 ml straws, round 0.5 ml straws and flattened 1.7 ml straws. Similarly Palacious *et al.* (1992) failed to demonstrate a difference between 4 ml straws and 0.5 ml straws in terms of motility or acrosome damage. However, other workers have demonstrated differences between different sized straws. Papa *et al.* (1991) demonstrated a difference in motility between 0.5 ml and 4 ml straws, in favour of the smaller straws. Similarly Advena (1990), Park *et al.* (1995) and Kneissl (1993) demonstrated an advantage, in terms of motility, in using 0.5 ml straws over 5 ml cryotubes. Heitland *et al.* (1996) demonstrated a similar effect of straw volume, specifically on curvilinear velocity, which was highest in spermatozoa stored in 0.5 ml straws compared with 2.5 ml straws. The reason for these discrepancies is unclear, and it is not apparent in all work how the dimensions of the straws change with volume, in addition to which different extenders were used and these would also be expected to have an effect. Advena (1990) also investigated the possible effect of two different methods of sealing the straws – namely, the use of aluminium plug or ball bearings – but no differences between the two methods were reported.

Aluminium packets have been used to store larger volumes of semen and are more popular in Russia and Eastern Europe. The aluminium packet is in the form of a pocket, usually 4 mm thick with an area of some 37 mm × 110 mm. Into these packets is placed 15 ml of extended semen. Once filled, the packet is sealed by folding over the open end several times. It is then held in the refrigerator at 2–4°C for 90 min prior to freezing, by suspension over liquid nitrogen (Naumenkov and Romankova, 1979; Muller, 1982, 1987;

Love *et al.*, 1989). The development of a slotted copper holder allows the packets to be immersed in the liquid nitrogen up to 1 cm, achieving a more controlled rate of freezing. After a specified period they are immersed in liquid nitrogen, for long-term storage at $-196°C$ (Muller, 1982, 1987; Love *et al.*, 1989). A comparison between the use of aluminium packets and straws for semen storage was carried out by Love *et al.* (1989): their work demonstrated no difference in conception rates between the use of 4 ml straws and 10–12 ml aluminium packets.

Lastly, plastic bags have been tried as storage containers for frozen semen. However, they have proved to be not as durable as other methods (Ellery *et al.*, 1971).

Comparisons between three of the main methods of storage – namely, pellets, straws and aluminium packets – have been carried out using similar semen doses. One-cycle conception rates of 53% for semen stored in 0.2 ml pellets, compared with 62% for aluminium packets, were reported by Naumenkov and Romankova (1979). Pregnancy rates of 53–55%, 47–73% and 60–64% have been reported for semen stored in 10 ml straws, pellets and aluminium packets, respectively (Aliev, 1981b). Love *et al.* (1989) reported no significant difference in fertility rates, at 46% and 55%, respectively, for semen collected as a spermatozoa-rich fraction and stored in either 4 ml plastic straws or aluminium packages after extension with EDTA–lactose extender. Once frozen (regardless of the method), semen can be stored indefinitely in flasks of liquid nitrogen, some of which are portable (Fig. 7.6).

7.5.5. Cooling rates

The freezing rate or rate of cooling for semen depends upon the method of processing and of storage. The use of pelleted semen ensures a rapid but rather uncontrolled drop in temperature. It has been reported that, for 0.1 ml pellets, $-79°C$ is reached within 4 min (Krause and Grove, 1967). The cooling of straws and ampoules is easily and conveniently done by either initial suspension in racks over a tank of liquid nitrogen or a computer-controlled programmable freezer, followed by plunging into the liquid nitrogen for long-term storage. Alterations to the rate of cooling and final temperature using the tank-rack method can be achieved by changes to the height of the straw above the liquid nitrogen and the time for which it is suspended. Once the specified time of suspension has been reached, or the programme is completed, the straws are plunged into the liquid nitrogen at $-196°C$ for long-term storage. However, this system is still rather imprecise in its control. Aluminium packets are similarly cooled, initially by suspension over liquid nitrogen, followed by plunging them into the nitrogen. The development of new copper holders has been reported, the use of which results in just a 7 min time span between $4°C$ and $-196°C$.

It is apparent that the rate of temperature drop does have an effect upon success rates. Using ampoules, Schafer and Baum (1964) recommended a

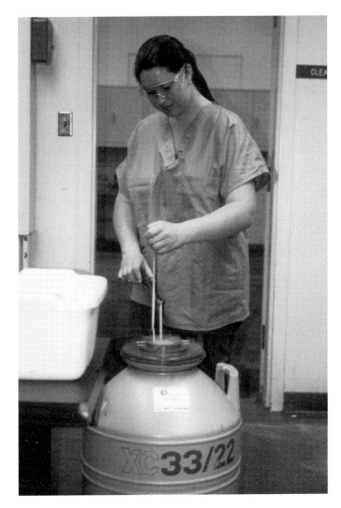

Fig. 7.6. Semen may be conveniently stored indefinitely in liquid nitrogen flasks. (Photograph by Julie Baumber and Victor Medina.)

rapid temperature drop of 8–10°C min^{-1} in the range from 6°C down to −79°C. More recent work suggested that a slower rate of decline (0.5°C min^{-1} in the temperature range 5–20°C) followed by a faster rate of decline (3°C min^{-1} in the temperature range −20°C down to −196°C) was the most appropriate, resulting in a low incidence of acrosome damage (Chao and Chang, 1962; Advena, 1990; Heitland *et al.*, 1996).

A significant amount of semen, especially in the UK and Europe, is stored in straws. In work comparing different cooling rates for 0.5 ml straws, semen was cooled at either 60°C min^{-1} throughout, by placing the straws horizontally over liquid nitrogen, or via a two-step cooling process. The two-step cooling process involved an initial rate of temperature drop of 10°C min^{-1} for the temperature range 20°C to −15°C followed by a 25°C min^{-1} drop for the remainder of the cooling process. No difference in motility was evident between the two treatments post thaw (Cochran et al., 1984; Christanelli et al., 1985). Borgel (1994) suggested that the rate of cooling does affect success rates: he demonstrated that a slower cooling rate of 0.3°C min^{-1} for the temperature range 20°C down to −15°C followed by 25°C min^{-1} from −15°C to −150°C was more advantageous (as regards motility and acrosome integrity) than a rate of decline of 10°C min^{-1} in the range 20°C to −15°C followed by 25°C min^{-1} down to −150°C. Further work investigated a more simplified cooling system, comparing a rate of temperature drop of 5°C min^{-1} with one of 25°C min^{-1}, both consistent throughout the cooling process. Surprisingly the results indicated that the faster freezing protocol yielded better post-thaw motility rates than the slower freezing rate (Robbelen, 1993). A typical recommended equine cooling curve for a programmable freezer starting at 4°C is 10°C min^{-1} to −10°C, 20°C min^{-1} to −100°C, and 60°C min^{-1} to −140°C, followed by plunging into liquid nitrogen (−196°C).

The use of a floating freezing rack, into which straws are placed and then floated on liquid nitrogen, has been reported to be successful (Hurtgen, 1997).

7.5.6. Thawing rates and extenders

The thawing rate can be manipulated by several factors, including the temperature and nature of the environment (air or water bath) and the thermal conductivity of the packaging (especially the diameter of the lumen). The easiest method available for thawing semen is to place the packaged semen in a warm water bath. Water bath temperatures of between 4°C and 75°C have been used successfully. The temperature chosen for the water bath depends on the desired rate of warming. For example, rapid thawing is reported to be achieved by immersing aluminium packages of semen in a water bath at 40–50°C for 20–50 s (Tischner, 1979; Muller, 1982, 1987; Love et al., 1989). Similar treatment of semen in polypropylene bags also ensures rapid thawing (Amann and Pickett, 1984; British Equine Veterinary Association, 1991). Large 4 ml straws are normally thawed in a water bath at 50°C for 40 s (Martin et al., 1979). Variations in the length of immersion and the water bath temperature (50°C for 40, 50 or 60 s, versus 60°C for 30, 40 or 50 s) have been shown to have no effect on motility or acrosin activity post thaw (Viera et al., 1980).

Smaller 0.5 ml straws are more readily affected by environmental temperature on thawing and can be thawed easily by immersion in a water bath at 38°C for 30 s (Cochran et al., 1983; Loomis et al., 1984). However, more recent work by Cochran et al. (1984) demonstrated that immersion of 0.5 ml straws in a water bath at 75°C for 7 s, followed by immediate immersion in a water bath at 37°C for longer than 5 s, resulted in better motility rates (34%) post thaw than the more conventional immersion in a 37°C water bath for 30 s (29%). Such treatment resulted in a much more rapid thaw, but it is most important that the 7 s immersion is adhered to strictly. Delay in removing the straws from the 75°C water bath resulted in overheating (temperatures in excess of 40°C) of the semen and a resultant risk to pregnancy rates. Further work directly comparing thawing rates of 7 s at 75°C with 30 s at 37°C failed to show a consistent detrimental effect of the greater thawing rate (Borg *et al.*, 1997). Similarily, when comparing large (4 ml) straws and small (0.5 ml) straws, Frey and Bernal (1984) demonstrated no differences in the motility of spermatozoa thawed at 45°C for 1 min or 5°C for 4 min.

The thermal conductivity and hence the composition of the packaging affects thawing rates. Cochran *et al.* (1984) illustrated a difference in thawing rate (warming curves) for two types of straw: 0.5 ml polypropylene straws and 0.5 ml polyvinyl chloride straws. The later required just 7 s immersion in the 75°C water bath, whereas the former required 10 s.

Some protocols for thawing semen involve the addition of warmed thawing extender to aid the thawing process. The extender may also increase the volume of inseminate and aid preservation of spermatozoan viability. Thawing extenders may be used for semen stored in pellets, vials or straws, and are added as part of the thawing process. The addition of 25 ml of HF-20 (see Table 7.8) or 20 ml of a sucrose milk solution (Table 7.17) at 42°C as a thawing extender has been reported to be advantageous to semen stored in straws and pellets or vials, respectively (Nishikawa *et al.*, 1972a, 1976; Nishikawa, 1975; Piao and Wang, 1988). Similarly the use of 25 ml of milk extender at thawing has been advocated as advantageous (Veselinovic *et al.*, 1980). Ringer's solution, lactose diluent, EDTA or Merck diluent plus milk have all been compared as thawing media. The conception rate reported for the four diluents was 67.4%, 75.0%, 60.0% and 30.0%, respectively, illustrating an advantage in using the lactose diluent (Papa, 1989).

Some work has been carried out into the benefit of oxygenation of semen during the thawing process, but results proved inconclusive (Matveev, 1986).

7.5.7. Dilution rates

Factors influencing dilution rates for frozen semen are similar to those for chilled semen and so have been discussed under the section dedicated to the chilling of semen. Most spermatozoa are frozen in a concentrated form

Table 7.17. A sucrose and milk diluent that may be used as a thawing extender (Piao and Wang, 1988).

Component	Quantity
Sucrose	6.0 g
Powdered skimmed milk	3.4 g
Distilled water	100.0 ml

– hence the use of thawing extenders as discussed in the previous section. As with chilled semen, the volume of inseminate should not exceed 100 ml (Rowley *et al.*, 1990) and concentrations of 20–400 × 10^6 spermatozoa ml^{-1} are most successful (Jasko *et al.*, 1992c; Heitland *et al.*, 1996). As with chilled semen, insemination of 100–600 × 10^6 total spermatozoa per dose per insemination is reported to be the most successful as far as pregnancy rates are concerned (Pace and Sullivan, 1975; Householder *et al.*, 1981).

7.5.8. Conception rates

Significant variation in the success rates for frozen semen has been reported. There is apparently a much wider variation in the performance of semen post freezing and thawing than is evident with bovine semen (S. Revell, Wales, 1997, personal communication). In general, pregnancy rates for frozen semen at best reach values approaching those of natural service but may also result in complete failure, despite the protocol for freezing apparently being unchanged (Palmer *et al.*, 1984; Julienne, 1987; Muller, 1987; Piao and Wang, 1988; Jernstrom, 1991; Thomassen, 1991; Pickett and Amann, 1993). It is impossible, therefore, to predict likely conception rates with any certainty and it is also very difficult to compare, with any accuracy, the pregnancy rates obtained in different research work using frozen semen. In most of the research carried out there are many variables, the details of which are often not specified. The results reported for pregnancy rates depend, among other things, upon: the stallion and the quality of semen produced; the minimum standard set for semen quality prior to acceptance for freezing; the number of spermatozoa per straw and straws per insemination; the freezing protocol; the thawing protocol; post-thaw semen quality control; the numbers and reproductive ability of the mares used for insemination; the timing of the insemination; the number of inseminations per cycle; and the number of cycles and inseminations per pregnancy. Lastly, but by no means least, a large number of experimental trials fail to use any form of control (ideally there should be coincidental insemination of an identical group of mares with fresh semen). As a result no absolute figure for pregnancy rates using frozen semen can be arrived at.

However, Pickett and Amann (1993) attempted to determine a reasonably accurate pregnancy rate by the amalgamation of the results from 5 years of work within their laboratory. In this they were able to pick out a number of stallions each year which had been used for both frozen and fresh semen insemination, between which valid comparisons could be made. These stallions all passed a semen evaluation assessment prior to their use (500 × 10^6 spermatozoa ml^{-1} ejaculate with 50% or greater motility). Between late April and the end of June they were collected from every third day for freezing, and every other day for use with fresh insemination. The frozen semen was stored in 0.5 ml and 1 ml straws and inseminated, after extension to 12–18 ml, at a dose rate of 100–340 × 10^6 motile spermatozoa per insemination. The fresh semen was extended in 10 ml of skimmed milk extender and inseminated at a dose of 200–300 × 10^6 spermatozoa per insemination. The stallions chosen for the comparisons all yielded 40% motility in thawed samples. Insemination was performed on one cycle only and the pregnancy rates were reported at rectal palpation on day 50. The results obtained are given in Table 7.18, which shows that pregnancy rates to frozen semen were significantly lower than those obtained as a result of using fresh semen. If the results are broken down on a year-by-year basis, the significantly better performance of fresh semen is apparent in both 1981 and 1982. The difference is not so evident in 1983 and this is reported to be due to a change in the processing protocol for freezing semen (Christenelli *et al.*, 1984; Cochran *et al.*, 1984). However, results using a similar freezing protocol in 1984 are not so encouraging.

Similar pregnancy rates have been reported by other workers. Palmer *et al.* (1984) report disappointingly low conception rates at 25%. Muller (1987), in work carried out over a 5-year period (1981–1985), reported one-cycle pregnancy rates of 32–51% based upon 69–413 cycles. One-cycle pregnancy rates of 38–40% are reported (Pickett and Amann, 1993) to have been obtained in commercial AI work in France between 1989 and 1990 based upon 690 mares, with an average of 3.8 and 3.9 doses, respectively, of semen required per pregnancy. Similar figures were reported by Julienne (1987), again working in France with French Saddle horses. Further work in Northeastern China, an area that traditionally inseminates many mares per year, was reported by Pickett and Amann (1993). An average one-cycle pregnancy rate of 44% in 1979 was reported, with 1030 mares inseminated with frozen semen; this figure compares quite well with the 50% pregnancy rates reported for insemination with fresh semen. In the same area in 1985, 31,832 mares were inseminated using frozen semen and a pregnancy rate of 68% was obtained. Piao and Wang (1988) reported upon the insemination of 89,176 mares and gave a one-cycle conception rate of 53%, which was similar to conception rates of 48% reported in China by Yi *et al.* (1983). Work in

Table 7.18. Comparison of 50-day conception rates obtained in mares inseminated with either frozen or fresh semen, collected from the same stallion, at the same time of year (mean pregnancy rates for frozen semen were less than for fresh; $P < 0.05$) (Amann and Pickett, 1987).

Stallion	Pregnancy rates		
	Fresh	Frozen	Frozen as % of fresh
1	67	50	75
2	67	56	84
3	61	61	100
4	77	8	10
5	70	21	30
6	79	48	61

Norway in 1991 indicated that conception rates of up to 70% for frozen semen are obtainable (Thomassen, 1991). In Germany, averaged over 2 years, insemination of 980 mares resulted in a conception rate of 47.5% (Haring, 1985). In Sweden, over a 9-year period, data from 1629 mares inseminated with frozen semen from 68 stallions indicated an overall conception rate of 60% (34.4–65.8%); further work for 1 year based upon 339 mares resulted in a conception rate of 63.1% (Jernstrom, 1991). Finally, based on 390 mares in Brazil, single-cycle conception rates of 54.7–61.8% were reported by Papa (1989, 1991).

It is evident that pregnancy rates as a result of using frozen semen vary enormously and the prediction of success is inaccurate. It is also apparent that, in general, pregnancy rates from frozen semen are not as high as those expected from using fresh semen (Wockener and Colenbrander, 1993): a difference of 5–30% is often quoted.

7.5.9. Factors affecting success rates of frozen semen

As indicated earlier, many factors affect the reported success rates in using frozen semen. The two major areas (excluding handling and storage techniques) that are normally considered to have the greatest effect are the timing of collection within the year and the inherent individual variation evident between stallions.

The effects of season

One of the major sources of variation which is reputed to have a considerable effect upon post-thaw spermatozoan viability, regardless of all other factors, is season. For reasons of management, it is much more convenient to collect semen destined for freezing outside the normal covering season. In the northern hemisphere this means collection from August to November, and for the southern hemisphere collection from February to May. This is practised because the increase in ejaculation frequency required if collection for insemination is to be carried out during the normal covering season may compromise both pregnancy rates for natural service and the quality of spermatozoa collected. Collection of semen for freezing, especially if it is intended for export, will require veterinary regulations to be met; these normally involve a series of blood tests, swabs and possibly semen culture. There may also be a period of quarantine for the stallion prior to collection. Satisfying these criteria is often not feasible during a busy covering season.

However, as discussed in Chapter 4, season is known to have a detrimental effect on reproductive efficiency in the stallion: a lack of libido, low spermatozoan concentrations and low total spermatozoa counts are evident during the non-breeding season. Stallions can be collected from during the non-breeding season provided they are given enough encouragement and time to compensate for their naturally lowered libido but, regardless of any effect on spermatozoan quality, the number of spermatozoa ejaculated (and, therefore, collected and available for freezing) is reduced during the non-breeding season. Nishikawa *et al.* (1972a) and Oshida *et al.* (1972) suggested that a detrimental effect of the non-breeding season may be evident in the percentage of motile spermatozoa post thaw and the pregnancy rates obtained. The differences were not consistent for all stallions, with some showing no effect at all. In practice, many workers encounter few problems with out-of-season semen collection, which is widely carried out (S. Revell, Wales, 1998, personal communication; J.M. Parlevliet, The Netherlands, 1998, personal communication).

Pickett *et al.* (1980) collected semen from five stallions over a 13-month period and failed to produce evidence to support the suggestion that a difference in motility might be apparent with time of year. Assessments were made either immediately post collection or after freezing in TRIS-based extender and storage in glass ampoules. Amann and Pickett (1987) also failed to show an effect of month of collection (October or January) in two successive years on spermatozoan motility post thaw. Magistrini *et al.* (1987) investigated the effect of season on spermatozoan characteristics post freezing and used three ponies and three horse stallions, which were collected from at either high frequency (five collections per week) or low frequency (three collections per week). The semen, post collection, was extended in a 2% egg yolk freezing extender and stored in 0.5 ml straws. Surprisingly, their results indicated that post-thaw motility was depressed in the breeding season, compared with the non-breeding season. This is opposite to the trend reported for motility in fresh semen samples collected. These authors suggested that the differences may be due to variations in seminal plasma. Accessory gland function is known to vary with the season, being depressed during the non-breeding season and resulting in a lower volume of ejaculate, especially a reduction in the volume of gel produced. It has been suggested that the secretions of the seminal vesicles, from which the gel originates, may be responsible for the possible adverse effects of seminal plasma on spermatozoa during storage (Varner *et al.*, 1987; Webb *et al.*, 1990). It has also been postulated that season may affect the components of seminal plasma, making it more or less suitable for freezing – for example, the changes to the volume of the secretions originating from the seminal vesicles. Magistrini *et al.* (1987) did not remove seminal plasma prior to freezing and it would be interesting to know if such a difference would also be apparent in semen samples which have had the seminal plasma removed prior to freezing. However, these workers did demonstrate a difference in the number of spermatozoa with intact acrosomes, in favour of the breeding season. Significant individual differences were reported throughout this work, which again highlights the problem of using small experimental numbers in an area where individual animal variation is known to be considerable. Work on semen after chilled storage also demonstrated no effect of season upon motility (Witte, 1989).

In summary, no real evidence exists to suggest a significant detrimental effect of season on quality post thaw. It is routine practice to collect from stallions for semen freezing during the non-breeding season to avoid clashes with their natural covering season. The resultant decline in total spermatozoan numbers and concentration can be compensated for by appropriate adjustment of dilution rates.

Stallion variation

Significant variation is reported to exist in the ability of semen from different stallions to survive the freeze–thaw process (Back *et al.*, 1975; Loomis, 1993; Torres-Boggino *et al.*, 1995). These differences are apparently not due to alterations in seminal plasma protein levels (Amann *et al.*, 1987) but may, in part, be due to the differing susceptibilities of spermatozoa to acrosome damage during the freezing and thawing process. Work using two small groups of stallions (one group whose semen habitually froze well and the other group whose semen did not) indicated that in the group whose semen froze well a 34.2% incidence of acrosome defects was apparent at thawing, compared with 44.0% for the second group (Grondahl *et al.*, 1993). It is apparent from work by Kuhne (1996) and Aurich *et al.* (1996) that seminal plasma is in part responsible for the ability of spermatozoa to survive freezing. In this work there was a reciprocal exchange of seminal plasma between samples from stallions whose spermatozoa habitually froze well and samples from stallions whose spermatozoa did not freeze well. As a result of this exchange, a significant improvement was observed in the motility and acrosome membrane integrity of spermatozoa from the normally poor sample. The converse was observed as regards motility with the spermatozoa from the stallion whose spermatozoa normally froze well. No effect was seen on acrosome integrity in these animals.

Breed differences in the ability of semen to survive the freeze–thaw process, manifest as differences in motility and the percentage of abnormal spermatozoa, have been reported (Yurdaydin *et al.*, 1985). It is also reported that the freezability (the effect of freezing on motility) varies between seasons. Vidament *et al.* (1997) indicated that only a 0.6 correlation existed between the freezability of semen collected from the same stallion in two successive years.

7.5.10. Conclusion

It is evident from a considerable amount of work that the biggest stumbling block to the development and widespread use of cryopreserved equine semen is the considerable variation in success apparent between stallions. The development or identification of some marker to indicate the ability of spermatozoa to survive the freeze–thaw process would be a significant step forward. Towards this end, Bittmar and Kosiniak (1992) attempted to identify whether biochemical parameters, such as asparagine aminotransferase activity or total protein, could be used as predictors of freezing success. They claimed to be able to use such markers to predict, with an 80% accuracy, the ability of a stallion's semen to survive the freeze–thaw process. If this proves to be repeatable, or an alternative can be identified, significant strides will have been made in advancing the popularity and commercial acceptability of cryopreservation of stallion semen.

7.6. Conclusion

It can be concluded that, though the storage and transportation of equine semen is a viable proposition, the sophistication of the techniques involved is nowhere near as advanced as those available for AI in other farm livestock, largely because of the high costs of equine research. Despite this significant strides have been made in the last few years, but the failure to identify a successful cryopreservation agent has proved to be a major stumbling block. Until consistently good results can be obtained from storing semen, especially via cryopreservation, it will be almost impossible to persuade the industry to support the development of equine AI to a level consistent with that of other farm livestock. In order for the industry to have faith in the techniques, we must be able to produce a consistently good quality product with some sort of quality assurance attached to it and ensure that the skills to use it properly exist in the field.

Mare Insemination

8.1. Introduction

The success of AI is measured by the ability of the procedure to result in a viable offspring. One of the major considerations is the placing of spermatozoa into the female tract in such a condition that fertilization will result. It is also essential that the timing of insemination is synchronized with ovulation. The synchronization of these two events is more difficult with AI than it is with natural service, as the signs of oestrus and ovulation are interpreted normally by man rather than by the stallion.

All the preceding careful preparation of the semen (that is, its collection, evaluation, extention and storage) is deemed useless if the final stage of insemination is not carried out appropriately. It is imperative that attention be given to the reproductive competence of the mares to be inseminated and to the detection of ovulation and the timing of insemination.

8.2. Selection of Suitable Mares for Insemination

In theory, any mare that would be presented for natural service can be presented for AI. In fact, AI can be used on mares that may be precluded from natural service – for example, mares with skeletal problems, behavioural problems, abnormalities and perineal operations (Caslick, 1937; Pouret, 1982). However, in such cases, careful consideration must be given to the possibility of perpetuating the abnormal trait in subsequent generations. AI can be successfully used on all categories of mares – maiden mares, barren mares and mares with foal at foot. Though some differences in conception rates to AI between these categories have been reported, no consistent differences are apparent (Abel, 1984; Merkt, 1984; Buttelmann, 1988). As with natural service, the best results are obtained when covering a mare with a history of high conception rates. AI will not improve upon nature in mares with inherently poor reproductive ability. It is, therefore, best to ensure reproductive competence before a mare is considered for AI, in order to avoid the disappointment of repeated returns that may be attributable not to the technique but rather to the innate inability of the mare to conceive. Reproductive assessment will allow not only assessment of the possible outcome of using AI but also the ease with which the technique may be carried out.

Reproductive assessment of the mare is a vast subject area and it is beyond the scope of this book to cover it comprehensively. The following sections will confine themselves to the assessment criteria and techniques that may be used in practice when considering AI.

8.2.1. History of the mare

The reproductive and general history of the mare should be investigated (Shideler, 1993a). Details on her past breeding performance should be noted, such as oestrous cycle frequency, length of oestrous cycle and period of oestrus, normal oestrous behaviour and signs of oestrus, synchrony of oestrus and ovulation, and incidences of dystocia and uterine infections. The most common cause of problems in mares presented unsuccessfully on numerous occasions for AI is persistent, unidentified or inappropriately treated uterine infection (McKinnon and Voss, 1993). The mare's general health should also be considered, and if problems have occurred their treatment and her subsequent reproductive performance should be scrutinized. Her general history as regards health and injury may not have a direct effect on the success of AI, provided she is not in pain at covering. However, consideration must be given to whether the mare's condition will allow her to carry a foal comfortably to term. This may be questionable in cases of skeletal injury, severe laminitis, navicular or tendonitis (Woods, 1989). Many mares presented for AI, by nature of the technique and the costs, are valuable ex-performance animals. Athletically fit animals will be disadvantaged due not only to the direct adverse effect of athletic performance on reproductive function, but also to the inherent hazards of such a career – for example, increased exposure to infections, the possible use of drugs (in particular, anabolic steroids), transportation and injury or pain (Shoemaker et al., 1989).

8.2.2. Temperament

The temperament of a mare must not be overlooked. Although there is no physical reason why mares with a poor temperament should not conceive via AI, an uncooperative and highly nervous mare makes the whole procedure hazardous for all concerned and introduces considerable stress. Unpredictable, aggressive or masculine behaviour in the mare may also be indicative of hormonal abnormalities such as granulosa cell tumours (Hinrichs and Hunt, 1990). It is not uncommon practice for such aggressive mares to be presented for AI to avoid the potential risk of damage to valuable stallions. This does not alleviate the problems associated with elevated stress levels nor the possibility that such temperament will be passed on to subsequent generations.

8.2.3. General condition

The general condition of a mare should be considered. There is conflicting evidence over whether body condition has a direct effect on fertilization rates (Van Niekerk and Van Hearden, 1972; Voss and Pickett, 1973; Kubiak *et al.*, 1987; Morris *et al.*, 1987). Nevertheless, it is generally believed that a mare in a good, fit body condition (that is, condition score 3 on a scale of 0–5), will be more likely to breed successfully than one in an over-thin or over-fat condition. As with natural service, considerable thought must be given to the suitability of inseminating mares with conditions such as umbilical hernias, chronic obstructive pulmonary disease (COPD), vaginal prolapse and cardiac disease (Turner and McIlwraith, 1982; Shideler, 1993a,b).

8.2.4. Age

Age may be an additional consideration, but within the range of 5–12 years no significant effect of age on fertility is observed. Mares either side of these limits may show a reduction in fertility to AI, especially older mares: they may have a build-up of infections, scar tissue, adhesions, etc. which may be the real cause of the reduction in fertility rather than age *per se* (Pascoe, 1979; Katila *et al.*, 1996).

8.2.5. Competence of the reproductive tract

Ideally the competence of the reproductive tract should also be assessed. This may be done by both external and internal examination.

External examination

When assessing the competence of the reproductive tract externally, the major area for consideration is the perineum. Poor external conformation is often indicative of additional internal problems and may have severe implications for the mare's ability to conceive (Easley, 1993). For ease of description, the mare's tract may be considered to include the outer protective structures and the inner functional (as far as conception and embryo development are concerned) structures. The outer protective structures, including the vulva, vagina and cervix (Fig. 8.1), form three protective seals.

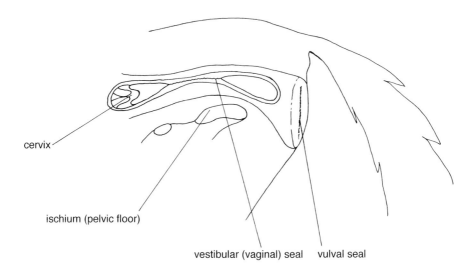

Fig. 8.1. The three seals providing protection for the inner reproductive tract of the mare.

The outer vulval seal is formed by the vulva constrictor muscle running down either side of the vulva lips. This forms the dorsal commissure at the upper junction of the labia lips, and the ventral commissure at the lower junction. In the ideally conformed mare these lips should be full, meet evenly along their length and form an efficient barrier against environmental contamination. It is clear that deficiencies in this seal will affect the protection of the reproductive tract. The major reasons for incompetence in the vulva seal are inherited malconformation of the vulva and perineal area, a vulva angle of greater than 10° from the vertical, or problems due to old injuries, lacerations, scar tissue, etc. One of the major causes of vulva sloping is inappropriate pelvic conformation.

The second seal of the tract is the vestibular or vaginal seal that is formed by the apposition of the walls of the posterior vagina as it sits on the floor of the pelvic girdle. The hymen is an added protection, if still present. A low pelvic floor predisposes to gaping of the vestibular seal and a sloping of the vulva. Mares with such conformation commonly suffer from the condition of pneumovagina – the passing of air, and with it bacteria, in and out of the vagina, so significantly increasing the bacterial challenge to the last seal, the cervix (Fig. 8.2). Figure 8.3 illustrates the various scenarios evident when perineal and pelvic conformation are inappropriate.

The ideal height for the pelvic floor is such that 80% of the vulva lies below the floor of the pelvic girdle and 20% above it (Fig. 8.4). As this proportion changes, the angle of the vulva slopes and allows faeces to collect in the ventral part of the vulva. This, along with the normally accompanying condition of pneumovagina, results in significant challenge to the internal reproductive system, invariably resulting in uterine infections.

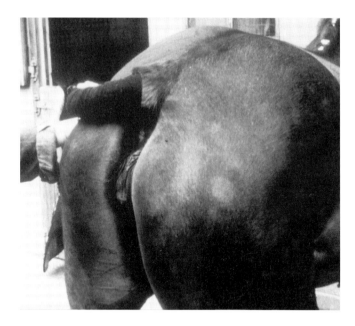

Fig. 8.2. A mare with poor perineal conformation. Note the distinct sloping of the vulva and the ease with which faeces could collect on the vulval lips.

Based on this conformation, mares may be classified as one of three types, and allocated a Caslick index, from which it is possible to predict the likelihood that the mare will suffer from uterine endometritis (Pascoe, 1979) and so not be a suitable candidate for AI (Fig. 8.5). Using such criteria, it is evident that Type II and III mares may not be expected to conceive as successfully as Type I. However, a Caslick operation (the suturing together of the vulva lips: Fig. 8.6) (Caslick, 1937) or a Pouret operation (realignment of the anus and vulva) (Pouret, 1982) may be performed to improve the competence of the mare's natural vulva seal.

The presentation for AI of mares that have had vulval suturing raises three areas of concern. Firstly, the necessity for the operation is a strong reason to suspect that the mare may have a history of uterine infections. Secondly, there is a limit to the number of times that repeated episiotomies and Caslick resuturing can be carried out; both are required at parturition as well as at natural service. The use of AI halves the number of times these operations need to be performed, as a mare can usually be inseminated through the hole left between the vulva lips for urination, but it does not negate the need for these procedures at parturition. Finally, it is increasingly evident that such poor conformation is heritable, and its prevalence in certain horse populations has significantly increased in the last few decades. All these factors must be considered when selecting mares as suitable candidates for AI.

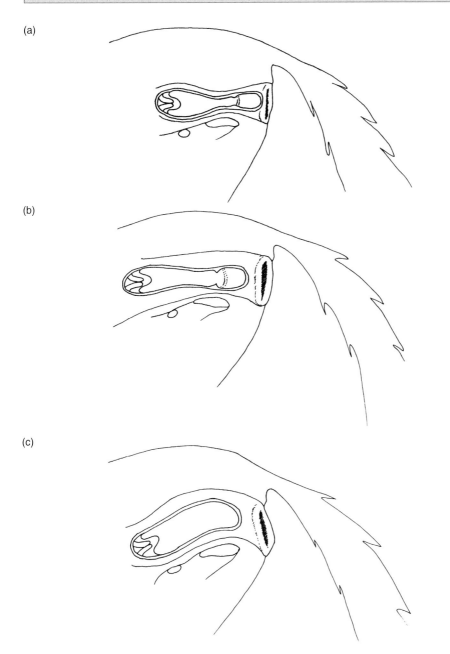

Fig. 8.3. The effect of perineal and pelvic conformation on the competence of the vulva, vestibular (vaginal) and cervical seals in the mare. (a) A low ischium (pelvic floor) results in an incompetent vestibular seal. In this case the vulval seal is still competent; therefore infection risk is limited. (b) A low ischium results in an incompetent vestibular seal. In this case the vulval seal is also incompetent; therefore infection risk is increased. (c) An incompetent vestibular and vulval seal plus a sloping perineal area results in a significant infection danger, especially from faecal contamination.

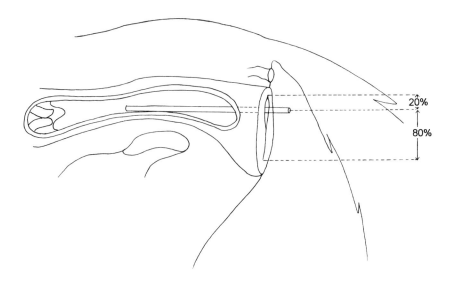

Fig. 8.4. Ideal height of the pelvic floor in relation to the proportion of the vulva that lies above and below.

Internal examination

Internal examination of the reproductive tract may be taken to some considerable depth if required. However, when considering selecting mares for AI it is very unlikely that use of all the techniques for such detailed examination is justified. The more standard assessment procedures include rectal palpation, ultrasonography and bacterial culturing; further more detailed examinations may include endoscopy and uterine biopsy.

RECTAL PALPATION

Rectal palpation allows a tactile impression to be gained of the inner, less accessible parts of the reproductive tract and is a relatively efficient means of identifying structural abnormalities. The procedure involves inserting a well-lubricated arm into the rectum of the mare. The rectal wall, being relatively thin, allows the practitioner to feel through it for the reproductive tract that lies below (Greenhof and Kenney, 1975; Shideler, 1993c). Rectal palpation allows the tone, size and texture of the uterine body, uterine horn, fallopian tubes and ovaries to be assessed, and can be used to ascertain the stage of the mare's cycle, whether she is cycling correctly and the correct time for insemination. Abnormalities such as ovarian and uterine cysts, tumours and neoplasms, stretched broad ligaments, uterine pyrometra, saculations, adhesions, lacerations, scars and delayed involution may be diagnosed via rectal palpation. The presence of any of these conditions would reduce the mare's suitability as a broodmare and may preclude her from use on an AI programme.

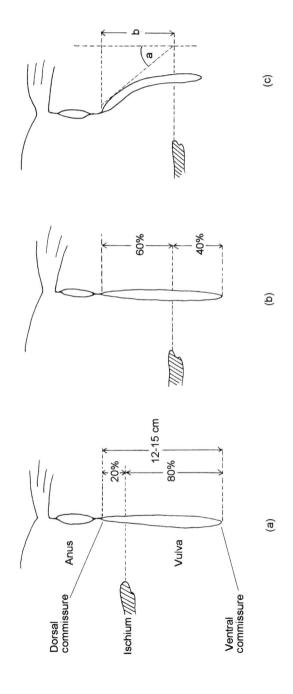

Fig. 8.5. Lateral view of the anatomical arrangement of the anus, pelvic floor and vulva in three types of mare to illustrate: (a) good conformation; (b) poor conformation; (c) very poor conformation (Caslick operation is required). The likelihood of requiring a Caslick operation is indicated by the Caslick index, which is calculated as angle a × distance b. Type I mare: Caslick index < 50 – Caslick operation is not required. Type II mare: Caslick index 50–150 – Caslick operation may be required later in life. Type III mare: Caslick index > 150 – Caslick operation is required immediately.

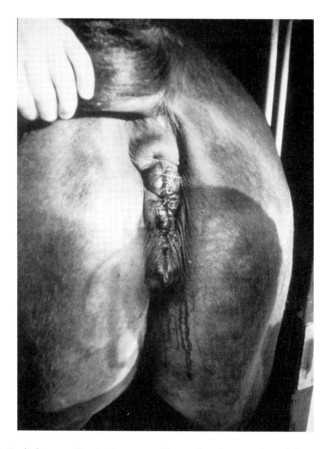

Fig. 8.6. A Caslick operation in the mare, illustrating the suturing of the upper vulva.

ULTRASONOGRAPHY

Ultrasonography offers a visual impression of the upper reproductive tract (Carnevale, 1998; Sertich, 1998; Turner, 1998). The principle on which it works is the deflection or absorption of high-frequency sound waves. The proportion of the sound waves reflected back and picked up by the transducer is recorded as an image on a visual display unit. Solid structures appear white and the fluid areas are black. For the evaluation of the reproductive tract in the mare, the transducer is placed in the rectum and angled ventrally so that the sound waves are emitted towards the reproductive tract (Ginther and Pierson, 1983; McKinnon and Carnevale, 1993; Le Blanc, 1995). The resultant image can be used to identify abnormalities similar to those assessed by rectal palpation, but particularly the presence of intrauterine fluid, cysts, air, debris, neoplasms, etc. (McKinnon *et. al.*, 1987). The technique can also be used to assess the size and shape of the uterus and the ovaries and so can be used to determine whether the mare is cycling and if she is at an appropriate stage to inseminate (Adams *et al.*, 1987; Perkins, 1996).

BACTERIAL CULTURING

The mare's reproductive tract is highly susceptible to infections, the major result of which is endometritis (Ricketts and Mackintosh, 1987). Endometritis has for many years been known to be a prime cause of infertility (Dimmock and Snyder, 1923). Complications arising from endometritis, such as exudate, adhesions, uterine endometrial degeneration and necrosis, may be identified by using rectal palpation, ultrasonography or endoscopy. However, endometritis has to be severe and chronic to result in such complications. Identification of the problem prior to its development to such a stage is obviously to be preferred, and allows confirmation of low-grade infection, which can itself result in depressed fertility. The most common causative bacterial agents are *Klebsiella pneumoniae*, *Pseudomonas aeroginosa*, *Haemophillus equigenitalis*, *Streptococcus zooepidemicus*, *Escherichia coli* and *Staphylococcus*.

Bacterial infections may be identified by swabbing of the reproductive tract. It is normal and recommended practice (Horse Race Betting Levy Board, 1997) that swabs be taken of the uterus, cervix, clitoris and urethra opening. Uterine swabbing should be carried out through an open cervix during oestrus, using a guarded swab to prevent contamination *en route* (Greenhof and Kenney, 1975). Other swabs may be taken throughout the mare's cycle (Ricketts, 1981). The resultant swabs are plated out and incubated under varying conditions, normally aerobic and microphillic, in order to identify any bacteria that might be present (Ricketts *et al.*, 1993). Fungal infections may be identified by similar means. The use of swabbing is a widespread and often compulsory practice. With some breed societies it has been successfully used to eradicate specific causes of infection in many areas worldwide. In particular, within the UK, the Horse Race Betting Levy Board publishes annual codes of practice which are used worldwide and have resulted in a near eradication of *Haemophillus equigenitalis*, the causative agent for contagious equine metritis (CEM) from the UK, as well as significantly reducing the incidence of venereal diseases caused by *Klebsiella pneumoniae* and *Pseudomonas aeruginosa* (Horse Race Betting Levy Board, 1997).

Uterine aspirations and washings may also be collected, especially if purulent material and fluid are present within the uterine lumen. Identification of bacteria is by culturing (Freeman and Johnston, 1987).

A positive identification of infection would preclude the mare from natural service with many breed societies. The use of AI overcomes the risk of disease transmission but does not overcome the hurdles that an infected uterus presents to a developing conceptus. The presence of infection often precludes the use of AI, as pregnancy rates are significantly reduced.

ENDOSCOPY

Endoscopy is a far more invasive technique than rectal palpation or ultrasonography, and has only limited applicability in general stud management for AI. Endoscopy involves the insertion of a movable light source and camera on the end of a series of optic fibres, termed an endoscope, into the uterus of a mare per vagina. It enables an illuminated image of the internal structures of the reproductive tract to be transmitted back up the optical fibres for display on a visual display unit (Threlfall and Carleton, 1996). The endoscope is designed to allow the tip to be rotated within the uterus, so that all internal structures may be viewed. This visual image will allow all the abnormalities identified with rectal palpation and ultrasound to be seen as a real image and, therefore, aid diagnosis. It also allows the colours of the internal structures to be viewed and the presence of uterine endometrial erosion and cloudy mucus or inflammation to be easily identified (Mather *et al.*, 1979; Le Blanc, 1993).

The use of the endoscope enables a biopsy sample of the uterine endometrium to be taken. This allows the histology of the uterine wall to be examined in detail (Ricketts, 1975; Doig and Waelchii, 1993).

8.3. Preparation of the Mare for Insemination

Once the intended mare has been proved appropriate for AI, her preparation must be geared towards achieving maximum conception rates. The correct timing of insemination relative to ovulation is of paramount importance. Insemination may be carried out in naturally cycling mares that have been identified as being in oestrus by observation, by teasing with a stallion or by the use of ultrasonography and rectal palpation. However, the period of oestrus is rather extended and ill-defined in the mare (Ginther *et al.*, 1972; Faulkner and Pineda, 1980). Synchronization of insemination and ovulation is difficult, and several inseminations throughout the oestrous period may be required. In such circumstances a mare would be inseminated at the same time as she would be covered naturally – that is, within 24–48 h of a 3 cm diameter follicle being identified – and reinseminated until she is no longer in oestrus. In most situations it is more convenient to manipulate the oestrous cycle of the mare, as this allows the arrival of the semen to be synchronized with the most opportune time for insemination and the use of fixed-time AI.

8.3.1. Manipulation of the oestrous cycle

It is beyond the scope of this book to provide a comprehensive review of the methods available to time oestrus and ovulation. Some of the methods have been mentioned in section 5.6.1, and other texts are available (Meyers, 1997). The following section will confine itself to further discussion of the methods normally recommended for use.

The natural oestrous cycle

The events within and control of the mare's oestrous cycle are a highly complex interrelated series of events, the explanation of which is beyond the scope of this book. Figure 8.7 illustrates the major hormonal events that occur within the normal 21-day cycle. There are many good reviews describing the hormonal changes and control of the oestrous cycle in the mare (Hafez, 1980; Rossdale and Ricketts, 1980; Daels and Hughes, 1993; Davies-Morel, 1993; Carleton, 1995).

The use of prostaglandins

Prostaglandin $F_{2\alpha}$, or one of its analogues, provides a successful means of timing oestrus and ovulation in the mare. Prostaglandin $F_{2\alpha}$ ($PGF_{2\alpha}$) is luteolytic in nature; that is, it destroys the corpus luteum and terminates the luteal phase of the cycle. In the natural cycle, in the absence of a pregnancy, prostaglandin is produced at a set time (12–14 days) post ovulation. As such it both marks and causes the termination of the luteal phase and the commencement of the endogenous hormone changes associated with oestrus and ovulation (Allen and Rowson, 1973; Loy et al., 1979; Neely et al., 1979; Cooper, 1981; Savage and Liptrap, 1987). Administration of exogenous prostaglandins, provided it is within certain time limits within the luteal phase, allows its termination to be controlled, and with it the timing of oestrus and ovulation.

Prostaglandins may be administered directly into the uterus or systemically. They may be used in their natural form (prostin $F_{2\alpha}$ and dinoprostromethamine

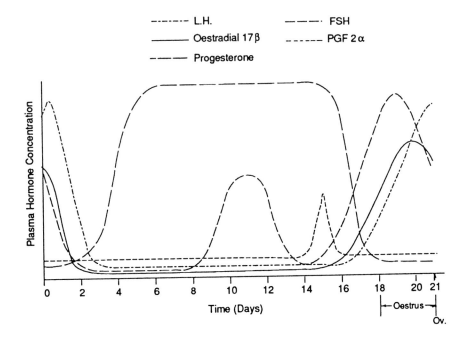

Fig. 8.7. A summary of the major plasma hormone concentration changes during the oestrous cycle of the non-pregnant mare.

5–10 mg i.m.) or as an analogue (α-prostol, 3 mg i.m.; fluprostenol, 0.25 µg i.m.; prostalene, 2 mg s.c.; or frenprostalene, 250 mg i.m.). The dose rates given are those appropriate for an average 400–500 kg mare (Bristol, 1987; Le Blanc, 1995).

The success of prostaglandin in timing oestrus in the mare is variable and depends upon the stage of the cycle. The corpus luteum of most mares is refractory to prostaglandin treatment prior to day 5 (Douglas and Ginther, 1972; Loy *et al.*, 1979). A good response is normally obtained when treatment is given between days 6 and 9 (Loy *et al.*, 1979). To be successful, the treatment must not only terminate the luteal phase but also induce ovulation. Considerable variation exists between the time of prostaglandin treatment and ovulation: a range of 24 h to 10 days is reported (Loy *et al.*, 1979). The time interval is determined by the stage of follicular development at treatment. Follicles at 4 cm in diameter, or greater, ovulate on average within 6 days, though again considerable variation is reported. If the follicle ovulates within 72 h it is often accompanied by an abbreviated oestrus or no oestrus at all. Occasionally, when a large follicle is present, prostaglandin treatment results in the regression of that follicle and the development and subsequent ovulation of another follicle; hence there is a longer time interval between treatment and ovulation (longer than 8 days) (Loy *et al.*, 1979). The most consistent results are obtained when treating mares earlier in the cycle, with small follicles (less than 4 cm in diameter), as less variation exists and the interval to ovulation is on average 6 days (Table 8.1).

As well as terminating the luteal phase, fenprostalene (250 mg), a long-acting prostaglandin, may be used in the mare during early oestrus (within 60 h of commencement) to hasten and synchronize ovulation. Ovulation is reported to occur in 81% of mares within 48 h of treatment, compared with 31% in untreated controls (Savage and Liptrap, 1987). Such treatment can, therefore, be used to improve the success of fixed-time AI carried out at 48 h after prostaglandin treatment.

The previously described use of prostaglandin relies upon a single injection, the major disadvantage of which is that, in order to optimize success rates, the stage of the mare's oestrous cycle must be known. In smaller

Table 8.1. Timing of oestrus and ovulation in the mare, using a single injection of prostaglandin, with suggested timing of AI.

Time	Drug to be administered/event
Day 0	Oestrus
Day 7	Prostaglandin
Day 9	Oestrus commences
Day 11	Ovulation may occur – AI
Day 13	AI – continue every 48 h until oestrus ceases or ovulation has occurred

(Note that considerable variation in an individual mare's response may be observed.)

intensive studs, where individual mares are monitored, this may present no problems. However, if AI is to be used on large groups of mares, kept in herd situations, or on mares whose stage of the cycle is unknown, a double injection of prostaglandin is required (Hyland and Bristol, 1979). The two prostaglandin injections need to be administered 14–18 days apart (Table 8.2).

The timing of the onset of oestrus with such treatment is quite successful: it has been reported that 60% of mares commence oestrus within 4 days of the second injection and 92% show oestrus within 6 days (Hyland and Bristol, 1979; Voss et al., 1979; Squires et al., 1981c; Squires, 1993a,b). Nevertheless, the synchrony and timing of ovulation is still very variable. Ovulation may occur anywhere between 2 and 12 days after the second injection. Although fixed-time AI at 4 days post second injection followed by three more inseminations at 48 h intervals results in normal conception rates, the multiple inseminations significantly increase costs (Bristol, 1993).

Further refinement of this protocol is the use of 1500–3700 IU of human chorionic gonadotrophin (hCG), a human placental gonadotrophin with properties similar to those of luteinizing (LH) and follicle-stimulating hormone (FSH). As such it enhances and supplements the natural release of gonadotrophins, which drive follicular development and, more specifically, ovulation. Its use is advocated to hasten ovulation and reduce the duration of oestrus, hence allowing just two inseminations to result in normal conception rates (Table 8.3) (Voss et al., 1975; Harrison et al., 1991).

Several timings for the injection of hCG have been advocated; most of them are between 4 and 6 days post second prostaglandin injection (Palmer and Jousett, 1975; Douglas and Ginther, 1972; Hyland and Bristol, 1979; Voss et al., 1979; Bristol, 1981; Squires et al., 1981c). Palmer (1976, 1979) reported that its use on day 6 post $PGF_{2\alpha}$ injection did help to alleviate the problem of variability in synchrony between ovulation and oestrus. Palmer and Jousett (1975) reported that 75.8% of mares ovulated within 72 h of an hCG injection which was given 6 days post second prostaglandin injection. Yurdaydin et al. (1993) achieved similar success using hCG on day 5 post prostaglandin, reporting that 80% of mares started oestrus within 24–36 h, and ovulated 5–6.5 days post hCG administration. When used on day 8,

Table 8.2. Timing of oestrus and ovulation in the mare, using two injections of prostaglandin, with suggested timing of AI.

Time	Drug to be administered/event
Day 0	Prostaglandin
Day 16	Prostaglandin
Day 20	Oestrus commences – AI
Day 22	Ovulation may occur – AI
Day 24	AI – continue every 48 h until oestrus ceases or ovulation has occurred

(Note that considerable variation in an individual mare's response may be observed.)

Table 8.3. Timing of oestrus and ovulation in the mare, using two injections of prostaglandin and a single injection of hCG, with suggested timings for AI.

Time	Drug to be administered/event
Day 0	Prostaglandin
Day 15	Prostaglandin
Day 19	Oestrus commences
Day 20	hCG
Day 21	Ovulation may occur – AI
Day 23	AI

(Note that considerable variation in an individual mare's response may be observed.)

oestrus synchronization rates of 90% post $PGF_{2\alpha}$ injection have been reported (Holtan *et al.*, 1977). Other work has demonstrated a more variable reaction or no significant improvement with the use of hCG (Holtan *et al.*, 1977; Squires *et al.*, 1981b). It has been advocated that hCG be used twice, on day 7 (7 days after the first $PGF_{2\alpha}$) and on day 21 (7 days after the second $PGF_{2\alpha}$ injection) (Table 8.4). The aim of this is to encourage the development of competent corpus lutea from the first prostaglandin injection, which would then react with less variation to the second prostaglandin injection. This regime is reported to result in up to 95% of mares ovulating on either day 22 or 23 (Allen *et al.*, 1974; Palmer and Jousett, 1975; Voss, 1993).

It is apparent from this work that the reaction of the follicle to hCG depends upon the stage of the breeding season. During the middle of the season, many mares may spontaneously ovulate before hCG is administered on day 6 post prostaglandin. It is advocated that, during the height of the breeding season, the interval from the second prostaglandin injection to hCG is reduced to less than 6 days so that a larger proportion of follicles react and ovulate within 48 h.

Table 8.4. Timing of oestrus and ovulation in the mare, using two injections of prostaglandin and two injections of hCG, with suggested timings for AI.

Time	Drug to be administered/event
Day 0	Prostaglandin
Day 7	hCG
Day 14	Prostaglandin
Day 18	Oestrus commences
Day 21	hCG
Day 22	Ovulation may occur – AI
Day 24	AI

(Note that considerable variation in an individual mare's response may be observed.)

Though reasonably successful in inducing ovulation, hCG has a major disadvantage. With repeated administration mares become refractory to hCG because of the development of antibodies (Roger *et al.*, 1979; Wilson *et al.*, 1990). Gonadotrophin-releasing hormone (GnRH) and its analogues have been advocated for use in its place. GnRH acts to stimulate the natural release of LH and FSH from the anterior pituitary. As such, its administration as a series of multiple injections (four at 12 h intervals) or via subcutaneous implants has been demonstrated to advance significantly the onset of ovulation in mares with follicles of greater than 3 cm in diameter (Table 8.5).

Successes rates of 88–100% of mares ovulating within 48 h of treatment (with 1.5–2.25 mg of deslorelin) have been reported (Johnson, 1986c; Harrison *et al.*, 1991; Meinert *et al.*, 1993; Jochle and Trigg, 1994; Mumford *et al.*, 1995). It has been suggested that GnRH may be more successful that hCG in inducing ovulation in larger, thicker-walled follicles. Also, GnRH does not have the disadvantage of inducing refractoriness of response due to antibody formation (Mumford *et al.*, 1995). However, much of the work to date on using GnRH has been in mares during their natural oestrous period, rather than with a synchronized oestrus regime. The limited work on using GnRH with prostaglandin to time oestrus and ovulation has suggested that there is no significant change in the timing of ovulation as a result of treatment, compared with the use of prostaglandin alone (Voss *et al.*, 1979; Booth *et al.*, 1980; Squires *et al.*, 1981b). Therefore, though the regime suggested in Table 8.5 would be feasible, it has yet to be proved that it will significantly improve the variation in the timing of oestrus and ovulation.

Prostaglandin does have some side effects, largely relating to its ability to activate smooth muscle contractions. Its use may be linked to increased gastrointestinal activity (manifest as diarrhoea), sweating and possibly slight caudal ataxia (Le Blanc, 1995). These side effects vary with the analogue used and with the individual mare. Provided the recommended dose rate is not exceeded, they are not serious.

Table 8.5. Timing of oestrus and ovulation in the mare, using two injections of prostaglandin and GnRH implants, with suggested timings for AI.

Time	Drug to be administered/event
Day 0	Prostaglandin
Day 15	Prostaglandin
Day 19	Oestrus commences
Day 21	GnRH implant
Day 22	Ovulation may occur – AI
Day 24	AI

(Note that considerable variation in an individual mare's response may be observed.)

The use of progesterone

Progesterone supplementation and subsequent withdrawal may also be used to time oestrus and ovulation. The use of progesterone, or one of its analogues, works on the principle of imitating the mare's natural dioestrous or luteal phase. This is achieved by mimicking the natural progesterone production during the luteal phase, by the administration of exogenous progestagens. Termination of this induced luteal state, achieved by cessation of treatment, acts like the end of the natural luteal phase and so induces the changes in the mare's endogenous hormones that are a prerequisite for oestrus and accompanying ovulation.

Progesterone may be used in its natural form, suspended in oil or in propylene glycol, or as progestagens – for example, altrenogest (allyl trenbolone). It may be administered orally [0.044 mg of altrenogest (Regumate) kg^{-1} body weight day^{-1}], intramuscularly (100–300 mg of progesterone day^{-1} suspended in oil, or 50–100 mg of progesterone $week^{-1}$ suspended in propylene glycol), or via intravaginal sponges (impregnated with 0.5–1.0 g of altrenogest and inserted for 20 days) (Loy and Swann, 1966; Allen, 1977; Hughes and Stabenfeldt, 1977; Palmer, 1979; Draincourt and Palmer, 1982; Squires, 1993a,b). Other workers have investigated, or are currently investigating, the use of PRIDs (progesterone-releasing intervaginal devices) in mares, along the lines of those available for cattle (Rutten *et al.*, 1986; Newcombe and Wilson, 1997; S. Revell, Wales, 1997, personal communication) or CIDRs (controlled internal drug-releasing dispensers) containing 1.9 g of progesterone (Arbeiter *et al.*, 1994). Whatever method of administration is chosen, the treatment must be long enough (15–18 days) to ensure that any natural corpus luteum has time to regress prior to withdrawal of exogenous progesterone. The subsequent termination of progesterone treatment will, therefore, remove all progesterone domination from the system.

Within 2–3 days of progesterone supplementation a mare will normally cease all oestrous activity, which will remain suppressed until treatment is terminated (Loy and Swann, 1966). After 15 days of treatment, oestrous behaviour is apparent 3–7 days after the withdrawal of progesterone, with ovulation at 8–15 days post progesterone withdrawal (Table 8.6) (Van Niekerk *et al.*, 1973). Squires *et al.* (1983) reported similar results but indicated that the timing of ovulation was more advanced, at 5.4 days on average. Progesterone or progestagen analogue supplementation via intramuscular injection has been reported to result in inconsistent synchronization of ovulation (Loy and Swann, 1966; Squires *et al.*, 1979b). However, other workers report that conception rates following 15 days of progesterone treatment are comparable with those associated with naturally occurring oestrus (Van Niekerk *et al.*, 1973; Squires *et al.*, 1979b, 1983).

Traditionally, long periods of progesterone supplementation were used (up to 20 days) and, although oestrus was suppressed, ovulation was not necessarily suppressed. Hence the timing of ovulation was not that successful (Loy and Swann, 1966). As a result the use of hCG after the termination of progesterone supplementation was investigated, with the idea that it might

Table 8.6. Timing of oestrus and ovulation in the mare, using progesterone supplementation, with suggested timings for AI.

Time	Drug to be administered/event
Day 0–16	Progesterone supplementation (intervaginal sponges or PRID, i.m. injection, oral)
Day 19 onwards	Oestrus
Day 21 onwards	Ovulation may occur – AI
Day 23	AI – continue every 48 h until oestrus ceases or ovulation has occurred

(Note that considerable variation in an individual mare's response may be observed.)

act to tighten up the synchrony of oestrus and ovulation, as it appears to do when used with prostaglandins. However, only limited success was reported (Holtan et al., 1977; Palmer, 1976, 1979).

Shorter periods of progesterone supplementation are now being used. As the period of supplementation may not be long enough to ensure that the natural corpus luteum has regressed, then a combination of progesterone supplementation and prostaglandin treatment is used. This is discussed in the following section.

The use of progesterone does have some disadvantages, mainly in its association with a decrease in neutrophil production in response to a bacterial challenge, which may well be significant in mares with poor perineal conformation or a history of uterine infections.

Combination treatments

There are several combination treatments, some of which have already been mentioned. The two most commonly used and of current interest – progesterone with prostaglandin, and progesterone with oestradiol – are specifically discussed in the following sections.

PROGESTERONE AND PROSTAGLANDINS
Combination treatments of progesterone and prostaglandin are increasingly popular as such treatments often improve the timing of ovulation and may reduce the length of progesterone supplementation. Progesterone can again be administered by one of the several methods already discussed. It may be administered for long periods, up to 20 days. Administration of 0.5 g or 1.0 g of altrenogest via intervaginal sponges for 20 days with a $PGF_{2\alpha}$ injection on the day of sponge removal was reported to result in oestrus at 1.8 (± 0.5) days and 2.2 (± 0.5) days, respectively, post $PGF_{2\alpha}$ injection, and by ovulation at 3.0 (± 0.7) and 5.4 (± 1.5) days, respectively, post $PGF_{2\alpha}$ injection (Palmer, 1979; Draincourt and Palmer, 1982). Today administration of progesterone is normally for only 7–9 days, with prostaglandin injection given on the day progesterone treatment ceases. Using this protocol and again with intervaginal

sponges containing 0.5 g altrenogest, Palmer (1979) demonstrated that on average oestrus occurred earlier (3.8 days) than figures suggested for progesterone treatment alone. The timing of ovulation was very variable, at 8–15 days post prostaglandin injection. Later work by Palmer *et al.* (1985), using the same sponges but inserted for 7 days, with 250 µg prostaglandin at sponge withdrawal, suggested better synchrony of ovulation, at 10.1–14.0 days post sponge removal (Table 8.7).

If the synchronization response to such treatment is assessed with respect to different times of the year, it is apparent that in April the average time from sponge removal to ovulation was 14 days but, during the period May to September synchrony was very good, with ovulation occurring between 10 and 10.7 days. This supported previous work that also indicated a seasonal effect in the response of mares to progesterone and prostaglandin treatment (Draincourt and Palmer, 1982). It is evident that, yet again, considerable variation in response is observed and that the variation may be greater with shorter-term progesterone supplementation during the early breeding season (Hughes and Loy, 1978; Palmer, 1979).

A further alternative is to use hCG with the aim of encouraging ovulation. It has been used on day 6, following 8 days of progesterone supplementation with a prostaglandin injection on the day of progesterone withdrawal (Table 8.8). This was reported to result in ovulation rates of 52.3% and 75% within 48 h and 96 h, respectively (Palmer, 1979), which is still quite variable.

PROGESTERONE AND OESTRADIOL

This combination treatment is increasingly popular. Both hormones may be administered daily via intramuscular injection for 10 days (R. Pryce-Jones and C. McMurchie, Wales, 1997, personal communication). Dose rates in the order of 150 mg of progestagen and 10 mg of oestradiol per day, followed, as in the previous protocols, with an injection of $PGF_{2\alpha}$ at the end of the treatment (Table 8.9), have proved successful, with a reported 81.3% of treated mares ovulating 10–12 days post $PGF_{2\alpha}$ injection (Loy *et al.*, 1981). Normal pregnancy rates have been reported to AI after such treatment (Jasko *et al.*, 1993b). PRIDs

Table 8.7. Timing of oestrus and ovulation in the mare, using progesterone supplementation, plus prostaglandin, with suggested timings for AI.

Time	Drug to be administered/event
Day 0–8	Progesterone supplementation (intervaginal sponges or PRID, i.m. injection, oral)
Day 8	Prostaglandin
Day 12	Oestrus
Day 16 onwards	Ovulation may occur – AI
Day 18	AI – continue every 48 h until oestrus ceases or ovulation has occurred

(Note that considerable variation in an individual mare's response may be observed.)

Table 8.8. Timing of oestrus and ovulation in the mare, using progesterone supplementation, prostaglandin and hCG, with suggested timings for AI.

Time	Drug to be administered/event
Day 0–8	Progesterone supplementation (intervaginal sponges or PRID, i.m. injection, oral)
Day 8	Prostaglandin
Day 12	Oestrus
Day 14	hCG
Day 16	Ovulation may occur – AI
Day 18	AI – continue every 48 h until oestrus ceases or ovulation has occurred

(Note that considerable variation in an individual mare's response may be observed.)

containing 2.3 mg progesterone and 10 mg oestradiol (held within a gelatin capsule) inserted for 10 days have been used with limited success (Rutten *et al.*, 1986). Further refinement of PRIDs or the development of slow-release subcutaneous capsules may enhance the use of such combination treatments, removing the need for time-consuming daily injections.

It is evident that the use of oestrus manipulation does not allow precise and accurate timing of ovulation to be predicted in the mare. However, there is no evidence to suggest that the use of such treatment significantly affects conception rates to AI (Palmer *et al.*, 1985), and some techniques are significantly better than others at allowing the time of ovulation to be predicted. As such, one of the major advantages of AI, that of fixed-time insemination, may be realized but with lower success than might have been hoped for judging from the success obtained with cows. Many of these techniques really only reduce the random spread of ovulations within a population, rather than allowing the exact timing to be predicted. Even so, such an effect allows ovarian examination or teasing to be concentrated into a shorter period, thus reducing labour costs, etc. Manipulation of the oestrous cycle is regularly practised with the additional use of ultrasonography and rectal palpation, which can allow a more accurate and precise prediction of ovulation.

Table 8.9. Timing of oestrus and ovulation in the mare, using progesterone and oestradiol treatment followed by prostaglandin, with suggested timings for AI.

Time	Drug to be administered/event
Day 0–10	Progesterone and oestradiol treatment
Day 10	Prostaglandin
Day 20	Ovulation may occur – AI
Day 22	AI – continue every 48 h until oestrus ceases or ovulation has occurred

(Note that considerable variation in an individual mare's response may be observed.)

8.3.2. Veterinary examination

The two major techniques used to assess the ovarian activity of a mare, and so predict the timing of ovulation, are rectal palpation and ultrasonography. Both techniques are widely used within stud management for infertility investigations and pregnancy diagnosis as well as ovarian assessment. Some detail on these two techniques has already been given earlier on in this chapter and it is beyond the scope of this book to detail them any further. Several good reviews of the techniques and their use are available (Greenhof and Kenney, 1975; Ginther and Pierson, 1983, 1984a,b; Squires *et al.*, 1984; Adams *et al.*, 1987; McKinnon and Carnevale, 1993; Shideler, 1993c; Le Blanc, 1995; Threlfall and Carleton, 1996; Carnevale, 1998; Sertich, 1998; Turner, 1998). In using either of these two techniques, two major features are looked for on the mare's ovary: the presence of a follicle greater than 3 cm in diameter, or the occurrence of ovulation. These two parameters are the major milestones considered when calculating the most appropriate time for insemination. Details on the exact timing of insemination will be discussed later.

8.3.3. Physical preparation of the mare

Once the timing of insemination has been determined, any necessary swabs have been taken, all paperwork has been completed or is ready for completion and the semen is available, the mare should be prepared for insemination largely as for natural covering and according to the minimum contamination techniques (Kenney *et al.*, 1975). Ideally, she should be bridled and restrained in stocks. If stocks are not available she may be backed up into a stable doorway, though this gives less protection to the practitioner. The mare's perineal area should be thoroughly washed and rinsed to remove any antibacterial agents that may be spermicidal or act as an irritant of the genital area (Fig. 8.8). The area should then be dried with a clean, dry disposable towel. During washing, particular attention should be paid to the removal of all faecal contamination. The mare's tail should be bandaged and covered with a sterile wrap, and tied or held out of the way to prevent contamination (British Equine Veterinary Association, 1991). All equipment should be non-toxic and sterile and ideally disposable.

8.4. Preparation of Semen for Insemination

Any requirements for preparation of the semen are normally determined by its means of storage. Most semen is processed and stored in a form that can be used directly for insemination with minimum preparation or handling. The major consideration is normally the thawing of frozen semen.

Mare Insemination

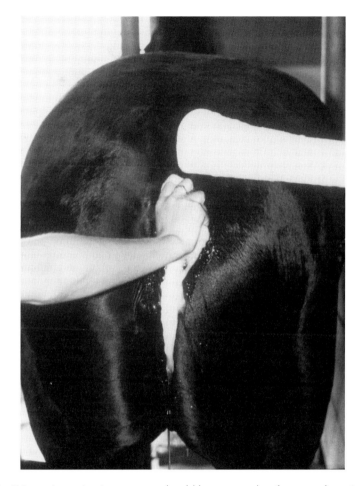

Fig. 8.8. Prior to insemination, a mare should be prepared as for natural service. This should include washing the perineal area. (Photograph by Julie Baumber and Victor Medina.)

8.4.1. Thawing or warming semen

Normally the establishment or laboratory providing the semen will also give recommendations regarding the thawing procedure required. This will largely depend upon its packaging. The thawing of frozen semen and the use of thawing extenders have already been discussed in some detail in section 7.5.6. In summary, frozen semen straws (the most common method of storage) (Figs 8.9 and 8.10) may be thawed in a variety of ways, either at a lower temperature for a longer period or at a higher temperature for a shorter period. For example, as a general guide, 0.5 ml straws may be thawed, usually in a water bath, at 5°C for 4 min, at 38°C for 30 s, or most rapidly at 75°C for 7 s

Fig. 8.9. Semen frozen in straws should be removed from the liquid nitrogen and thawed prior to insemination. (Photograph by Julie Baumber and Victor Medina.)

(Cochran *et al.*, 1983, 1984; Loomis *et al.*, 1984; Borg *et al.*, 1997). Larger straws do not thaw as effectively at high temperatures for short periods; therefore, thawing at 50°C for 40–55 s is recommended (Martin *et al.*, 1979). It must be remembered that if groups of straws are being thawed they need to be separated, otherwise differential thawing will occur; such straws are normally thawed at 37°C for 30 s.

Frozen semen is also available in ampoules or pellets, which can be thawed similarly. Occasionally, large doses of frozen semen are available in polypropylene bags or aluminium packages, which may be thawed by immersion in water at room temperature or by holding in the palm of the hand (Tischner, 1979; Muller, 1982, 1987; Amann and Pickett, 1984; Love *et al.*, 1989; British Equine Veterinary Association, 1991).

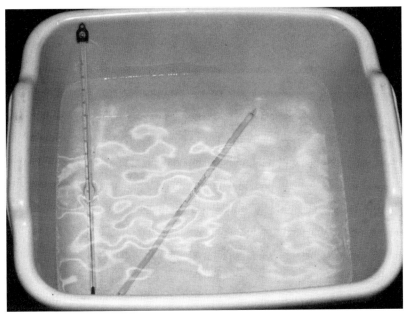

Fig. 8.10. Semen stored in straws can be conveniently thawed prior to insemination, by placing in a warm water bath. (Photograph by Julie Baumber and Victor Medina.)

Some protocols for thawing semen involve the inclusion of a thawing extender to aid the thawing process, as the extender is added warm. The extender may also increase the volume of inseminate and aid retention of spermatozoan viability. The addition of 25 ml HF-20 (see Table 7.8) or 20 ml of a sucrose milk solution (see Table 7.17) at 42°C as a thawing extender has been reported to be advantageous to semen stored in straws and pellets or vials, respectively (Nishikawa et al., 1972a, 1976; Nishikawa, 1975; Piao and Wang, 1988).

Once the sample has thawed it may be assessed prior to (or sometimes immediately after) insemination. As discussed in some detail in Chapter 7, freezing is known to have an adverse affect on fertility rates: some stallions' semen suffers considerably more than others. In addition, there is no simple test that will give an accurate indication of spermatozoan viability post thaw. Motility, which in fresh semen is the parameter with the best, but not ideal, correlation to fertility, is an even less accurate predictor in post-thaw spermatozoa, but in the absence of any other parameter it is often used as a guide to viability. Due to the relative inaccuracy of such prediction in many circumstances, semen is not evaluated prior to insemination. It is assumed, rightly or wrongly, that if the semen was of good enough quality to allow it to be frozen in the first place it will be of adequate quality for insemination. This rather hit-and-miss method underlines the real need for some system of quality assurance based upon conception rates for stallions whose semen has been frozen and inseminated previously.

Chilled semen requires no warming prior to insemination and should remain chilled until immediate use. It will be warmed quickly by the body heat of the mare. Fresh semen is presented at body temperature straight from an incubator.

If the semen to be used is from a number of sources (that is, straws or ampoules, whether frozen or chilled), it should be thoroughly mixed immediately post thaw and prior to use. Regardless of the container in which the semen arrives, it is invariably drawn up into a clean, dry and sterile syringe, once the insemination pipette is in place within the mare's uterus. Drawing up from vials or ampoules is relatively easy, as is drawing up from thawed pelleted frozen semen. The use of straws for freezing makes the system slightly more complicated. The contents of the straws either can be drained directly into the pipette and expelled by means of an air-filled syringe; or they may be drained into the syringe, which is then attached to the pipette and emptied in the normal way; or the straw may be designed to be placed directly into the insemination pipette and the semen emptied by means of a long plunger running the length of the pipette (as used for cow insemination) (Curnow, 1993) (Figs 8.11 and 8.12).

8.4.2. Insemination dose

One of the major considerations is the dose of semen to be inseminated. The dilution of semen and to some extent the dose required for insemination have already been discussed in Chapter 7 when considering the storage of semen. If the number of spermatozoa required per insemination for different farm livestock is compared (Table 8.10), it is evident that significantly greater numbers of spermatozoa are required per insemination in the mare than in the cow. In the cow, insemination doses of $2.5-15 \times 10^6$ result in good fertility rates. In contrast, in the mare, insemination doses of $250-500 \times 10^6$ are normally used, and these may vary within the range of $100-500 \times 10^6$, depending upon the method of storage, the normal fertility of the stallion and the time of year (Cooper and Wert, 1975; Demick et al., 1976; Boyle, 1992). Doses of 100×10^6 progressively motile spermatozoa have been reported to produce acceptable pregnancy rates with stallions of good fertility (Pickett et al., 1975d; Demick et al., 1976). Householder et al. (1981) and Pace and Sullivan (1975) demonstrated pregnancy rates of 37% when doses of 50×10^6 spermatozoa were used, compared with 75% when doses of 500×10^6 were inseminated. Jasko et al. (1992c) demonstrated a significant decline in embryo recovery rates between mares inseminated with 250×10^6 (70% embryo recovery) and 25×10^6 (35.3% embryo recovery) spermatozoa. Some of this difference, though it is unlikely to be all of it, may be accounted for by the difference in volume inseminated (250×10^6 in 10 ml, 25×10^6 in 50 ml). Volkmann and Van Zyl (1987) showed a significant increase in pregnancy rate when doses of inseminate were increased from 175×10^6 (44%) to 249×10^6 (73%). Investigations into a

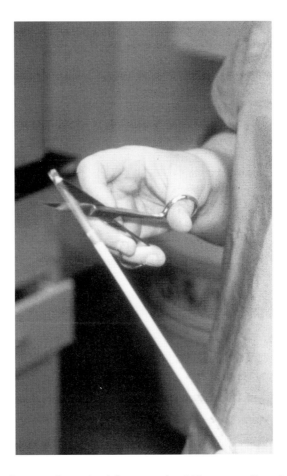

Fig. 8.11. After thawing, the ends of the straw should be cut to allow the semen to drain into a syringe ready for insemination. (Photograph by Julie Baumber and Victor Medina.)

narrower range of doses indicated that no significant difference exists in pregnancy rates between the insemination of 250×10^6 and 500×10^6 spermatozoa (Witte, 1989), between 150×10^6 and $200-250 \times 10^6$ (Piao and Wang, 1988) or between 150×10^6 and 300×10^6 (Vidament *et al.*, 1997). Based upon this, it is normal for doses of 250×10^6 to be inseminated when using fresh or chilled semen.

The insemination of frozen semen normally requires an increase in dose rate in order to compensate for spermatozoan mortality during the freezing and thawing process. Dose rates of 800×10^6 spermatozoa are used, which have been demonstrated to improve pregnancy rates with frozen semen (Samper, 1995b).

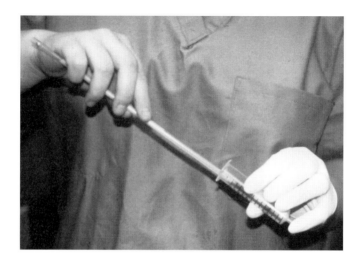

Fig. 8.12. A semen sample, either from a single straw or from a number of straws, may be drained directly into a syringe prior to insemination. This syringe is then attached to an insemination pipette for insemination. (Photograph by Julie Baumber and Victor Medina.)

8.4.3. Insemination volume

The volume of inseminate used appears to have less of an effect on pregnancy rates than the number of spermatozoa (Papa, 1989). The volumes of inseminate used may vary from 0.5 to 100 ml, though commonly they range

Table 8.10. A summary of the comparative insemination procedures used in farm livestock (Demick *et al.*, 1976; Garner, 1991; Boyle, 1992).

Animal	Method	Site of semen deposition	Insemination time	Oestrus detection	Sperm numbers ($\times 10^6$)	Insemination volume (ml)
Horse	Vaginal Rectovaginal	Uterus	Every 48 h from day 2 of oestrus until ovulation	Cycle history Observe signs Teasing	250–500	0.5–5
Cattle	Rectovaginal	Uterus	Onset of oestrus and at 9 h	Cycle history Observe signs Tail/bull markers	2.5–15	0.25–1.0
Sheep	Speculum and pipette (fresh)	Vagina	Onset of oestrus and 10–20 h	Cycle history Vasectomized marker ram	> 50	0.05–0.2
	Laparoscopy (frozen)	Uterus	Onset of oestrus and 10–20 h		200	0.05–0.2
Pig	Pipette	Cervix/uterus	Onset of oestrus and 10–30 h	Cycle history Observe signs Lordosis	> 50	50

from 10 to 30 ml for fresh semen, 30–60 ml for chilled semen and 0.5–5 ml for frozen semen (British Equine Veterinary Association, 1997). As regards pregnancy rates, it has been suggested that volumes in excess of 100 ml may be detrimental (Rowley *et al.*, 1990), but Bedford and Hinrichs (1994) indicated no significant differences in pregnancy rates between insemination volumes of 30 ml or 120 ml. It has also been reported that no effect of volume on fertility is evident within the range of 0.6–26 ml (Pickett *et al.*, 1987). Other work indicates that some decline in embryo recovery rates is seen when 50 ml is used rather than 10 ml (60% compared with 80%) (Jasko *et al.*, 1992c).

Once the desired dose is known, the insemination volume may be calculated as follows (PMS = progressively motile spermatozoa):

insemination volume (ml) = number of PMS required/number of PMS ml^{-1}

where PMS ml^{-1} = % PMS × number of sperm ml^{-1}.

8.5. Methods of Insemination

Once the mare and semen have been prepared there are two main methods of insemination, which involve the guiding of the insemination pipette through the cervix into the uterus via an arm placed either in the vagina (per vagina) or in the rectum (per rectum) of the mare. Whichever system is used, it is very important that all the equipment is warm, clean, dry and sterile. It is also important, throughout the whole process of preparing the semen and during insemination, that all equipment is maintained at 37°C to avoid cold shock. Semen should also be protected from exposure to air and sunlight (Coulter and Foote, 1977; Yates and Whitacre, 1988). Concern has recently been expressed over the contact of spermatozoa with rubber equipment, as discussed in Chapter 7; it is normally advised that such contact should be avoided, and in particular that syringes with plastic rather than rubber plungers are used.

8.5.1. Insemination per vagina

The inseminator's arm should be covered in a sterile insemination glove. The insemination pipette (Fig. 8.13) is then placed in the centre of the hand.

A small amount of lubricant may be placed on the knuckle of the hand, avoiding contact with the insemination pipette (Fig. 8.14). The amount of lubricant should be minimized, as evidence suggests that some kinds of lubricant may be detrimental to motility and hence viability (Fromann and Amann, 1983; Lampinen and Kattila, 1994). The inseminator then places the hand, plus pipette, into the vagina (Fig. 8.15) and proceeds up towards the cranial part of the vagina near the cervix. The opening of the cervix is located using the index finger, and the insemination pipette is then guided in through the cervix and into the uterus. The pipette should ideally be pushed into the uterus far enough to allow at least 2 cm to protrude into the lumen. This may

Fig. 8.13. In readiness for insemination, the filled syringe is attached to the end of the insemination pipette. (Photograph by Julie Baumber and Victor Medina.)

be ascertained if a small elastic band has been placed around the insemination pipette about 5 cm from the end. Estimating the size of the cervix to be 2–3 cm, then the pipette will be appropriately placed within the uterus when the elastic band can be felt against the cervix (Fig. 8.16).

8.5.2. Insemination per rectum

An alternative but less popular method of insemination is to guide the pipette per rectum. (This is the preferred method in cattle AI.) The inseminator

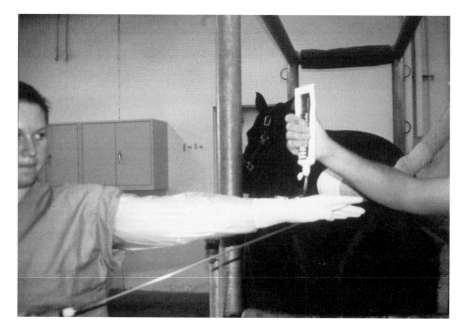

Fig. 8.14. Prior to insemination, a small amount of lubricant should be placed on the knuckle of the hand, which is positioned to cover the end of the insemination pipette, thus avoiding contamination. (Photograph by Julie Baumber and Victor Medina.)

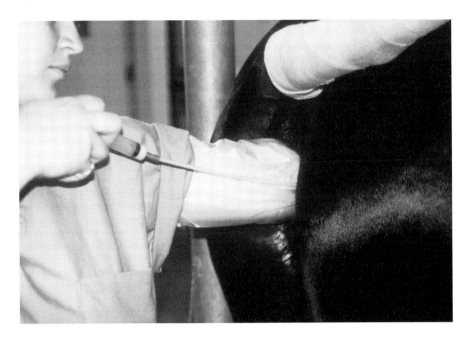

Fig. 8.15. Once the hand is lubricated it should be introduced slowly into the vagina along with the insemination pipette. when the pipette is in place, the plunger of the syringe should be slowly depressed to expel all the semen into the uterus. (Photograph by Julie Baumber and Victor Medina.)

places the gloved hand and arm into the rectum of the mare, as for rectal palpation. Once the inseminator has located the cervix by feeling through the rectum wall, the insemination pipette is placed into the mare's vagina and guided up and through the cervix by means of the hand within the rectum.

Both methods have their merits and disadvantages. Insemination per rectum reduces the risk of contamination of the reproductive tract as the arm is not introduced into the vagina and only a relatively small breach of the natural physical seals occurs, due to the small size of the insemination pipette. However, it is more difficult to locate and manipulate the cervix per rectum. It is largely for this reason that the preferred method of insemination in the mare is guiding the pipette per vagina. Some pipettes have a flexible tip, which allows direction into either of the uterine horns in the belief that deposition of the semen into the horn ipsilateral to the ovulating ovary will improve success rates.

Laparoscopic AI, though in theory possible in the mare, is not practised except for research, as the cervix and uterus are easily accessible via the vagina (J. Baumber, California, 1998, personal communication).

Once the pipette is in place, the syringe, pre-loaded with semen, is attached to the end of the pipette and the plunger is slowly depressed, pushing the semen out into the uterus. It is important that all the inseminate is deposited:

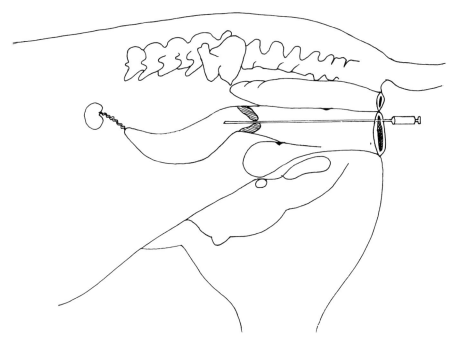

Fig. 8.16. Correct positioning of the insemination pipette for deposition of semen.

in the preparation of the pipette, it is advantageous to ensure that a small volume of air is included, which can be pushed out after the semen to ensure that the pipette is fully emptied. If inseminating per vagina it is suggested that manual closing of the cervix for a few seconds after insemination will help to reduce any loss of semen back through into the vagina (S. Revell, Wales, 1999, personal communication).

Once inseminated, the mare should be removed from the stocks and returned to her normal environment. Mares who have been inseminated require no different or special care than mares that have been covered naturally. Due to the often high value of such mares and the semen inseminated, close monitoring post insemination is often practised. An inseminated mare might be examined at 24 h post insemination to ensure that she has indeed ovulated. If she has not, and enough semen remains, she is often re-inseminated. A Caslick operation may be carried out if the mare's perineal conformation indicates that one is necessary. There is some evidence to suggest that the extent of neutrophil production in response to covering is related to the number of spermatozoa. Hence, monitoring of the inflammatory response in mares may be particularly beneficial after insemination with frozen semen (Kotilainen *et al.*, 1994; Katila, 1995).

8.6. Factors Affecting the Success of AI

Several factors affect the success of AI, including the method of storage, and the volume and dose rate of semen inseminated. Two other considerations have a significant effect on pregnancy rates: the timing and frequency, along with the total number of inseminations per oestrous cycle.

8.6.1. Timing and frequency of insemination

As with natural service, the longevity of the spermatozoa within the female tract and the duration of the viability of the ovum determine the optimum timing of insemination. Significant variation exists in spermatozoan longevity (Watson and Nikolakopoulos, 1996). It has been reported to vary from 24 h to 7 days, whereas ovum viability is much shorter (in the region of 8 h) (Hunter, 1990). Synchronization of the arrival of the spermatozoa at the site of fertilization and the timing of ovulation is critical if fertilization rates are to be maximized. Bearing this in mind, as with natural service, good success rates are normally achieved with insemination every 48 h while oestrus lasts, or until the mare has ovulated (Watson and Nikolakopoulos, 1996). However, due to the characteristically long oestrus of the mare this may involve numerous, potentially expensive, inseminations. It is, therefore, recommended that mares be monitored, by scanning or rectal palpation, from the first signs of oestrus or the expected time of oestrus. This will allow identification of a soft follicle of 3 cm or more in diameter. Such mares are likely to ovulate within the next 48 h and so immediate insemination is recommended, followed by another at 48 h. Pregnancy rates using this regime are reported to be good, provided the stallion is of good fertility. In many cases the mare will continue to be monitored to ascertain that ovulation has occurred. If ovulation has not occurred 48 h after the second insemination, it is recommended that another insemination is given immediately.

Acceptable fertilization rates are achieved with insemination (using stallions of good fertility) between 48 and 72 h prior to ovulation (Watson and Nikolakopoulos, 1996). Work by Woods *et al.* (1990) indicated that a single insemination at 24 h prior to ovulation gave the best pregnancy rates, at 88%, though pregnancy rates of 75% were reported with insemination at 72 h prior to ovulation. Insemination of mares within 6 h of ovulation gave comparable results and proved to be more successful than insemination 18–24 h post ovulation. Insemination at later than 30 h post ovulation resulted in no pregnancies (Woods *et al.*, 1990). Katila *et al.* (1996) obtained in general lower pregnancy rates, but supported the evidence that insemination at less than 48 h prior to ovulation resulted in the highest pregnancy rates. With a stallion of unknown fertility, it is normally recommended that insemination within the 24 h period preceding ovulation should be practised. Similarly, when using chilled semen (even from stallions of known good fertility), due to the potential adverse affect of chilling, it is again recommended that insemination is carried out in the 24 h prior to ovulation (British Equine Veterinary Association, 1997).

The timing of insemination when using frozen semen is more critical, as it is apparent that the longevity of frozen semen in the female tract is reduced. This may in part be accounted for by the occurrence of an acrosome-type reaction during freezing, which subsequently reduces the time that the spermatozoa requires to be in contact with the secretions of the female tract in order to undergo capacitation. *In vitro* studies estimate that the time for capacitation is 18–29 h: the time required *in vivo* is as yet unknown. It is likely that frozen semen needs to be inseminated closer to the time of ovulation than fresh and chilled semen, and it is normally advised that insemination should take place not sooner than 24 h, and ideally within the 12 h period, preceding ovulation and be repeated again after 24 h if ovulation has not been confirmed. Scanning at intervals of 12 or 24 h is used to ascertain the timing of ovulation. Some work still demonstrates good conception rates with frozen semen inseminated every other day during oestrus, as advised for fresh or chilled semen (Volkmann and Van Zyl, 1987). Such intensive monitoring and insemination regimes are potentially very time consuming, expensive and wasteful of precious semen, though it does ensure that semen is deposited during that critical 24 h period prior to ovulation.

The relative synchrony between ovulation and insemination can be improved by the use of hCG, GnRH or LH once a follicle with a 3 cm diameter has been identified. The use of hCG or GnRH as an agent to induce ovulation has been discussed in section 8.3.1. However, again, significant mare variation exists and there is the risk that mares may be inseminated post ovulation.

Work has been conducted recently into the use of post-ovulatory insemination. In theory, if insemination was successful post ovulation, then more accurate timing of insemination relative to ovulation could be achieved. Good conception rates have been reported for insemination up to 6 h post ovulation, and conception is reported to be possible with insemination up to 24 h post ovulation (Belling, 1984; Kloppe *et al.*, 1988; Woods *et al.*, 1990). Pregnancy rates of up to 33% have been reported for insemination up to 12 h post ovulation (Heiskanen *et al.*, 1993). Further work by Heiskanen *et al.* (1993, 1994a) indicated that better conception rates were achieved when mares were inseminated both before and after ovulation (100% conception) compared with insemination just before ovulation (58% conception). Koskinen *et al.* (1990) demonstrated that insemination up to 24 h post ovulation could be successful, but that embryo mortality rates might be increased. This supported work by Belling (1984) which also indicated a problem of embryo mortality with post-ovulation insemination. It has been suggested that the equine ovum is capable of being fertilized up to 18 h post ovulation, but increased senescence leads to an increased risk of embryo mortality (Koskinen *et al.*, 1990). Although fertilization may result from insemination up to 24 h post ovulation, it is apparent that the best time as regards pregnancy rates is within the first 18 h. Within this time frame, pregnancy rates of up to 100% are reported and embryo mortality is minimized (Katila *et al.*, 1988).

One particular area that may benefit most from post-ovulatory insemination is in the use of frozen semen. As discussed previously, frozen spermatozoa are likely to require less time for capacitation. They are, therefore, capable of fertilizing an ovum sooner than fresh or chilled semen, hence reducing the senility of the ovum at fertilization. It is suggested that there is no difference in the pregnancy rates in mares inseminated with frozen semen shortly before or after ovulation (British Equine Veterinary Association, 1997).

The aim of all insemination programmes should be a single insemination at the most opportune time; this minimizes the risk of uterine infections and inflammatory response and increases the efficiency of the use of semen. The exact timing of ovulation is the stumbling block. Single inseminations shortly before or after ovulation are reported to result in similar pregnancy rates (33% and 41%, respectively) (British Equine Veterinary Association, 1997) and so may prove to be the way ahead. However, until an accurate, cheap and effective method can be developed either to predict or to time ovulation precisely, most inseminations consist of a double dose inseminated at a 48 h interval. Increased frequencies and numbers of inseminations can be used but there is no reason why these should result in any better pregnancy rates than a single, appropriately timed insemination. Silva Filho *et al.* (1991b) demonstrated no advantage in inseminating every 24 h compared with every 48 h throughout oestrus post identification of a follicle 3 cm in diameter. Conception rates of 95% were reported for both.

As far as the effect of timing inseminations within the breeding season is concerned, Vidament *et al.* (1997) reported that conception rates are significantly increased ($P < 0.001$) if the mare is first inseminated before May 1 (58%) compared with mares inseminated later (37%).

8.7. Conception Rates to AI

Conception rates to AI have been discussed in some detail with relation to the means of storage of semen in Chapter 7. It is evident from this and the discussions in this chapter that so many variables can effect conception rates that to draw any conclusions would be meaningless. However, provided all the necessary criteria are met, there is no reason why conception rates to AI should not be as good as those to well-managed natural service.

8.7. Conclusion

Tables 8.10 and 8.11 summarize and note the success of comparable procedures involved in AI in a variety of farm livestock. Many similarities may be seen. As with cattle, the procedure in horses is relatively simple and uses a

Table 8.11. Some reported fertility rates in farm livestock after natural service and AI (Amann, 1984; Evans and Maxwell, 1987; Garner, 1991).

	Pregnancy rates (%)		
		Artificial insemination	
Animal	Natural mating	1 insemination fresh/cooled semen	1 insemination frozen semen
Horses	40–75	50–65	35–50
Cattle	60–75	50–65	50–65
Sheep	80–90	60–80	5–70
Pigs	85–95	65–90	62–64

largely non-invasive technique. The biggest problem as far as equine insemination is concerned is the variability in natural length of the oestrous period and relative synchrony of oestrus and ovulation. Even the manipulation of the oestrous cycle does not allow the precise timing of ovulation to be predicted. Insemination remains a relatively hit-and-miss affair with regards to its synchrony with ovulation, often necessitating detailed monitoring of ovarian activity and multiple, potentially wasteful inseminations. One of the most promising developments is the apparently good conception rates being obtained with insemination post ovulation. Such a system would not remove the need for detailed monitoring, but would allow the timing of insemination to be more accurate and so increase the chances of success with single dose inseminations.

9 Conclusion

It can be concluded that the use of equine AI has progressed significantly since the resurgence of interest in the technique in the mid 1970s. However, it still has some considerable way to go to match the sophistication of bovine, ovine and porcine AI. Its scientific development has met no greater challenges than were faced by researchers developing the technique for use in cattle; in fact, considerable benefit was derived from the cattle work, with much having a direct application to the horse. The major handicap to the development of techniques for use in equines has been the equine industry itself: many of the notable leading breed societies have been very reluctant to accept such technological advances. It is really only a matter of time before all equine breed societies are forced to accept AI and the technique will then become truly worldwide. Towards this end, the major pressure is likely to be an economic one, and this is already evident in the UK. As a result of the advent of free trade within the European Community, the UK now has to compete directly with other European countries. The importance of a 'level playing-field' is becoming increasingly evident, forcing the UK breed societies to fall in line with their European counterparts and permit the use of AI.

The future development of equine AI will present some interesting dilemmas or concerns. Many of these are ethical considerations and there is much concern over the ethical and welfare consequences of several aspects of AI. Artificial insemination allows the mating of a mare and a stallion that under normal circumstances could not be covered naturally. There are many reasons that may preclude an animal from natural service, including inherited abnormalities, both physical and psychological. While it is now possible to mate such animals, is it right to do so and thus perpetuate an abnormality in subsequent generations? This is of particular concern when AI also increases the breeding potential or life span of individuals with genetic abnormalities, so that the problem is perpetuated even further in future generations. Mares with poor perineal conformation may be considered as just such a type of animal. They routinely undergo Caslick operations which alleviate the problem but as a result of which they are not able to be covered naturally without an episiotomy being performed, followed shortly after by the resuturing of the Caslicks. The dilemma as to whether to breed from such mares applies to any form of breeding, but in this context it is of particular concern as AI allows these mares to be bred from a greater number of times: the limit on the number of pregnancies is normally the physical restriction placed on opening and resuturing the vulva. The incidence of poor perineal conformation within the population is known to have increased since the widespread use of Caslick operations. There is the possibility, therefore, that this increase in incidence within the population, especially with athletically built performance mares, may be further accelerated with the widespread use of AI.

A further ethical concern, expressed by many, is the potential to use the semen of a stallion long after his death. Some breed societies therefore restrict the use of semen after a stallion's death, usually by way of a storage time limit. Other breed societies refuse to allow semen to be used after a stallion's death. However, such restrictions limit one of the major potential advantages of AI: that of setting up a gene bank, composed of samples collected from superior stallions and stored for prosperity, for judicious future reintroduction. This would allow the breadth of genetic base to be restored in the future as a safeguard against overconcentration on a few fashionable traits at the expense of some of the more traditional types of animal.

There are still many breeders (and hence breed societies) who are worried by the concept of 'interfering with nature', especially if it involves (or contorts) the selection of animals for breeding. The argument is that selection should be dominated by survival of the fittest. However, since 1400 BC when humans first discovered how to determine the characteristics of a foal by appropriate selection of a mare and a stallion, we have interfered with natural selection. In this context the effect of AI will be minimal.

Our ability to collect and store semen for long periods presents considerable scope for manipulation of spermatozoa or the semen sample as a whole, prior to insemination. This is commonly carried out to improve semen from stallions that habitually produce poor samples, often of low concentrations. Such samples can be concentrated by centrifugation and resuspension, prior to the insemination of an appropriate number of spermatozoa. Bearing in mind previous comments, it is questionable whether such stallions should be bred from at all. Our ability to manipulate spermatozoa begs the question as to how far is it acceptable for us to go in order to achieve fertilization. This issue is of considerable current interest, especially with reference to humans; and it may well not be long before such biotechnology and gamete manipulation is possible in horses.

The sexing of embryos during early pregnancy is now possible (Curran and Ginther, 1989, 1991, 1993; Peippo *et al.*, 1995). This could have serious consequences. The only reason why it would be desirable to know the sex of the embryo is in order that the information may be acted upon. The major demanded action, as a result of sexing embryos, is the elective abortion of embryos of the non-desired sex. This not only raises significant ethical concerns but will distort the population ratio of mares to stallions and affect the markets, resulting in major oscillations in one sex or another being presented for sale. In many breeds it is of greater advantage to have fillies born, as these can serve the dual purpose of being a breeding animal and also a riding animal. In other levels of the industry it is most advantageous to have a colt foal, especially if it is well bred and is destined to remain entire, as his potential earning power can be considerably greater than that of an equivalent mare. The sexing of spermatozoa is the next possibility: work on this has been carried out for many years – largely in other mammals, but some work has been done specifically in horses (Bhattacharya *et al.*, 1966; Johnson, 1991a; Hafez, 1993b). It is apparent that the sexing of spermatozoa would be possible and, through this, sex determination and selection may be carried out. Colts or fillies could then be produced to order to satisfy demand. It could be argued that such selection is an improvement on the elective abortion of embryos, but it will have similar detrimental effects on the equine population and markets.

Before any of these ideas can be taken beyond the experimental stages there must be a demand from the industry. At present it appears unlikely that the rather traditional equine industry will accept such techniques as *in vitro* fertilization, spermatozoa and embryo sex determination, embryo splitting or even cloning. Judging from the reluctance of some to accept the vast array of current techniques, any acceptance of more advanced techniques will take some time. However, similar comments to these were made about AI 15 years ago.

If these developments are to be allowed and regulations and restrictions on their use laid down, the question arises as to who will police them. In countries that currently have strict control and regulation over their equine breeding, such as France, Germany and The Netherlands, it is relatively easy to see that such policing could be incorporated into the present system. It is much harder to envisage how it would be achieved in countries which have a much less controlled system, such as the UK and the USA. It is likely that in these countries it will fall into the hands of the breed societies, which no doubt will pass the responsibility and judgement on to the veterinary profession – a potentially expensive and controversial move. It is also questionable whether in a so-called free society we have the right to restrict such freedom of choice.

Despite all its advantages and potential, equine AI is not universally accepted. One of the major bodies still holding out against the technique is the English Thoroughbred which quotes many of the concerns given previously as reasons for its reluctance to accept the technique. Before the technique can become truly worldwide in its use, such breed societies have to be persuaded of the benefits of the technique and assured that their concerns can be adequately addressed by the use of feasible regulations. The failure of the Thoroughbred industry to accept AI is one of the major reasons for its relatively poor acceptance rate, until recently, within the UK, especially when compared with the acceptance rate for other new technologies that help equine management and welfare.

For many of the disadvantages, concerns or problems relating to the use of AI to be addressed, significant specialist research is required. The major problem with this recommendation is the lack of financial backing in many of the major equine breeding nations for such research. There is a particular shortfall in financial backing in countries that do not have state control over their industry – for example, the USA and the UK. In both these countries the respective governments have decided that equine research does not warrant significant funding, if any at all. Research funding, therefore, has to be obtained from the industry or from charitable organizations, a far from ideal situation if truly unbiased and widely accessible research is to be conducted. Limited funding also ensures that those experiments that do run tend (through necessity) to involve minimal expenditure; this is reflected in the small number of animals used, casting doubt on the validity of many of the results obtained. Countries such as Germany and France, whose equine industries are under state control, often have well-funded research as a spin-off from the financial input into maintaining the central control and administration, often via national studs. In such circumstances financial backing is available but, by the nature of the establishment, it tends to have a greater commercial application rather than research which 'pushes forward the frontiers of science'.

The particular areas of research that are required, which would also go some way to allaying some of the fears over the use of AI, include improvements to spermatozoan evaluation, extenders and freezing protocols. At present there is no reliable method or parameter by which the fertilizing potential of a certain semen sample can be ascertained. In the absence of anything more accurate, spermatozoan motility is used as the indicator as it is the parameter with the best correlation to fertilizing ability. This is obviously not satisfactory. As a result many samples are discarded, on the basis of poor motility, which might well have proved most successful when inseminated into a mare, and vice versa. It is vital that research is conducted into identifying a simple, easy and accurate test that can, ideally, be done on the farm, to indicate the fertilizing capability of a semen sample. This test can then be used to not only assess accurately the quality of a semen sample prior to its use or storage, but it will also allow accurate monitoring of factors that might have an adverse affect on fertility. This information can then be used in the development and evaluation of different semen extenders.

At present, using motility as a guide, it is concluded that the freezing of spermatozoa has a significantly detrimental effect on their fertilizing capabilities. However, these are *in vitro* indications: they are not invariably reflected in the results of inseminations, and it is apparent that the correlation between motility and fertilizing potential is reduced in semen samples post thaw. Another major area of concern is the freezing process itself, in particular the identification of a suitable cryoprotectant. Glycerol, which is used traditionally, along with many of the others investigated, apparently has a detrimental effect on spermatozoan viability. The effect that freezing has on spermatozoa is very variable, some stallion's semen being apparently much more able to withstand the process than others. Without an accurate *in vitro* test for fertilizing potential, it is not possible to confirm whether this observation is reflected *in vivo*, and if so which stallions suffer the most, or which extenders or components cause the detrimental effects. Lacking this information, it is very difficult to diagnose the problem and hence provide a solution. Solving the problem would allow individual extending protocols to be developed for different individuals or groups of stallions. Once the problems and the reasons for the variability of success achieved when using frozen semen are understood, attempts may be made to reduce the problems and the variability. Then much of the disappointment currently associated with the use of frozen semen and the negative impressions of the whole technique can be addressed.

On a more commercial basis it is evident that some form of quality control is urgently required. This again will enhance the acceptance of AI. Many countries have addressed this problem by applying rigorous standards that have to be met for all stallions standing for use with AI and for all samples they produce for insemination, for use both at home and abroad. However, many countries have little or no control over the quality of semen sold. Many of the health issues are now legislated for but there is little regulation on the quality of the semen sample itself. Many studs fall foul of this and spend considerable amounts of money importing semen either of unknown quality or of such poor quality that it is of little use. The identification of a test of fertilizing potential would address many of these problems, allowing individuals to request test results prior to semen exportation/importation and allowing the test to be repeated when the sample arrives. This would reduce much wasted time and effort. Some form of universally accepted quality assurance standard would be ideal.

Many of the points raised here are specific examples of more major issues that the industry will have to address in the future. Despite this and the associated concerns, it is evident that the use of equine AI is here to stay and will continue to expand, opening up with it exciting opportunities in the selection and breeding of equine species.

References

Abel, W. (1984) Collection and evaluation of data relevant to fertilisation of trotter and warmblood mares at an AI centre during 1983 breeding season. Thesis, Tierärztliche Hochschule Hannover, Germany, 110 pp.

Abou-Ahmed, M.M., El Belely, M.S., Ismail, S.T., El-Baghdady, Y.R.M. and Hemeida, N.A. (1993) The influences of age and season on certain biochemical constituents of seminal plasma of Arabian horses. *Animal Reproduction Science* 32(3–4), 237–244.

Adams, C.E. (1962) Artificial insemination in rodents. In: Maule, J.P. (ed.) *The Semen of Animals and AI.* Commonwealth Agricultural Bureau, Farnham Royal, UK, pp. 316–330.

Adams, G.P., Kastelic, J.P., Bergfelt, D.R. and Ginther, O.J. (1987) Effect of uterine inflammation and ultrasonically detected uterine pathology on fertility in the mare. *Journal of Reproduction and Fertility, Supplement* 35, 444–454.

Advena, I. (1990) Deep frozen storage of stallion semen: comparison of the flat straw and macrotube methods in relation to their effects on the motility and acrosome integrity of the spermatozoa. Thesis, Tierärztliche Hochschule Hannover, Germany, 93 pp.

Ahlemeyer, B. (1991) Freezing of stallion semen. Effects of seminal plasma on sperm motility and acrosome integrity. Thesis, Tierärztliche Hochschule Hannover, Germany, 89 pp.

Alexander, S.L. and Irvine, C.H.G. (1993) Follicle stimulating hormone and luteinising hormone. In: McKinnon, A.O. and Voss, J.L. (eds) *Equine Reproduction.* Lea and Febiger, Philadelphia, pp. 45–57.

Aliev, A.I. (1981a) The effect of ovaritropin on reproductive function of mares, and the optimum time of insemination with frozen–thawed semen. *Sbor Nauch Trudov* 115, 88–92. *Animal Breeding Abstracts* 49(805), 6815.

Aliev, A.I. (1981b) A new method of freezing horse semen in polypropylene tubes. *Sbor Nauch Trudov* 115, 92–97. *Animal Breeding Abstracts* 49(805), 6816.

Allen, W.R. (1977) Artificial control of the mare's reproductive cycle. *Veterinary Record* 100, 68–71.

Allen, W.R. and Cooper, M.J. (1993) Prostaglandins. In: McKinnon, A.O. and Voss, J.L. (eds) *Equine Reproduction.* Lea and Febiger, Philadelphia, pp. 69–80.

Allen, W.R. and Rowson, L.E.A. (1973) Control of the mare's oestrus cycle by prostaglandins. *Journal of Reproduction and Fertility* 33, 539–543.

Allen, W.R., Stewart, F., Cooper, N.J., Crowhurst, R.C., Simpson, D.J., McEnry, R.J., Greenwood, R.E.S., Rossdale, P.D. and Ricketts, S.W. (1974) Further studies on the use of synthetic prostaglandin analogue for inducing luteolysis in mares. *Equine Veterinary Journal* 6, 31–35.

Allen, W.R., Stewart, F., Trounson, A.O., Tishner, M. and Bielanski, W. (1976a) Viability of horse embryos after storage and long distance transport in the rabbit. *Journal of Reproduction and Fertility* 47, 387–390.

Allen, W.R., Bowen, J.M., Frank, C.J., Jeffcote, L.B. and Rossdale, P.D. (1976b) The current position of artificial insemination in horse breeding. *Equine Veterinary Journal* 8 (2), 72–74.

Almahbobi, G., Papadopoulas, V., Carreaus, V. and Silberzahn, P. (1988) Age related morphological and functional changes in the leidig cells of the horse. *Biology of Reproduction* 38, 653–665.

Almquist, J.O. (1949a) A comparison of penicillin, streptomycin and sulphonilamide for improving the fertility of semen from relatively infertile bulls. *Journal of Dairy Science* 32, 722.

Almquist, J.O. (1949b) The effect of penicillin upon the fertility of semen from relatively infertile bulls. *Journal of Dairy Science* 32, 950–954.

Althouse, G.C. and Hopkins, S.M. (1995) Assessment of boar sperm viability using a combination of two fluorophores. *Theriogenology* 43, 595–603.

Althouse, G.C., Seager, S.W.J., Varner, D.D. and Webb, G.W. (1989) Diagnostic aids for the detection of urine in the equine ejaculate. *Theriogenology* 31, 1141–1148.

Alvarez, J.G. and Storey, B.T. (1984a) Assessment of cell damage caused by spontaneous lipid peroxidation in rabbit spermatozoa. *Biology of Reproduction* 30, 323–331.

Alvarez, J.G. and Storey, B.T. (1984b) Lipid peroxidation and the reactions of superoxide and hydrogen peroxide in mouse spermatozoa. *Biology of Reproduction* 30, 833–841.

Amann, R.P. (1981a) A review of anatomy and physiology of the stallion. *Equine Veterinary Science* 1(3), 83–105.

Amann, R.P. (1981b) Spermatogenesis in the stallion. A review. *Equine Veterinary Science* 1(4), 131–139.

Amann, R.P. (1984) Effects of extender, storage temperature and centrifugation on stallion spermatozoal motility and fertility. *Proceedings of the 10th International Congress on Animal Reproduction and Artificial Insemination*, Urbana, Paper No. 186.

Amann, R.P. (1988) Computerised evaluation of stallion spermatozoa. *Proceedings of the 14th Annual Convention of the American Association of Equine Practice*, pp. 453–473.

Amann, R.P. (1989a) Treatment of sperm to predetermine sex. *Theriogenology* 31, 49–60.

Amann, R.P. (1989b) Spermatogenesis. In: Pickett, B.W., Amann, R.P., McKinnon, A.O., Squires, E.L. and Voss, J.L. (eds) *Management of the Stallion for Maximum Reproductive Efficiency*, Vol. II. Colorado State University, Fort Collins, Colorado, pp. 27–39.

Amann, R.P. (1993a) Functional anatomy of the adult male. In: McKinnon, A.O. and Voss, J.L. (eds) *Equine Reproduction*. Lea and Febiger, Philadelphia, pp. 645–657.

Amann, R.P. (1993b) Physiology and endocrinology. In: McKinnon, A.O. and Voss, J.L. (eds) *Equine Reproduction*. Lea and Febiger, Philadelphia, pp. 658–685.

Amann, R.P. (1993c) Effects of drugs or toxins on spermatogenesis. In: McKinnon, A.O. and Voss, J.L. (eds) *Equine Reproduction*. Lea and Febiger, Philadelphia, pp. 831–839.

Amann, R.P. and Ganjam, V.K. (1981) Effects of hemicastration or hCG treatment on steroids in testicular vein and jugular vein blood of stallions. *Journal of Andrology* 3, 132–139.

Amann, R.P. and Graham, J.K. (1993) Spermatozoal function. In: McKinnon, A.O. and Voss, J.L. (eds) *Equine Reproduction*. Lea and Febiger, Philadelphia, pp. 715–745.

Amann, R.P. and Hammerstedt, R.H. (1980) Validation of a system for computerised measurements of spermatozoal velocity and percentage of motile sperm. *Biology of Reproduction* 23, 647–656.

Amann, R.P. and Pickett, B.W. (1984) An overview of frozen equine semen: procedures for thawing and insemination of frozen equine spermatozoa. *Experimental Station Animal*

Reproductive Laboratory, Special Series No. 33. Colorado State University, Fort Collins, 30 pp.

Amann, R.P. and Pickett, B.W. (1987) Principals of cryopreservation and a review of cryopreservation of stallion spermatozoa. *Journal of Equine Veterinary Science* 7, 145–173.

Amann, R.P., Johnson, L. and Pickett, B.W. (1977) Connection between the seminiferous tubules and efferent ducts in the stallion. *American Journal of Veterinary Research* 38, 1571–1579.

Amann, R.P., Thompson, D.L. Jr, Squires, E.L. and Pickett, B.W. (1979) Effects of age and frequency of ejaculation on sperm production and extragonadal sperm reserves in stallions. *Journal of Reproduction and Fertility, Supplement* 27, 1–6.

Amann, R.P., Loomis, P.R. and Pickett, B.W. (1983) Improved filter system for an equine Artificial Vagina. *Equine Veterinary Science* 3(4), 120–125.

Amann, R.P., Cristanelli, M.J. and Squires, E.L. (1987) Proteins in stallion seminal plasma. *Journal of Reproduction and Fertility, Supplement* 35, 113–120.

Amann, R.P., Squires, E.L. and Pickett, B.W. (1988) Effects of sample thickness and temperature on spermatozoal motion. *Proceedings of 11th International Congress on Animal Reproduction and AI*, Vol. 3. Dublin, Republic of Ireland, pp. 221a–221c.

Anderson, J. (1945) The semen of animals and its use for artificial insemination. *Technical Communication of the Bureau of Animal Breeding Genetics* 6, viii.

Anwar, M.M., Megahed, G.A., El-Deeb, T.S. and Shehata, H.S. (1996) The effect of melatonin on bull liquid semen and enzymatic release in seminal plasma. *Assiut Veterinary Medical Journal* 35(69), 42–63.

Arbeiter, K., Barth, U. and Jochle, W. (1994) Observations on the use of progesterone intra-vaginally and of desorelin in acyclic mares for induction of ovulation. *Journal of Equine Veterinary Science* 14(1), 21–25.

Argo, C.M., Cox, J.E. and Gray, J.L. (1991) Effect of oral melatonin treatment on the seasonal physiology of pony stallions. *Journal of Reproduction and Fertility, Supplement* 44, 115–125.

Arhipov, G. (1957) O dozirovke semeni pri iskusstvennom osemenenii kobyl (The dosage of semen in the artificial insemination of mares). *Konevodstvo* 27(5), 33–34. *Animal Breeding Abstracts* 25, 1690.

Arighi, M., Singh, A. and Bosu, W.T.K. (1987) Histology of the normal and retained equine tesis. *Acta Anatomica* 129, 127–139.

Arnold, R. (1992) Biological investigations on thermally exposed stallion semen. Laboratory studies on diluted semen stored for short periods. Thesis, Tierärztliche Hochschule Hannover, Germany, 116 pp.

Arns, M.J., Webb, G.W., Kreider, J.L., Potter, G.D. and Evans, J.W. (1987) Use of diluent glycolysable sugars to maintain stallion sperm viability when frozen or stored at 37°C and 5°C in bovine serum albumin. *Journal of Reproduction and Fertility, Supplement* 35, 135–141.

Arras, N. (1994) A semen pooling model for AI in horses. Thesis, Tierärztliche Hochschule Hannover, Germany.

Arriola, J. and Foote, R.H. (1982) Effects of amikacin sulphate on the motility of stallion and bull spermatozoa at different temperatures and intervals of storage. *Journal of Animal Science* 54, 1105–1110.

Ashdown, R.R. (1963) The anatomy of the inguinal canal in the domesticated mammal. *Veterinary Record* 75, 200–233.

Ashley, K.B., Thompson, D.L. Jr, Garza, F. Jr and Weist, J.J. (1986) Testosterone proprionate treatment of stallions: effects on secretion of LH and FSH in daily samples and after administration of GnRH. *Domestic Animal Endocrinology* 3, 295–299.

Atherton, J.G. (1975) The identification of equine genital strains of *Klebsiella* and *Enterobacter* species. *Equine Veterinary Journal* 7, 207–209.

Augero, A., Miragaya, M.H., Mora, N.G., Chavas, M.G., Neild, D.M., Trinchero, G.D. and Beconi, M.T. (1994) A study of the dynamic, metabolic and morphological parameters of equine semen subjected to different treatments. *Proceedings of the 6th International Symposium on Equine Reproduction*, pp. 191–192.

Augusto, C., Gacek, F. and Artes, R. (1992) Seasonal changes in the osmolarity of stallion seminal plasma. *Revista Brasileira de Reproducae Animal* 16(1–2), 13–22.

Aurich, J.E., Kuhne, A., Hoppe, H. and Aurich, C. (1996) Seminal plasma affects membrane integrity and motility of equine spermatozoa after cryopreservation. *Theriogenology* 46(5), 791–797.

Aurich, J.E., Schoumlautnherr, U., Hoppe, H. and Aurich, C. (1997) Effects of antioxidants on motility and membrane integrity of chilled-stored stallion semen. *Theriogenology* 48(2), 185–192.

Austin, J.W., Hupp, E.W. and Murphree, R.L. (1961) Effect of scrotal insulation on semen of Hereford bulls. *Journal of Animal Science* 20, 307–310.

Autorino, G.L., Careleti, G., Rosati, R., Ferrari, G., Vulcano, G., Ammaddeo, D., McCollum, W.H. and Timoney, P.J. (1994) Pathogenic and virological properties of an Italian strain of equine arthritis virus in stallions following intranasal challenge with infective raw semen. In: Nakajima, H. and Plowright, W.J. (eds) *Equine Infectious Diseases. VIII. Proceedings of the 7th International Conference.* Tokyo, Japan, 8–11 June, 1994. R. and W. Publications, Newmarket, UK, pp. 335–336.

Back, D.G., Pickett, B.W., Voss, J.L. and Seidel, G.E. Jr (1975) Effect of antibacterial agents on motility of stallion spermatozoa at various storage times, temperature and dilution ratios. *Journal of Animal Science* 41, 137–143.

Bader, H. and Huttenrauch, O. (1966) Fraktionierte Ejakulatgewinnung beim Hengst. (Collection of stallion ejaculates in fractions). *Deutsche Tierärztliche Wochenschrift* 73, 547–548.

Baemgartl, C., Bader, H., Drommer, W. and Luning, I. (1980) Ultrastructural alterations of stallion spermatozoa due to semen conservation. *Proceedings of the International Congress of Animal Reproduction and Artificial Insemination* 5, 134–137.

Bailey, M.T., Bott, R.M. and Gimenez, T. (1995) Breed registries regulations on artificial insemination and embryo transfer. *Journal of Equine Veterinary Science* 15(2), 60–66.

Bain, A.M. (1966) The role of infection in infertility in the Thoroughbred mare. *Veterinary Record* 78, 168–175.

Ball, B.A. and Mohammed, H.O. (1995) Morphometry of stallion spermatozoa by computer-assisted image analysis. *Theriogenology* 44(3), 367–377.

Ball, B.A., Fagnan, M.S. and Dobrinski, V. (1997) Determination of acrosin amidase activity in equine spermatozoa. *Theriogenology* 48(7), 1191–1198.

Baranska, K. and Tischner, M. (1995) Evaluating capacitation of stallion spermatozoa obtained from the mare's reproductive tract. *Biology of Reproduction, Monograph Equine Reproduction*, VI, 707–712.

Bardin, C.W. (1989) Inhibin structure and function in the male. *Annals of the New York Academy of Sciences* 564, 102–123.

Barker, C.A.V. and Gandier, J.C.C. (1957) Pregnancy in a mare resulted from frozen epididymal spermatozoa. *Canadian Journal Compendium on Medical Veterinary Science* 21, 47–51.

Barrelet, F.E. (1992) Artificial insemination in European horse breeding. *Equine Veterinary Education* 4(5), 225–228.

Bartlett, D.E. (1973) Use of artificial insemination for horses. In: *Stud Manager's Handbook*, Vol. 9. Agriservices Foundation, Clovis, CA.

Bataille, B., Magistrini, M. and Palmer, E. (1990) Objective determination of sperm motility in frozen–thawed stallion semen. Correlation with fertility. In: *Quoi de Neuf en Matière d'Études et de Recherches de la Cheval? 16ème Journée d'Étude, Paris, 7 Mars 1990.* CEREOPA, Paris, France, pp. 138–141.

Batellier, F., Magistrini, M., Fauquant, J. and Palmer, E. (1997) Effect of milk fractions on survival of equine spermatozoa. *Theriogenology* 48(3), 391–410.

Bearden, H.J. and Fuquay, J.W. (1992) Environmental management. In: Bearden, H.J. and Fuquay, J.W. (eds) *Applied Animal Reproduction*, 3rd edn. Prentice Hall International, London, pp. 273–282.

Beatty, R.A., Bennett, G.H., Hall, J.G., Hancock, J.L. and Stewart, D.L. (1969) An experiment with heterospermic insemination in cattle. *Journal of Reproduction and Fertility* 19, 491–496.

Beckett, S.D., Hudson, R.S. and Walker, D.F. (1973) Blood pressures and penile muscle activity in the stallion during coitus. *American Journal of Physiology* 225, 1072–1075.

Beckett, S.D., Walker, D.F. and Hudson, R.S. (1975) Corpus spongiosus penis pressure and penile muscle activity in the stallion during coitus. *Australian Journal of Veterinary Research* 36, 431–436.

Bedford, J.M. and Hoskins, D.D. (1990) The mammalian spermatozoon. Morphology, biochemistry and physiology. In: Lamming, G.E. (ed.) *Marshall's Physiology of Reproduction, Vol. 2. Male Reproduction*. Churchill Livingston, London, pp. 379–568.

Bedford, J.M. and Overstreet, J.W. (1972) A method for objective evaluation of the fertilising ability of spermatozoa irrespective of genetic character. *Journal of Reproduction and Fertility* 31, 407–411.

Bedford, S.J. and Hinrichs, K. (1994) The effect of insemination volume on pregnancy rates of pony mares. *Theriogenology* 42(4), 571–578.

Bedford, S.J., Jasko, D.J., Graham, J.K., Amann, R.P., Squires, E.L. and Pickett, B.W. (1995a) Use of two freezing extenders to cool stallion spermatozoa to 5°C with and without seminal plasma. *Theriogenology* 43(5), 939–953.

Bedford, S.J., Jasko, D.J., Graham, J.K., Amann, R.P., Squires, E.L. and Pickett, B.W. (1995b) Effect of seminal extenders containing egg yolk and glycerol on motion characteristics and fertility of stallion spermatozoa. *Theriogenology* 43(5), 955–967.

Belling, T.M. Jr (1984) Post ovulation breeding and related reproductive phenomena in the mare. *Equine Practice* 6, 12–19.

Bennett, D.G. (1986) Therapy of endometritis in mares. *Journal of the American Veterinary Medical Association* 188, 1390–1392.

Bergin, W.C., Gier, H.T., Marion, G.B. and Coffman, J.R. (1970) A development concept of equine cryptochidism. *Biology of Reproduction* 3, 82–92.

Berliner, V.R. (1940) An improved artificial vagina for the collection of stallion and jack semen. *Journal of the American Veterinary Medical Association* 96, 667.

Berliner, V. (1947) Horses and jackstock. In: Perry, E.J. (ed.) *The Artificial Insemination of Farm Animals*. Chapman & Hall, London, pp. 99–132.

Berndtson, W.E. and Desjardins, C. (1974) The cycle of the seminiferous epithelium and spermatogenesis in the bovine testis. *American Journal of Anatomy* 140, 167–180.

Berndtson, W.E. and Foote, R.M. (1969) The survival of bovine spermatozoa following minimum exposure to glycerol. *Cryobiology* 5, 398–400.

Berndston, W.E. and Jones L.S. (1989) Relationship of intratesticular testosterone content to age, spermatogenesis, sertoli cell distribution and germ cell:sertoli cell ratio. *Journal of Reproduction and Fertility* 85, 511–518.

Berndtson, W.E., Pickett, B.W. and Nett, T.M. (1974) Reproductive physiology of stallion. IV. Seasonal changes in the testosterone concentration of peripheral plasma. *Journal of Reproduction and Fertility* 39, 115–118.

Berndtson, W.E., Hoyer, J.H., Squires, E.L. and Pickett, B.W. (1979) Influences of exogenous testosterone on sperm production, seminal quality and libido of stallions. *Journal of Reproduction and Fertility, Supplement* 27, 19–23.

Berndtson, W.E., Squires, E.L. and Thompson, D.L. Jr (1983) Spermatogenesis, testicular composition and the concentration of testosterone in equine testis as influenced by season. *Theriogenology* 20, 449–457.

Bertrand, M., Wierzbowski, S. and Zakrzewska, G. (1957) Wyniki masowych badan nasiena I odruchow plciowych ogierow. (The results of large scale investigations on the semen and sexual reflexes of stallions.) *Zesz. Nauk. Wyzsz. Szkot. Roln. Krakowie, Zootech.* Z. 1, 1957, 3, 97–114. *Animal Breeding Abstracts* 26, 1170.

Bertrand, M., Ferney, J., Biron, M. and Bind, J.P. (1959) Le sperme de l'étalon. Appréciation du pouvoir fecondant. (Evaluation of the fertility of stallion semen.) *Revue de Médecine de Toulouse* 110, 373–391. *Animal Breeding Abstracts* 27, 1722.

Betsch, J.M., Hunt, P.R., Spalart, M., Evenson, D. and Kenney, R.M. (1991) Effects of chlorhexidene penile washing on stallion semen parameters and sperm chromatin structure assay. *Journal of Reproduction and Fertility, Supplement* 44, 655–656.

Bhattacharya, B.C., Bangham, A.D., Cro, R.J., Keynes, R.D. and Rowson, L. (1966) An attempt to determine the sex of calves by artificial insemination with spermatozoa separated by sedimentation. *Nature* 211, 863.

Bhattacharya, B.C., Shome, P. and Gunther, A.H. (1977) Successful separation of X and Y spermatozoa in human and bull semen. *International Journal of Fertility* 22, 30–35.

Bielanski, W. (1951) Characteristics of semen of stallions. Macro and microscopic investigations with estimations of fertility. *Mem. Acad. Pol. Sci. Let. B.* No. 16.

Bielanski, W. (1975) The evaluation of stallion's semen in aspects of fertility control and its use for Artificial Insemination. *Journal of Reproduction and Fertility, Supplement* 23, 19–24.

Bielanski, W. and Kaczmarski, F. (1979) Morphology of spermatozoa in semen from stallions of normal fertility. *Journal of Reproduction and Fertility, Supplement* 27, 39–45.

Binor, Z., Sokoloski, J.E. and Wolf, P.P. (1980) Penetration of the zona free hamster egg by human sperm. *Fertility and Sterility* 33, 321–327.

Bishop, M.W.H., David, J.S.E. and Messervey, A. (1964) Some observations on cryptorchidism in the horse. *Veterinary Record* 76, 1041–1048.

Bishop, M.W.H., David, J.S.E. and Messervey, A. (1966) Cryptorchidism in the stallion. *Proceedings of the Royal Society for Medicine* 59, 769–774.

Bittmar, A. and Kosiniak, K. (1992) The role of selected biochemical components of equine seminal plasma in determining suitability for deep freezing. *Polskie Archiwum Weterynaryjne* 32(1–2), 17–29.

Blach, E.L., Amann, R.P., Bowen, R.A., Sawyer, H.R. and Hermenet, M.J. (1988) Use of a monoclonal antibody to evaluate integrity of the plasma membrane of stallion sperm. *Gamete Research* 21(3), 233–241.

Blach, E.L., Amann, R.P., Bowen, R.A. and Frantz, D. (1989) Changes in quality of stallion spermatozoa during cryopreservation: plasma membrane integrity and motion characteristics. *Theriogenology* 31(2), 283–298.

Blanc, G., Magistrini, M., Noue, P. and Palmer, E. (1989) Analysis of cryopreservation of horse spermatozoa. *Centre d'Étude et de Recherche sur l'Économie et l'Organisation des Productions Animales*, pp. 48–61.

Blanc, G., Magistrini, M. and Palmer, E. (1991) Use of concanavalin A for coating the membranes of stallion spermatozoa. *Journal of Reproduction and Fertility, Supplement* 44, 191–198.

Blanchard, T.L. (1981) Comparison of two techniques for obtaining endometrial bacteriological cultures in mares. *Theriogenology* 16, 85–93.

Blanchard, T.L. and Johnson, L. (1997) Increased germ cell degeneration and reduced germ cell:sertoli cell ratio in stallions with low sperm production. *Theriogenology* 47(3), 665–677.

Blanchard, T.L. and Varner, D.D. (1993) Testicular degeneration. In: McKinnon, A.O. and Voss, J.L. (eds) *Equine Reproduction.* Lea and Febiger, Philadelphia, pp. 855–860.

Blanchard, T.L., Elmore, R.G., Youngquist, R.S., Loch, W.E., Hardin, D.K., Bierschwal, C.J., Ganjam, V.K., Balke, J.M., Ellersiek, M.R., Dawson, L.J. and Miner, W.S. (1983) The effects of stanozolol and boldenone undecylenate on scrotal width, testis weight and sperm production in pony stallions. *Theriogenology* 20, 121–131.

Blanchard, T.L., Varner, D.D., Love, C.C., Hurtgen, J.P., Cummings, M.R. and Kenney, R.M. (1987) Use of semen extender containing antibiotic to improve the fertility of a stallion with seminal vesiculitis due to *Pseudomonas aeruginosa. Theriogenology* 28, 541–546.

Blanchard, T.L., Varner, D.D., Hurtgen, J.P., Love, C.C., Cummings, M.R., Strezmienski, P.J., Benson, C. and Kenney, R.M. (1988) Bilateral seminal vesiculitis and ampullitis in a stallion. *Journal of the American Veterinary Medical Association* 192, 525–526.

Blanchard, T.L., Evans, J.W., Varner, D.D., Mollett, T.A., Hardin, D.K., Elmore, R.G. and Youngquist, R.S. (1990) Pulsatile release of gonadotrophins in young pony stallions. *Theriogenology* 34(6), 1087–1095.

Blanchard,T.L., Jorgensen, J.B., Varner, D.D., Forrest, D.W. and Evans, J.W. (1996) Clinical observations on changes in concentrations of hormones in plasma of two stallions with thermally-induced testicular degeneration. *Journal of Equine Veterinary Science* 16(5), 195–201.

Blue, B.J., McKinnon, A.O., Squires, E.L., Seidel, G.E. and Muscari, K.T. (1989) Capacitation of stallion spermatozoa and fertilization of equine oocytes *in vitro. Equine Veterinary Journal, Supplement* 8, 111–116.

Blue, B.J., Pickett, B.W., Squires, E.L., McKinnon, A.O., Nett, T.M., Amann, R.P. and Shiner, K.A. (1991) Effect of pulsatile or continuous administration of GnRH on reproductive function of stallions. *Journal of Reproduction and Fertility, Supplement* 44, 145–154.

Bogart, R. and Mayer, D.T. (1950) The effects of egg yolk on the various physical and chemical factors detrimental to spermatozoal viability. *Journal of Animal Science* 9, 143–152.

Bonadonna, T. and Caretta, A. (1954) Indagini sui nemaspermi degli animali domestici mediante il microscopio elettronico Iia nota. Delle microstrutture dei nemaspermi di "*Equus caballus*" e di "*Ovis sudanica*". (Examination of the spermatozoa of domestic animals by means of the electron microscope. II. The microstructure of the spermatozoa of *Equus caballus* and *Ovis sudanica.*) *Zootec. E Vet.* 9, 65–69. *Animal Breeding Abstracts* 22, 847.

Booth, L.C., Oxender, W.D., Douglas, T.H. and Woodley, S.L. (1980) Estrus, ovulation and serum hormones in mares given prostaglandin F2α, estradiol and gonadotrophin releasing hormone. *American Journal of Veterinary Research* 41, 120–122.

Borg, K., Colenbrander, B., Fazeli, A., Parlevliet, J. and Malmgren, L. (1997) Influence of thawing method on motility, plasma membrane integrity and morphology of frozen–thawed stallion spermatozoa. *Theriogenology* 48(4), 531–536.

Borgel, C. (1994) Studies on cryopreservation of horse spermatozoa in relation to different diluents, sperm concentrations and freezing programmes. Thesis, Tierärztliche Hochschule Hannover, Germany, 104 pp.

Bostock, D.E. and Owen, L.N. (1975) *Neoplasms in the Cat, Dog and Horse.* Wolfe Medical Publishers, London.

Bowen, J.M. (1969) Artificial insemination in the horse. *Equine Veterinary Journal* 1, 98–100.

Bowen, J.M. (1986) Venereal diseases of stallions. In: Robinson N.E. (ed.) *Current Therapy in Equine Medicine,* 2nd edn. W.B. Saunders, Philadelphia, pp. 508–511.

Bowen, J.M., Tobin, N., Simpson, R.B., Ley, W.B. and Ansari, M.M. (1982) Effects of washing on the bacterial flora of the stallion's penis. *Journal of Reproduction and Fertility, Supplement* 32, 41–45.

Boyle, M.S. (1992) Artificial Insemination in the horse. *Annales de Zootechnie* 41(3–4), 311–318.

Boyle, M.S., Skidmore, J., Zhange, J. and Cox, J.E. (1991) The effects of continuous treatment of stallions with high levels of a potent GnRH analogue. *Journal of Reproduction and Fertility, Supplement* 44, 169–182.

Brachen, F.K. and Wagner, P.C. (1983) Cosmetic surgery for equine pseudohermaphrodism. *Veterinary Medicine and Small Animal Clinician* 78, 879–884.

Brackett, B.G., Cofone, M.A., Boice, M.L. and Bousquet, D. (1982) Use of zona-free hamster ova to assess sperm fertilising ability of bulls and stallions. *Gamete Research*, 5, 217–227.

Braun, J., Schfels, W. and Stolla, R. (1990) Successful use of frozen stallion semen stored for 18 years. *Berliner und Münchener Tierärztliche Wochenschrift* 103(6), 211–212.

Braun, J., Oka, A., Sato, K. and Oguri, N. (1993) Effect of extender, seminal plasma and storage temperature on spermatozoal motility in equine semen. *Japanese Journal of Equine Science* 4(1), 25–30.

Braun, J., Torres-Boggino, F., Hochi, S. and Oguri, N. (1994a) Effect of seminal plasma on motion characteristics of epididymal and ejaculated stallion spermatozoa during storage at 5°C. *Deutsche Tierärztliche Wochenschrift* 101(8), 319–322.

Braun, J., Sakai, M., Hochi, S. and Oguri, N. (1994b) Preservation of ejaculated and epididymal stallion spermatozoa by cooling and freezing. *Theriogenology* 41(4), 809–818.

Bretschneider, L.H. (1949) An electron-microscopical study of sperm. IV. The sperm tail of bull, horse and dog. *Proceedings of the Academy of Science Amst.* 52, 526–534. *Animal Breeding Abstracts* 17, 1226.

Brinsko, S.P. and Varner, D.D. (1992) Artificial Insemination and preservation of semen. *Veterinary Clinics of North America Equine Practice* 8, 205–218.

Brinsko, S.P. and Varner, D.D. (1993) Artificial insemination. In: McKinnon, A.O. and Voss, J.L. (eds) *Equine Reproduction*. Lea and Febiger, Philadelphia, pp. 790–797.

Bristol, F. (1981) Studies of oestrus synchronisation in mares. *Proceedings of the Society of Theriogenology 1981*, pp. 258–264.

Bristol, F. (1987) Fertility of pasture bred mares in synchronised oestrus. *Journal of Reproduction and Fertility, Supplement* 35, 39–43.

Bristol, F. (1993) Synchronization of ovulation In: McKinnon, A.O. and Voss, J.L. (eds) *Equine Reproduction*. Lea and Febiger, Philadelphia, pp. 348–352.

British Equine Veterinary Association (1991) *Codes of Practice for 1) Veterinary Surgeons and 2) Breed Societies in the United Kingdom and Ireland Using AI for Breeding Equids*. British Equine Veterinary Association, London, 24 pp.

British Equine Veterinary Association (1997) *Equine AI: Course for Technicians*. British Equine Veterinary Association, London.

Brook, D. (1993) Uterine cytology. In: McKinnon, A.O. and Voss, J.L. (eds) *Equine Reproduction*. Lea and Febiger, Philadelphia, pp. 246–254.

Brooks, D.E. (1971) Nucleotides in spermatozoa. *Journal of Reproduction and Fertility* 25, 302–303.

Broussard, J.R., Roussel, J.D., Hibbard, M., Thibodeaux, J.K., Moreau, J.D., Goodeaux, S.D. and Goodeaux, L.L. (1990) The effect of Monojet and Rit-Tite syringes on equine spermatozoa. *Theriogenology* 33, 200.

Broussard, J.R., Goodeaux, S.D., Goodeaux, L.L., Thibodeaux, J.K., Moreau, J.D., Godke, S. and Roussel, J.D. (1993) The effect of different types of syringe on equine spermatozoa. *Theriogenology* 39(2), 389–399.

Brown, B.W. (1994) A review of nutritional influences on reproduction in boars, bulls and rams. *Reproduction, Nutrition, Development* 34, 98–114.

Brown, D.V. and Senger, P.L. (1980) Influence of homologous blood serum on motility and head to head agglutination in non motile ejaculated bovine spermatozoa. *Biology of Reproduction* 23, 271–275.

Buell, J.R. (1963) A method for freezing stallion semen and tests of its fertility. *Veterinary Record* 75, 900–902.

Buiko-Rogalevic, A.N. (1949) Dliteljnoe hranenie spermy zerebca. (Storage of stallion semen for a long period.) *Konevodstvo*, 1949, 5, 27–34. *Animal Breeding Abstracts* 18, 41.

Burgos, M.H., Vitale-Calpe, R. and Aoki, A. (1970) Fine structure of the testis and its functional significance. In: Johnson, A.D., Gromes, W.R. and Van Denmark, N.L. (eds) *The Testis*, Vol. 1. Academic Press, New York, pp. 551—580.

Burkman, L.J., Overstreet, J.W. and Katz, D.F. (1984) A possible role for potassium and pyruvate in the modulation of sperm motility in the rabbit oviductal isthmus. *Journal of Reproduction and Fertility* 71, 367–376.

Burns, K.A. and Casilla, E.R. (1990) Partial purification and characterisation of an acetylcarnitine hydrolase from bovine epididymal spermatozoa. *Archives of Biochemistry and Biophysics* 277, 1–7.

Burns, P.J. and Douglas, R.H. (1985) Reproductive hormone concentrations in stallions with breeding problems: case studies. *Journal of Equine Veterinary Science* 5, 40–42.

Burns, P.J. and Reasner, D.S. (1995) Computerized analysis of sperm motion: effects of glycerol concentration on cryopreservation of equine spermatozoa. *Journal of Equine Veterinary Science* 15(9), 377–380.

Burns, P.J., Kumaresan, P. and Douglas, R.H. (1981) Plasma oxytocin concentrations in cyclic mares and sexually aroused stallions. *Theriogenology* 16, 531–539.

Burns, P.J., Jaward, M.J., Edmundson, A., Cahill, C., Boucher, J.K., Wilson, E.A. and Douglas, R.H. (1982) Effect of increased photoperiod on hormone concentrations in Thoroughbred stallions. *Journal of Reproduction and Fertility, Supplement* 32, 103–111.

Burns, S.J. and Reasner, D.S. (1995) Computerised analysis of sperm motion: effect of glycerol concentration on cryopreservation of equine spermatozoa. *Journal of Equine Veterinary Science* 15(9), 377–380.

Burns, S.J., Simpson, R.B. and Snell, J.R. (1975) Control of microflora in stallion semen with a semen extender. *Journal of Reproduction and Fertility, Supplement* 23, 139–142.

Butler, W.J. and Roberts, T.K. (1975) Effect of some phosphatidyl compounds on boar spermatozoa following cold shock or cooling. *Journal of Reproduction and Fertility* 43, 183–187.

Buttelmann, A. (1988) Evaluation of results for a large AI centre for horses. Thesis, Tierärztliche Hochschule Hannover, Germany.

Byers, S.W., Dowsett, K.F. and Glover, T.D. (1983) Seasonal and circadian changes of testosterone levels in peripheral blood plasma of stallion and their relation to semen quality. *Journal of Endocrinology* 99, 141–150.

Calvete, J.J., Nessau, S., Mann, K., Sanz, L., Sieme, H., Klug, E. and Topfer-Petersen, E. (1994) Isolation and biochemical characterisation of stallion seminal plasma proteins. *Reproduction in Domestic Animals* 29(6), 411–426.

Calvete, J.J., Mann, K., Schafer, W., Sanz, L., Reinert, M., Nessau, S., Raida, M. and Topfer-Petersen, E. (1995) Amino acid sequence of HSP-1 a major protein of stallion seminal plasma: effect of glycosylation on its heparin- and gelatin-binding capabilities. *Biochemistry Journal (London)* 310(2), 615–622.

Cameron, R.D.A. and Blackshaw, A.W. (1980) The effect of elevated ambient temperature on spermatogenesis in the boar. *Journal of Reproduction and Fertility* 59, 173–179.

Campo, M.R. del, Donoso, M.X., Parrish, J.J. and Ginther, O.J. (1990) *In vitro* fertilisation of *in vitro*-matured equine oocytes. *Journal of Equine Veterinary Science* 10(1), 18–22.

Canchie, J. and Mann, T. (1957) Glycosidases in mammalian sperm and seminal plasma. *Nature* (London) 179, 1190–1191.

Card, C.E., Manning, S.T., Bowman, P. and Leibel, T. (1997) Pregnancies from imipramine and xylazine-induced ex copula ejaculation in a disabled stallion. *Canadian Veterinary Journal* 38(3), 171–174.

Carleton, C.L. (1995) Pathophysiology and diagnosis of reproductive disease – mare physiology. In: Kobluk, C.N., Ames, T.R. and Goer, R.J. (eds) *The Horse – Diseases and Clinical Management*, Vol. 2. Saunders and Company, Philadelphia, pp. 897–901.

Carnevale, E.M. (1998) Folliculogenesis and ovulation. In: Rantanen, N.W. and McKinnon, A.O. (eds) *Equine Diagnostic Ultrasonography*. Williams and Wilkins, Baltimore, pp. 201–213.

Caron, J., Barber, S. and Bailey, J. (1985) Equine testicular neoplasia. *Compendium of Continuing Education for Practising Vets* 6, 5296.

Casey, P.J., Robertson, K.R., Lui, I.K.M., Botta, E.S. and Drobnis, E. (1991) Separation of motile spermatozoa from stallion semen. *Proceedings of the 17th Annual Convention of the American Association of Equine Practitioners*. San Francisco, pp. 203–210.

Casey, P.J., Hillman, R.B., Robertson, K.R., Yudin, A.I., Lui, I.K. and Drobonis, E. (1993) Validation of an acrosomal stain for equine sperm that differentiates between living and dead sperm. *Journal of Andrology* 14, 282–297.

Casey, P.J., Gravance, C.G., Davis, R.O., Chabot, D.D. and Liu, I. (1997) Morphometric differences in sperm head dimensions of fertile and subfertile stallions. *Theriogenology* 47(2), 575–582.

Caslick, E.A. (1937) The vulva and the vulval–vaginal orifice and its relationship to genital health of Thoroughbred mares. *Cornell Veterinarian* 27, 178–187.

Castro, T.A.M., Gastal, E.L., Castro, F.G. Jr and Augusto, C. (1991) Physical, chemical and biochemical traits of stallion semen. In: *Anais, IX Congresso Brasileiro de Reproducao Animal*. Belo Horizonte, Brazil, 22 a 26 de Junho de 1991. Vol II. Colegio Brasileiro de Reproducao Animal, Belo Horizonte, Brazil, pp. 443–447.

Cermak, O. (1975) Relationship between citric acid concentration in semen and some indices of ejaculation quality in stallions. *Veterinaria Medicina* 20(4), 223–226.

Chaechowski, F.I. (1894) *Sterilitat der pferede*, Ihre Ursachen und Behandlung. Oester, Monates. F. Therheilk.

Chakraborty, J., Constantinou, A. and McCorquodale, M. (1985) Monoclonal antibodies to bull sperm surface antigens. *Animal Reproduction Science* 9, 101–109.

Chang, M.C. and Walton, A. (1940) The effect of low temperature and acclimatisation on the respiratory activity and survival of ram spermatozoa. *Proceedings of the Royal Society of London (Biology)* 129, 517–527.

Chao, T. and Chang, P.-H. (1962) An experimental report of the low temperature storage of stallion semen. *Chinese Journal of Animal Husbandry, Veterinary Science* 1, 5–6. From English abstract in *Scientific Abstracts China Biological Science* 1964, 2, 1, No. 343. *Animal Breeding Abstract* 32, 445.

Chapin, R.E. and Foster, P.M.D. (1989) Testis en plastique: use and abuse of *in vitro* systems. In: Working, P.K. (ed.) *Toxicity of the Male and Female Reproductive Systems*. Hemisphere Publishers, New York, pp. 273–284.

Charneco, R., Pool, K.C. and Arns, M.J. (1993) Influence of vesicular gland rich and vesicular gland poor seminal plasma on the freezability of stallion spermatozoa. *Proceedings of the 13th Conference of the Equine Nutrition and Physiology Society*, pp. 385–386.

Cheng, P.L. (1962) The present situation of artificial insemination in horses in China and some investigations on increasing conception rates of mares and breeding efficiency of stallions. *Acta Veterinaria Zootechnologie Seneca* 5, 29–34.

Chevalier-Clement, F., Hochereau de Reviers, M.T., Perreau, C. and Magistrini, M. (1991) Alterations of the semen and genital tract of infertile stallions. *Journal of Reproduction and Fertility, Supplement* 44, 657–658.

Chirnside, E.D. (1992) Equine arteritis virus: an overview. *British Veterinary Journal* 148(3), 181–197.

Chow, P.Y.W., White, I.G. and Pickett, B.W. (1986) Stallion sperm and seminal plasma phospholipids and glycerylphosphorylcholine. *Animal Reproduction Science* 11(3), 207–213.

Christanelli, M.J., Squires, E.L., Amann, R.P. and Pickett, B.W. (1984) Fertility of stallion semen processed, frozen and thawed by a new procedure. *Theriogenology* 22(1), 39–45.

Christanelli, M.J., Amann, R.P., Squires, E.L. and Pickett, B.W. (1985) Effects of egg yolk and glycerol levels in lactose–EDTA–egg yolk extender and of freezing rate on the motility of frozen–thawed stallion spermatozoa. *Theriogenology* 24(6), 681–686.

Christensen, A.K. (1975) Endocrinology Section 7. In: Hamilton, D.W. and Greep, R.O. (eds) *Handbook of Physiology, Volume V, Male Reproductive Systems*. American Physiology Society, Washington, DC, pp. 57–94.

Christensen, P., Parlevliet, J.M., Van Buiten, A., Hyttel, P. and Colenbrander, B. (1995) Ultrastructure of fresh and frozen–thawed stallion semen. *Biology of Reproduction, Monograph Equine Reproduction VI* 1, 769–777.

Christensen, P., Whitfield, C.H. and Parkinson, T.J. (1996) *In vitro* induction of acrosome reactions in stallion spermatozoa by heparin and A23187. *Theriogenology* 45(6), 1201–1210.

Claus, R., Dimmick, M.A., Gimenez, T. and Hudson, L.W. (1992) Estrogens and prostaglandin F2α in the semen and blood plasma of stallions. *Theriogenology* 38(4), 687–693.

Clay, C.M. (1988) Influences of season and artificial photoperiod on reproduction in stallions. PhD thesis, Colorado State University, Fort Collins, USA.

Clay, C.M. and Clay, J.N. (1992) Endocrine and testicular changes with season, artificial photoperiod and the peri-pubertal period in stallions. *Veterinary Clinics of North America, Equine Practice* 8(1), 31–56.

Clay, C.M., Squires, E.L., Amann, R.P. and Pickett, B.W. (1984) Effect of dilution, polyvinyl alcohol (PVA) and bovine serum albumin (BSA) on stallion spermatozoal motility. *10th International Congress on Animal Reproduction and Artificial Insemination*. University of Illinois, Paper No. 187, 3 pp.

Clay, C.M., Squires, E.L., Amann, R.P. and Pickett, B.W. (1987) Influences of season and artificial photoperiod on stallion: testicular size, seminal characteristics and sexual behaviour. *Journal of Animal Science* 64(2), 517–525.

Clay, C.M., Squires, E.L., Amman, R.P. and Nett, T.M. (1988) Influences of season and artificial photoperiod on stallions: LH, FSH and testosterone. *Journal of Animal Science* 66, 1246–1255.

Clay, C.M., Squires, E.L., Amman, R.P. and Nett, T.M. (1989) Influences of season and artificial photoperiod on stallion's pituitary and testicular responses to exogenous GnRH. *Journal of Animal Science* 67, 763.

Clement, F., Guerin, B., Vidament, M., Diemert, S. and Palmer, E. (1993) Microbial quality of stallion semen. *Pratique Vétérinaire Equine* 25(1), 37–43.

Clement, F., Vidament, M. and Guerin, B. (1995) Microbial contamination of stallion semen. *Biology of Reproduction, Monograph Equine Reproduction VI* 1, 779–786.

Clemmons, A.J., Thompson, D.L. Jr and Johnson, L. (1995) Local initiation of spermatogenesis in the horse. *Biology of Reproduction* 52(6), 1258–1267.

Cochran, J.D., Amann, R.P., Squires, E.L. and Pickett, B.W. (1983) Fertility of frozen thawed stallion semen extended in lactose EDTA egg yolk extender and packaged in 1.0 ml straws. *Theriogenology* 20, 735–741.

Cochran, J.D., Amann, R.P., Froman, D.P. and Pickett, B.W. (1984) Effects of centrifugation, glycerol level, cooling to 5°C, freezing rate and thawing rate on the post thaw motility of equine spermatozoa. *Theriogenology* 22, 25–38.

Colas, G. (1979) Fertility in the ewe after artificial insemination with fresh and frozen semen at the induced oestrus, and influence of the photoperiod on semen quality of the ram. *Livestock Production Science* 6, 153–166.

Colborn, D.R., Merilan, C.P. and Loch, W.E. (1990) Temperature and pH effects on rubber toxicity for equine spermatozoa. *Journal of Equine Veterinary Science* 10(5), 343–347.

Colenbrander, B., Puyk, H., Zandee, A.R. and Parlevliet, J. (1992) Evaluation of the stallion for breeding. *Acta Veterinaria Scandinavica, Supplement* 88, 29–37.

Comhaire, F.H., Vermeulen, L. and Schoonjans, F. (1987) Reassessment of the accuracy of traditional sperm characteristics and adenosine triphosphate (ATP) in estimating fertilising potential of human semen *in vivo*. *International Journal of Andrology* 10, 653–662.

Conboy, H.S. (1992) Training the novice stallion for artificial breeding. *Veterinary Clinics of North America, Equine Practice* 8(1), 101–109.

Cooper, M.J. (1981) Prostaglandins in veterinary practice. *In Practice* 3(1), 30–34.

Cooper, W.C. (1979) The effect of rapid temperature changes on oxygen uptake by and motility of stallion spermatozoa. *Proceedings of the Annual Meeting of the Society of Theriogenology* 1979, pp. 10–13.

Cooper, W.L. and Wert, N. (1975) Wintertime breeding of mares using artificial light and insemination: six years experience. *Proceedings of the 21st Annual Convention of the American Association of Equine Practitioners*, p. 245.

Cornwall, J.C. (1972) Seasonal variation in stallion semen and puberty in the Quarter horse colt. MSc thesis, Louisiana State University.

Cornwall, J.C., Hauer, E.P., Spillman, T.E. and Vincent, C.K. (1973) Puberty in the Quarter Horse colt. *Journal of Animal Science* 36, 1215.

Correa, J.R. and Zavos, P.M. (1994) The hypoosmotic swelling test: its employment as an assay to evaluate the functional integrity of frozen–thawed bovine spermatozoa. *Theriogenology* 42, 351–360.

Correa, J.R., Heersche, G. Jr and Zavos, P.M. (1997) Sperm membrane functional integrity and response of frozen–thawed bovine spermatozoa during the hypoosmotic swelling test incubation at varying temperatures, *Theriogenology* 47, 715–721.

Coulter, G.H. and Foote, R.H. (1977) Effects of package, extender and light on stored frozen bull spermatozoa. *Journal of Dairy Science* 60(9), 1428–1432.

Courot, M., Hochereau-de Rivers, M.T. and Ortavant, R. (1970) Spermatogenesis. In: Johnson, A.D., Gromes, W.R. and Van Denmark, N.L. (eds) *The Testis*, Vol. 1. Academic Press, New York, pp. 339–431.

Couto, M.A. and Hughes, J.P. (1993) Sexually transmitted (venereal) diseases of horses. In: McKinnon, A.O. and Voss, J.L. (eds) *Equine Reproduction*. Lea and Febiger, Philadelphia, pp. 845–854.

Cox, J.E. (1975) Experiences with a diagnostic testosterone for equine cryptorchidism. *Equine Veterinary Journal* 7, 179–183.

Cox, J.E. (1988) Hernias and ruptures: words to the heat of deeds. *Equine Veterinary Journal* 20, 155–156.

Cox, J.E. (1993a) Developmental abnormalities of the male reproductive tract. In: McKinnon, A.O. and Voss, J.L. (eds) *Equine Reproduction*. Lea and Febiger, Philadelphia, pp. 895–906.

Cox, J.E. (1993b) Cryptorchid castration. In: McKinnon, A.O. and Voss, J.L. (eds) *Equine Reproduction*. Lea and Febiger, Philadelphia, pp. 915–920.

Cox, J.E., Edwards, G.B. and Neal, P.A. (1979) An analysis of 500 cases of equine cryptorchidism. *Equine Veterinary Journal* 11, 113–116.

Cran, D.G., Johnson, L.A. and Polge, C. (1995) Sex preselection in cattle: a field trial. *Veterinary Record* 136, 495–496.

Cross, N.L. and Meizel, S. (1989) Methods for evaluating the acrosome status of mammalian sperm. *Biology of Reproduction* 41, 635–641.

Crouch, J.R.F., Atherton, J.G. and Platt, H. (1972) Venereal transmission of *Klebsiella aerogenes* in a Thoroughbred stud from a persistently infected stallion. *Veterinary Record* 90, 21–24.

Crowhurst, R.C., Simpson, D.Y., Greenwood, R.E.S. and Ellis, D.R. (1979) Contagious equine metritis. *Veterinary Record* 104, 465.

Crump, J. Jr and Crump, J. (1989) Stallion ejaculation induced by manual stimulation of the penis. *Theriogenology* 31, 341–346.

Cueva, F.I. de la C., Pujol, M.R., Rigau, T., Bonet, S., Briz, M. and Rodriguez-Gill, J.E. (1997a) Resistance to osmotic stress of horse spermatozoa: the role of ionic pumps and their relationship to cryopreservation success. *Theriogenology* 48, 947–968.

Cueva, F.I. de la C., Rigau, T., Bonet, S., Briz, M and Rodriguez-Gill, J.E. (1997b) Subjecting horse spermatozoa to hypoosmotic incubation: effects of ouabain. *Theriogenology* 47, 765–784.

Cupps, P.T. (1991) *Reproduction in Domestic Animals*, 4th edn. Academic Press, New York.

Curnow, E.M. (1993) Tutorial article: AI and mare management. *Equine Veterinary Education* 5(1), 10–13.

Curran, S. and Ginther, O.J. (1989) Ultrasonic diagnosis of equine fetal sex by location of the genital tubercle. *Journal of Equine Veterinary Science* 9(2), 77–83.

Curran, S. and Ginther, O.J. (1991) Ultrasonic determination of fetal gender in horses and cattle under farm conditions. *Theriogenology* 36(5), 809–814.

Curran, S. and Ginther. O.J. (1993) Ultrasonic fetal gender diagnosis during months 5 to 11 in mares. *Theriogenology* 40(6), 1127–1136.

Curtis, S.K. and Amann, R.P. (1981) Testicular development and establishment of spermatogenesis in Holstein bulls. *Journal of Animal Science* 53, 1645–1657.

Daels, P.F. and Hughes, J.P. (1993) The normal estrus cycle. In: McKinnon, A.O. and Voss, J.L. (eds) *Equine Reproduction*. Lea and Febiger, Philadelphia, pp. 133–143.

Danek, J. and Wisniewski, E. (1992) Changes in semen quality in zinc-deficient stallions. *Medycyna Weterynaryjna* 48(12), 566–568.

Danek, J., Wisniewski, E. and Klimczak, J. (1993) Facultative pathogens in the semen of horses. *Bulletin of the Veterinary Institute in Pulawy* 37(2), 94–99.

Danek, J., Wisniewski, E. and Krumrych, W. (1994a) A case of urospermia in a stallion. *Medycyna Weterynaryjna* 50(3), 129–131.

Danek, J., Wisniewski, E., Krumrych, W. and Dabrowska, J. (1994b) Prevalence and sensitivity to antibiotics of facultative pathogenic bacteria isolated from semen of stallions. *Medycyna Weterynaryjna* 50(8), 385–387.

Danek, J., Wisniewski, E. and Krumrych, W. (1996a) Effects of dietary calcium excess on the quality of semen in stallions. *Medycyna Weterynaryjna* 52(7), 459–461.

Danek, J., Wisniewski, E., Krumrych, W. and Dabrowska, J. (1996b) Changes in qualities of stallion semen after incubation with *Streptococcus zooepidemicus*. *Bulletin of the Veterinary Institute in Pulawy* 40(2), 111–116.

David, J.S.E., Frank, G.J. and Powell, D.G. (1977) Contagious equine metritis. *Veterinary Record* 101, 189–190.

Davies Morel, M.C.G. (1993) *Equine Reproductive Physiology, Breeding and Stud Management*, 1st edn. Farming Press, Ipswich, UK.

Davis, I.S., Brattan, R.W. and Foote, R.H. (1963) Liveability of bovine spermatozoa at 5°C and −85°C in Tris buffered egg yolk glycerol extenders. *Journal of Dairy Science* 46, 333–336.

Davis, R.O., Gravance, C.G. and Casey, P.J. (1993) Automated morphometric analysis of stallion spermatozoa. *American Journal of Veterinary Research* 54(11), 1808–1811.

Dawson, R.M.C., Mann, T. and White, I.G. (1957) Glycerylphosphorylcholine and phosphorylcholine in semen and their relation to choline. *Biochemistry Journal* 65, 627–634.

Day, F.T. (1940) The stallion and fertility. The technique of sperm collection and insemination. *Veterinary Record* 52, 597–602.

De Vries, P.J. (1987) Evaluation of the use of fresh, extended, transported stallion semen in the Netherlands. *Journal of Reproduction and Fertility, Supplement* 35, 641.

De Vries, P.J. (1993) Diseases of the testis, penis and related structures. In: McKinnon, A.L. and Voss, J.L. (eds) *Equine Reproduction*. Lea and Febiger, Philadelphia, pp. 878–884.

Debuff, Y. (1991) *The Veterinary Formulary*, 1st edn. The Pharmaceutical Press, London.

Delius, D. (1985) Investigations on the freezing of stallion semen in units of 0.5 ml by means of a computerised freezing programme. Thesis, Tierärztliche Hochschule Hannover, Germany, 161 pp.

Demick, D.S., Voss, J.L. and Pickett, B.W. (1976) Effect of cooling, storage, glycerolization and spermatozoal numbers on equine fertility. *Journal of Animal Science* 43, 633–637.

Denniston, D.J., Graham, J.K., Squires, E.L. and Brinsko, S.P. (1997) The effect of liposomes composed of phosphatidylserine and cholesterol on fertility rates using thawed equine spermatozoa. *Journal of Equine Veterinary Science* 17(12), 675–676.

Dimmock, W.W. and Snyder, E.M. (1923) Bacteria of the genital tract of mares and the semen of stallions and their relation to breeding efficiency. *Journal of the American Veterinary Medical Association* 64, 288–298.

Dinger, J.E. and Noiles, E.E. (1986) Prediction of daily sperm output in stallions. *Theriogenology* 26(1), 61–67.

Dinger, J.E., Noiles, E.E. and Hoagland, T.A. (1986) Effect of controlled exercise on semen characteristics in two year old stallions. *Theriogenology* 25(4), 525–535.

Dippert, K.D. (1997) Regulations on artificial insemination and embryo transfer in the equine industry. In: *Proceedings of the Fifteenth Equine Nutrition and Physiology Symposium*. Fort Worth, Texas, pp. 296–297.

Dixon, K.F., Spongy, E.A., Thrasher, D.M. and Kreider, J.L. (1980) Effect of bovine serum albumin in the evaluation of boar spermatozoa and their fertility. *Theriogenology* 13, 437–444.

Dobrinski, I., Ignotz, G.G., Fagnan, M.S., Yudin, A.I. and Ball, B.A. (1997) Isolation and characterisation of a protein with homology to angiotensin converting enzyme from the periacrosomal plasma membrane of equine spermatozoa. *Molecular Reproduction and Development* 48(2), 251–260.

Doig, P.A. and Waelchii, R.O. (1993) Endometrial biopsy. In: McKinnon, A.O. and Voss, J.L. (eds) *Equine Reproduction*. Lea and Febiger, Philadelphia, pp. 255–257.

Domerg, D. (1984) Artificial insemination in horses. Results of AI studs. *Insemination Artificielle et Amelioration Genetique: Bilan et Perspectives Critiques*. Toulouse-Auzeville, France, 22–24, November 1983, pp. 349–358.

Donaldson, L.E. (1976) Artificial insemination in beef cattle. *Australian Veterinary Journal* 52, 565–569.

Dorotle, J.M. (1955) *Insemination artificielle et reproduction des équidés. Recueil de notes et communications publiées au titre des recherches zootechniques en Algérie*. Annales de l'Institute d'Agriculture Algér 9(2), 3–13.

Dott, H.M. (1975) Morphology of stallion spermatozoa. *Journal of Reproduction and Fertility, Supplement* 23, 41–46.

Dott, H.M. and Foster, G.C. (1972) A technique for studying the morphology of mammalian spermatozoa which are eosinophilic in differential 'live/dead' stain. *Journal of Reproduction and Fertility* 29, 443–445.

Dott, H.M. and Foster, G.C. (1975) Preservation of different staining of spermatozoa by formal citrate. *Journal of Reproduction and Fertility* 45, 57–60.

Douglas, R.H. and Ginther, O.J. (1972) Effect of prostaglandin F2α on length of dioestrus in mares. *Research in Prostaglandins* 2, 265–268.

Douglas-Hamilton. D.H., Osol, R., Osol, G., Driscoll, D. and Noble, H. (1984) A field study of the fertility of transported equine semen. *Theriogenology* 22(3), 291–304.

Douglas-Hamilton, D.H., Burns, P.J., Driscoll, D. and Viale, K.M. (1987) Fertility characteristics of slow cooled stallion semen. *Journal of Reproduction and Fertility* 35, 649–650.

Dowsett, K.F. and Knott, L.M. (1996) The influence of age and breed on stallion semen. *Theriogenology* 46(3), 397–412.

Dowsett, K.F. and Pattie, W.A. (1982) Characteristics and fertility of stallion semen. *Journal of Reproduction and Fertility, Supplement* 32, 1–8.

Dowsett, K.F. and Pattie, W.A. (1987) Variation in characteristics of stallion semen caused by breed, age, season of year and frequency. *Journal of Reproduction and Fertility, Supplement* 35, 645–647.

Dowsett, K.F., Osbourne, H.G. and Pattie, W.A. (1984) Morphological characteristics of stallion spermatozoa. *Theriogenology* 22(5), 463–472.

Dowsett, K.F., Dunn, B.L., Knott, L.M. and Cooper, J.W. (1995) Observations on the use of deep frozen stallion semen. *Australian Equine Veterinarian* 13(3), 53–60.

Dowsett, K.F., Knott, L.M., Tshewang, V., Jackson, A.E., Bodero, D.A.V. and Trigg, T.E. (1996) Suppression of testicular function using two dose rates of a reversible water soluble gonadotrophin releasing hormone (GnRH) vaccine in colts. *Australian Veterinary Journal* 74(3), 228–235.

Draincourt, M.A. and Palmer, E. (1982) Seasonal and individual effects on ovarian and endocrine responses of mares to a synchronisation treatment with progestagens impregnated vaginal sponges. *Journal of Reproduction and Fertility* 32, 283–291.

Dzuik, P.J., Graham, E.F., Donker, J.D., Marion, G.B. and Petersen, W.E. (1954) Some observations in collection of semen from bulls, goats, boars and rams by electrical stimulation. *Veterinary Medicine* 49, 455.

Easley, J. (1993) External perineal conformation. In: McKinnon, A.O. and Voss, J.L. (eds) *Equine Reproduction*. Lea and Febiger, Philadelphia, pp. 20–24.

Eddy, E.M. (1988) The spermatozoon. In: Knobil, E. and Neill, J. (eds) *The Physiology of Reproduction*. Raven Press, New York, pp. 27–68.

Edgar, D.G. (1957) A comparison of methods of rectal electrical stimulation of rams for semen collection. *New Zealand Veterinary Journal* 5, 17.

Ellery, J.C. and Hall, A.C. (1987) Temperature effects on red cell membrane transport processes. In: Bowler, K. and Fuller, B.J. (eds) *Temperature and Animal Cells*. Company of Biologists, Cambridge, UK, pp. 53–66.

Ellery, J.C., Graham, E.F. and Zemjanis, R. (1971) Artificial insemination of pony mares with semen frozen and stored in liquid nitrogen. *American Journal of Veterinary Research* 32, 1693–1698.

Ellington, J.E., Ball, B.A., Blue, B.J. and Wilker, C.E. (1993) Capacitation-like membrane changes and prolonged viability *in vitro* of equine spermatozoa cultured with uterine tube epithelial cells. *American Journal of Veterinary Research* 54(9), 1505–1510.

Elliott, F.E., Sherman, J.K., Elliott, E.J. and Sullivan, J.L. (1973) A photographic method of measuring percentage of progressing motile sperm cells using dark field microscopy. *The 8th Symposium of Zootechnology*. Milan, p. 160.

England, G.C. and Plummer, J.M. (1993) Hypo-osmotic swelling of dog spermatozoa. *Journal of Reproduction and Fertility, Supplement* 47, 261–270.

England, G.C.W. and Keane W. (1996) The effect of X radiation upon the quality and fertility of stallion semen. *Theriogenology* 46, 173–180.

Ericsson, R.J. and Glass, R.H. (1982) Functional differences between sperm bearing the X- and Y-chromosome. In: Amann, R.P. and Seidel, G.E. (eds) *Prospects for Sexing Mammalian Sperm*. Colorado Association University Press, Boulder, pp. 201–211.

Erokhin, A.S., Nikiforov, G.A. and Epishina, T.M. (1996) Protective effect of new synthetic antioxidants in the cryopreservation of ram semen. *Sel'skokhozyaistvennaya Biologiya* 6, 62–67.

European Union Commission Decision of 24th July 1995 on determining the specimen animal health certificate for trade in ova and embryos of the equine species (95/294/EC). *Official Journal of the European Communities* No. L 182/27 1995.

European Union Commission Decision of 24 July 1995 on determining specimen animal health certificate for trade in semen of the equine species (95/307/EC). *Official Journal of the European Communities* No. L 185, 58 1995.

European Union Commission Decision of 4th September 1996 on animal health requirements and veterinary certification for imports into the Community of semen of the equine species (96/539/EC). *Official Journal of the European Communities* No. L 230, 23 1996.

European Union Commission Decision of 4th September 1996 on animal health requirements and veterinary certification for imports into the Community of ova and embryos of the equine species (96/540/EC). *Official Journal of the European Communities* No. L 230, 28 1996.

European Union Council Directive of 26th June 1990 on animal health conditions governing the movement and import from third countries of Equidae (90/426/EEC). *Official Journal of the European Communities* No. L 224/42 1990.

European Union Council Directive of 26th June 1990 on the zootechnical and genealogical conditions governing intra Community trade in equidae (90/427/EEC). *Official Journal of the European Communities* No. L 224/55 1990.

European Union Council Directive of 26th June 1990 on trade in equidae intended for competitions and laying down the conditions for participation therein (90/428/EC). *Official Journal of the European Communities* No. L 224/60 1990.

Evans, G. and Maxwell, W.M.C. (1987) *Salamon's Artificial Insemination of sheep and goats*, 2nd. edn. Butterworths, Sydney, Australia, 194 pp.

Evans, M.J., Hamer, J.M., Garson, L.M., Graham, C.S., Asbury, A.C. and Irvine, C.H.G. (1986) Clearance of bacteria and non-antigenic markers following intrauterine inoculation into maiden mares. Effect of steroid hormone environment. *Theriogenology* 25, 37–50.

Evans, M.J., Hamer, J.M., Garson, L.M. and Irvine, C.H.G. (1987) Factors affecting uterine clearance of inoculated materials in mares. *Journal of Reproduction and Fertility, Supplement* 35, 327–334.

Evans, M.J., Alexander, S.L., Irvine, C.H.G., Livesey, J.H. and Donald, R.A.S. (1991) *In vitro* and *in vivo* studies of equine prolactin secretion throughout the year. *Journal of Reproduction and Fertility, Supplement* 44, 27–35.

Evans, R. (1996) A possible role for the gel fraction of stallions ejaculate in the assurance of paternity. MSc Thesis, University of Wales, Aberystwyth.

Evenson, D.P. and Melamed, M. (1983) Rapid evaluation of normal and abnormal cell types in human semen and testis biopsies by flow cytometry. *Journal of Histochemistry and Cytochemistry* 31, 248–252.

Evenson, D.P., Sailer, B.L. and Josh, L.K. (1995) Relationship between stallion sperm deoxyribonucleic acid (DNA), susceptibility to denaturation *in situ* and presence of DNA strand breaks: implications for fertility and embryo viability. *Biology of Reproduction, Monograph Equine Reproduction VI*, 1, 655–659.

Everett, N.B. (1945) The present status of the germ-cell problem in vertebrates. *Biological Reviews* 20, 45–55.

Fahy, G.M. (1986) The relevance of cryoprotectant 'toxicity' to cryobiology. *Cryobiology* 23, 1–13.

Fallon, E.H. (1967) The clinical aspects of streptococci infections in horses. *Journal of the American Veterinary Medical Association* 155, 413–414.

Farlin, M.E., Jasko, D.J., Graham, J.K. and Squires, E.L. (1992) Assessment of *Pisum sativum* agglutinin in identifying acrosomal damage in stallion spermatozoa. *Molecular Reproduction and Development* 32(1), 23–27.

Faulkner, L.C. and Pineda, M.H. (1980) Artificial insemination. In: McDonnell, L.E. (ed.) *Veterinary Endocrinology and Reproduction*, 3rd edn. Lea and Febiger, Philadelphia, pp. 330–336.

Fawcett, D.W. (1970) A comparative review of sperm ultrastructure. *Biology of Reproduction, Supplement* 2, 90–127.

Fawcett, D.W. (1975) The mammalian spermatozoon. *Developmental Biology* 44, 394–436.

Fayrer-Hosken, R.A. and Caudle, A.B. (1989) Stallion fertility evaluation: part III. *Equine Practice* 11(4), 30–34.

Fayrer-Hosken, R.A., Caudle, A.B. and Shur, B.D. (1991) Galactosyltransferase activity is related to the plasma membranes of equine and bovine sperm. *Molecular Reproduction and Development* 34(1), 74–78.

Fazeli, A.R., Steenweg, W., Bevers, M.M., Broek van den, J., Bracher, V., Parlevliet, J.M. and Collenbrander, B. (1995) Relationships between stallion sperm binding to homologous hemizonae and fertility. *Theriogenology* 44, 751–760.

Feuchter, F.A., Vernon, R.B. and Eddy, E.M. (1981) Analysis of the sperm surface with monoclonal antibodies: topographically restricted antigens appearing in the epididymis. *Biology of Reproduction* 24, 1099–1110.

Finland, Central Association of AI Society (1985) *Artificial insemination in Finland 1984.*

Finland, Central Association of AI Society (1987) *Artificial insemination in Finland 1985.*

Finland, Central Association of AI Society (1988) *Artificial insemination in Finland 1987.*

Fiser, P.S., Hansen, C., Underhill, K.L. and Shrestha, J.N.B. (1991) The effect of induced ice nucleation (seeding) on the post thaw motility and acrosomal integrity of boar spermatozoa. *Animal Reproduction Science* 24, 293–304.

Flink, G. (1988) Gonadotrophin secretion and its control. In: Knobil, E. and Neill, J. (eds) *The Physiology of Reproduction.* Raven Press, New York, pp. 1349–1377.

Flipse, R.J., Potton, S. and Almquist, J.O. (1954) Dilutes for bovine semen. III. Effect of lactenin and of lactoperoxidase upon spermatozoan livability. *Journal of Dairy Science* 32, 1205–1211.

Fomina, E.L. and Miroshnikova, K.I. (1989) The duration of semen storage in horses. *Konevodstvo i konnyi Sport* 12, 25.

Foote, R.H. (1978) General principles and basic techniques involved in synchronisation of oestrus in cattle. *Proceedings of the 7th Technical Conference on AI and Reproduction.* NAAB, Madison, pp. 55–61.

Foote, R.H. (1980) Artificial insemination. In: Hafez, E.S.E. (ed.) *Reproduction in Farm Animals,* 4th edn. Lea and Febiger, Philadelphia, USA, pp. 521–545.

Foote, R.H. (1982) Cryopreservation of spermatozoa and AI, past, present and future. *Journal of Andrology* 3, 85–100.

Foote, R.H. (1988) Preservation and fertility prediction of spermatozoa. *Proceedings of the 11th International Congress on Animal Reproduction and AI.* Dublin, pp. 126–134.

Foote, R.H. and Oltenacu, E.A.B. (1982) Increasing fertility in AI by culling bulls or ejaculates within bulls. *Proceedings of the 8th Technical Conference of the National Association of Animal Breeders,* pp. 6–12.

Foote, R.H., Hauser, E.R. and Casida, L.E. (1960) Some causes of variation in post-partum reproductive activity in Hereford cows. *Journal of Animal Science* 19, 238.

Foulkes, J.A. (1977) The separation of lipoproteins from egg yolk and their effect on the motility and integrity of bovine spermatozoa. *Journal of Reproduction and Fertility* 49, 277–284.

Francavilla, S., Gabriele, A., Romano, R., Ginaroli, L., Ferraretti, A.P. and Francavilla, F. (1994) Sperm–zona pellucida binding of human sperm is correlated with immunocytochemical presence of proacrosin and acrosin in the sperm heads but not with proteolytic activity of acrosin. *Fertility and Sterility* 62(6), 1226–1233.

Francl, A.T., Amann, R.P., Squires, E.L. and Pickett, B.W. (1987) Motility and fertility of equine spermatozoa in milk extender after 12 or 24 hours at 20°C. *Theriogenology* 27, 517–525.

Fraser, M. (ed.) (1986) *Merk's Veterinary Manual,* 6th edn. Merk and Co. Inc., Rahway, New Jersey.

Frazelli, A.R., Steenweg, W., Bevers, M.M., van der Broek, J., Brecher, V., Parlevliet, Y. and Colenbrander, B. (1995) Relation between stallion sperm binding to homologous hemizonae and ferility. *Theriogenology* 44, 751–760.

Freeman, K.P. and Johnston, J.M. (1987) Collaboration of a cytopathologist and practitioner using equine endometrial cytology in a private broodmare practice. *Proceedings of the 13th Annual Convention of the American Association of Equine Practitioners,* pp. 629–639.

Frey, G.W. von and Bernal, S.A. (1984) Freezing and evaluation of stallion semen. *Investigaciones del Departamento de Produccion Animal 1981–1982.* Universidad de Chile, Santiago, Chile, pp. 176–180.

Friedman, R., Scott, M., Heath, S.E., Hughes, J.P., Daels, P.F. and Tran, T. (1991a) Seminal characteristics of stallions with transitory testicular impairment. *Proceedings of the Annual Convention of the American Association of Equine Practitioners* 36, 3–13.

Friedman, R., Scott, M., Heath, S.E., Hughes, J.P., Daels, P.F. and Tran, T.Q. (1991b) The effects of increased testicular temperature on spermatogenesis in the stallion. *Journal of Reproduction and Fertility, Supplement* 44, 127–134.

Friedman, R., Daels, P.F., Scott, M.A., Roser, J.F., Mohammed, H.O. and Hughes, J.P. (1995) Endocrine profiles of stallions with impaired testicular function due to increased testicular temperature. *Biology of Reproduction, Monograph Equine Reproduction VI,* 1, 641–646.

Fromann, D.P. and Amann, R.P. (1983) Inhibition of motility of bovine, canine and equine spermatozoa by AV lubricants. *Theriogenology* 20(3), 357–361.

Fromann, D.P., Amann, R.P., Riek, P.M. and Olar, T.T. (1984) Acrosin activity of canine spermatozoa as an index of cellular damage. *Journal of Reproduction and Fertility* 70, 301–308.

Fukunaga, Y., Wada, R., Matsumura, T., Anzai, T., Imagawa, H., Sugiura, T., Kumanomido, T., Kanemaru, T. and Kamada, M. (1992) An attempt to protect against persistent infection of equine viral arteritis in the reproductive tract of stallions using formalin-inactivated virus vaccine. In: Plowright, W., Rossdale, P.D. and Wade, J.F. (eds) *Equine Infectious Diseases VI: Proceedings of the Sixth International Conference.* R. and W. Publications, Newmarket, pp. 239–244.

Galli, A., Basetti, M., Balduzzi, D., Martignoni, M., Bornaghi, V. and Maffii, M. (1991). Frozen bovine semen quality and bovine cervical-mucus penetration test. *Theriogenology* 35(4), 837–844.

Gamzu, R., Yogev, L., Amit, A., Lessing, J., Hamonnai, Z.T. and Yavetz, H. (1994) The hemizona assay is good prognostic value for the ability of sperm to fertilise oocytes *in vitro. Fertility and Sterility* 62(5), 1056–1059.

Ganjam, V.K. and Kenney, R.M. (1975) Androgens and oestrogens in normal and cryptorchid stallions. *Journal of Reproduction and Fertility, Supplement* 23, 67–73.

Garcia, M.L., Ganjam, V.K., Blanchard, T.L., Brown, E., Hardin, K., Elmore, R.C., Youngquist, R.S., Loch, W.E., Ellersieck, M.R. and Balke, J.M. (1987) The effects of stanozolol and boldenene undecylenate on plasma testosterone and gonadotrophins and on testis histology in pony stallions. *Theriogenology* 28, 109–119.

Garner, D.L. (1991) Artificial insemination. In: Cupps, P.T. (ed.) *Reproduction in Domestic Animals,* 4th edn. Academic Press, San Diego, California, pp. 251–278.

Garner, D.L., Pinkel, D., Johnson, L.A. and Pace, M.M. (1986) Assessment of spermatozoal function using dual fluorescent staining and flow-cytometric analyses. *Biology of Reproduction* 34, 127–138.

Gastel, E.L., Augusto, C., Castro, T.A.M.G. and Gastel, M.O. (1991) Relationship between the quality of stallion semen and fertility. In: *Anais, IX Congresso Brasileiro de Reproducao Animal*. Belo Horizonte, Brazil, 22 a 26 de Junho de 1991. Vol II. Colegio Brasileiro de Reproducao Animal, Belo Horizonte, Brazil, p. 446.

Gather, C. (1994) Effect of a skim milk diluent modified by Tyrode's medium on the sperm motility, morphology and viability of stored fresh stallion semen. Thesis, Tierärztliche Hochschule Hannover, Germany.

Gebauer, M.R., Pickett, B.W. and Swierstra, E.E. (1974a) Reproductive physiology of the stallion, II. Daily production and output of sperm. *Journal of Animal Science* 39, 732–736.

Gebauer, M.R., Pickett, B.W. and Swierstra, E.E. (1974b) Reproductive physiology of the stallion, III. Extra gonadal transit time and sperm reserves. *Journal of Animal Science* 39, 737–742.

Gebauer, M.R., Pickett, B.W., Voss, J.L. and Swierstra, E.E. (1974c) Reproductive physiology of the stallion: daily sperm output in testicular measurements. *Journal of the American Veterinary Medical Association* 165, 711–713.

Gebauer, M.R., Pickett, B.W., Swierstra, E.E., Faulkner, L.C., Remmenga, E.E. and Berndtson, W.E. (1976) Reproductive physiology of stallions, VII. Chemical characteristics of seminal plasma and spermatozoa. *Journal of Animal Science* 43, 626–632.

Geisler, A. (1990) A comparison of vital staining methods for the evaluation of semen from domestic animals. Thesis, Ludwig-Maximilians-Universität, Munchen, Germany.

Ghei, J.C., Uppal, P.K. and Yaday, M.P. (1994) Prospects of AI in equines. *Cataur* XI, 1, July 1994.

Gibbons, B.H. and Gibbons, I.R. (1972) Flagella movement and adenosine triphosphatase activity in sea urchin sperm extracted with Triton X-100. *Journal of Cell Biology* 54, 74–80.

Gibbs, E.P.Y., Roberts, M.C. and Morris J.M. (1972) Equine coital exanthema in the UK. *Equine Veterinary Journal* 4, 74.

Ginther, O.J. (1979) *Reproductive Biology of the Mare*. McNaughton and Gunn, Ann Arbor, Michigan.

Ginther, O.J. (1992) Endocrinology of the ovulatory season. In: *Reproductive Biology of the Mare, Basic and Applied Aspects*, 2nd edn. Equiservices, Wisconsin, pp. 233–290.

Ginther, O.J. and Pierson, R.A. (1983) Ultrasonic evaluation of reproductive tracts of mares; principles, equipment and techniques. *Journal of Equine Veterinary Science* 3, 195–201.

Ginther, O.J. and Pierson, R.A. (1984a) Ultrasonic evaluation of reproductive tracts of the mare: ovaries. *Journal of Equine Veterinary Science* 4, 11–16.

Ginther, O.J. and Pierson, R.A. (1984b) Ultrasonic anatomy of equine ovaries. *Theriogenology* 21, 471–483.

Ginther, O.J., Whitmore, H.L. and Squires, E.L. (1972) Characteristics of estrus, diestrus and ovulation in mares and effects of season and nursing. *American Journal of Veterinary Research* 33, 1935–1939.

Glass, R.H., Drouin, M.T., Ericsson, S.A., Marcoux, L.J., Ericsson, R.J. and Sullivan, H. (1991) The resazurin reduction test provides an assessment of sperm activity. *Fertility and Sterility* 56(4), 743–746.

Golnik, W. and Cierpisz, J. (1994) Reproductive use of stallions naturally infected with equine arteritis virus. *Medycyna Weterynaryjna* 50(10), 482–483.

Golnik, W., Paweska, J. and Dzik, W. (1991) Stallion as a potential source of infection with equine arteritis virus. *Medycyne Weterynaryjna* 47(10), 459–461.

Gomes, W.R. (1977) Artificial insemination. In: Cole, H.H. and Cupps, P.T. (eds) *Reproduction in Domestic Animals*, 3rd edn. Academic Press, New York, pp. 257–284.

Gomes, W.R., Butler, W.R. and Johnson, A.D. (1971) Effects of elevated ambient temperature on testis and blood levels and *in vitro* biosynthesis of testosterone in the ram. *Journal of Animal Science* 33, 804–807.

Goodeaux, S.D. and Kreider, J.L. (1978) Motility and fertility of stallion sperm isolated in bovine serum albumin. *Theriogenology* 10, 405–414.

Goodpasture, J.C., Reddy, J.M. and Zanveld, L.J.D. (1981) Acrosin, proacrosin and acrosin inhibitor of guinea pig spermatozoa capacitated and acrosome reacted *in vitro*. *Biology of Reproduction* 25, 44–55.

Goodrowe, K.L. and Heath, E. (1984) Disposition of the manchette in the normal equine spermatid. *Anatomical Record* 209(2), 177–183.

Gordon, I. (1983) Control and manipulation of reproduction in horses. In: *Controlled Breeding in Farm Animals*, 1st edn. Pergamon Press, Oxford, pp. 379–419.

Gordon, I. (1997) Introduction to controlled reproduction in horses. In: Gordon, I. (ed.) *Controlled Reproduction in Horses, Deer and Camelids. Controlled Reproduction in Farm Animals Series*, Vol. 4, 1st edn. CAB International, Wallingford, Oxon, pp. 1–35.

Gotze, R. (1949) *Besamung und Unfruchtbarkeit der Haussaugetiere.* (Insemination and infertility in domestic mammals.) M. and H. Schaper, Berlin.

Graham, E.F., Crabo, B.G. and Pace, M.M. (1978) Current status of semen preservation in the ram, boar and stallion. Proceedings of the XII Biennial Symposium on Animal Reproduction. *Journal of Animal Science, Supplement* 11, 80–119.

Graham, E.F., Schmehl, M.LO. and Deyo, R.C.M. (1984) Effects of prostaglandin supplementation on frozen thawed ram spermatozoa. *Proceedings of the 10th Technical Conference on AI and Reproduction*. NAAB, Milwaukee, pp. 4–29.

Graham, J.K. (1996) Methods for induction of capacitation and the acrosome reaction of stallion spermatozoa. *Veterinary Clinics of North America, Equine Practice* 12(1), 131–147.

Gravance, C.G., Liu, I.K.M., Davis, R.O. Hughes, J.P. and Casey, P.J. (1996) Quantification of normal morphometry of stallion spermatozoa. *Journal of Reproduction and Fertility* 108(1), 41–46.

Gravance, C.G. Champion, Z. Liu, I.K.M. and Casey, P.J. (1997) Sperm head morphometry analysis of ejaculate and dismount stallion semen samples. *Animal Reproduction Science* 47, 149–155.

Greenhof, G.R. and Kenney, R.M. (1975) Evaluation of the reproductive status of non pregnant mares. *Journal of the American Veterinary Medical Association* 167, 449–458.

Grondahl, C., Grondahl, M.L., Hyttel, P. and Greve, T. (1993) Acrosome damage in fresh and frozen stallion semen. Evaluation by scanning electron microscopy. *Dansk Veterinaertidsskrift* 76(6), 225–230.

Grubaugh, W.R. (1982) The effect of pinealectomy in pony mares. *Journal of Reproduction and Fertility* 32, 293–295.

Gunn, R.M.C. (1936) Fertility in sheep. Artificial production of seminal ejaculation and the characteristics of spermatozoa contained therein. *Bulletin of the Council of Science and Industrial Research of Australia* No. 94.

Gunnarsson, V. (1997) A survey of the fertility of Icelandic stallions. MSc thesis, University of Wales, Aberystwyth.

Gunsalus, I.C., Salisbury, G.W. and Willet, E.L. (1941) The bacteriology of bull semen. *Journal of Dairy Science* 24, 911–919.

Haard, M.C. and Haard, M.G.H. (1991) Successful commercial use of frozen stallion semen abroad. *Journal of Reproduction and Fertility, Supplement* 44, 647–648.

Hafez, E.S.E. (1980) Reproductive cycles, horses. In: Hafez, E.S.E. (ed.) *Reproduction in Farm Animals*, 4th edn. Lea and Febiger, Philadelphia, pp. 387–409.

Hafez, E.S.E. (1986) Reproductive cycles, horses. In: Hafez, E.S.E. (ed.) *Reproduction in Farm Animals*, 5th edn. Lea and Febiger, Philadelphia.

Hafez, E.S.E. (1993a) Artificial insemination. In: Hafez, E.S.E. (ed.) *Reproduction in Farm Animals*, 6th edn. Lea and Febiger, Philadelphia, pp. 424–440.

Hafez, E.S.E. (1993b) X- and Y-chromosome bearing spermatozoa. In: Hafez, E.S.E. (ed.) *Reproduction in Farm Animals*, 6th edn. Lea and Febiger, Philadelphia, pp. 441–445.

Hagenas, L., Ritzen, E.M. and Suginami, H.L. (1978) Temperature dependence of sertoli cell function. *International Journal of Andrology, Supplement* 2, 449–458.

Hall, J.L. (1981) Relationship between semen quality and human sperm penetration of zona free hamster ova. *Fertility and Sterility* 35, 457–463.

Hammerstedt, R.H. and Graham, J.K. (1992) Cryopreservation of poultry sperm. The enigma of glycerol. *Cryobiology* 29, 26–38.

Hammerstedt, R.H., Graham, J.K. and Nolan, J.P. (1990) Cryopreservation of mammalian sperm: what we ask them to survive. *Journal of Andrology* 11, 73–88.

Hancock, J.L. (1957) The structure of spermatozoa. *Veterinary Record* 69, 996–997.

Haring, H. (1985) Artificial insemination in horses. *Tierzuchter* 37(11), 482–483.

Harris, J.M., Irvine, C.H.G. and Evans, M.J. (1983) Seasonal changes in serum levels of FSH, LH and testosterone and in semen parameters in stallions. *Theriogenology* 19, 311–322.

Harrison, L.A., Squires, E.L. and McKinnon, A.O. (1991) Comparison of hCG, buserelin and luprostiol for induction of ovulation in cycling mares. *Journal of Equine Veterinary Science* 11, 163–166.

Harrison, R.A.P. and Vickers, S.E. (1990) Use of fluorescent probes to assess membrane integrity in mammalian spermatozoa. *Journal of Reproduction and Fertility* 88, 343–352.

Harrison, R.A.P. and White, J.G. (1972) Glycolytic enzymes in the spermatozoa and cytoplasmic droplets of bull, boar and rams and their leakage after shock. *Journal of Reproduction and Fertility* 30, 105–115.

Harrison, R.A.P., Dott, H.M. and Foster, G.C. (1982) Bovine serum albumin, sperm motility and 'dilution effect'. *Journal of Experimental Zoology* 222, 81–88.

Harrison, R.G. (1975) Effect of temperature on the mammalian testis. In: Hamilton, D.W. and Greep, R.O. (eds) *Handbook of Physiology*, Vol. 5, Section 7. American Physiology Society, Washington, DC, pp. 219–223.

Hayes, H.M. (1974) Congenital umbilical and inguinal hernias in cattle, horses, swine, dogs and cats: risk by breed and by sex among hospital patients. *American Journal of Veterinary Research* 35, 839–842.

Heape, W. (1897) The artificial insemination of mammals and subsequent fertility on impregnation of their ova. *Proceedings of the Royal Society of London*, 61, 52.

Heffe, G. (1993) Investigations on the storage of fresh stallion semen. Effect of various centrifugation diluents on keeping properties of fresh semen. Thesis, Tierärztliche Hochschule Hannover, Germany.

Heiskanen, M.L. (1995) Hevosten keinosiemennys sperman kasittelyn ja sailytksen vaikutus tiinehtymiseen. (Artificial insemination of horse. The effect of semen handling and storage on insemination results.) PhD Thesis, Kuopio Univeristy (cited in Katila, 1997).

Heiskanen, M.L., Pirhonen, A., Koskinen, E. and Maenpaa, P.H. (1987) Motility and ATP content of extended equine spermatozoa in different storage conditions. *Journal of Reproduction and Fertility, Supplement* 35, 103–107.

Heiskanen, M.L., Huhtinen, M., Pirhonen, A. and Maenpaa, P.M. (1993) Insemination results with stallion semen stored for two days. *International Journal of Thymology* 1(1), 48–49.

Heiskanen, M.L., Huhtinen, M., Pirhonen, A. and Maenpaa, P.M. (1994a) Insemination results with slow-cooled stallion semen stored for approximately 40 hours. *Acta Veterinaria Scandinavica* 35(3), 257–262.

Heiskanen, M.L., Hilden, L., Hyyppa, S., Kangasniemi, A., Pirhonen, A. and Maenpaa, P.H. (1994b) Freezability and fertility results with uncentrifuged stallion semen. *Acta Veterinaria Scandinavica* 35(4), 377–382.

Heitland, A.V., Jasko, D.J., Graham, J.K., Squires, E.L., Amann, R.P. and Pickett, B.W. (1995) Motility and fertility of stallion spermatozoa cooled and frozen in a modified skim milk extender containing egg yolk and liposome. *Biology of Reproduction, Monograph Equine Reproduction VI* 1, 753–759.

Heitland, A.V., Jasko, D.J., Squires, E.L., Graham, J.K., Pickett, B.W. and Hamilton, C. (1996) Factors affecting motion characteristics of frozen–thawed stallion spermatozoa. *Equine Veterinary Journal* 28(1), 47–53.

Hejzlar, Z. (1957) Knekterym otazkam umeleho osemenovani koni. (The artificial insemination of horses.) *Sborn. Csl. Akad. Zemed. Ved. Zivoc. Vyr.* 2(30), 597–614. *Animal Breeding Abstracts* 26, 51.

Hendrikse, J. and Van Der Kaay, F.C. (1950) Devruchtingsresultaten met verschillende methoden van kunstmatige inseminatie. (Conception with different methods of artificial insemination.) *Tijdschrift Voor Diergeneeskunde* 74, 983–999.

Henri-Petri, I. (1993) Effect of 3 methods of treatment on the sperm motility and morphology of stallion semen after different periods of storage. Thesis, Tierärztliche Hochschule Hannover, Germany, 128 pp.

Hermenet, M.J., Sawyer, H.R., Pickett, B.W., Amann, R.P., Squires, E.L. and Long, P.L. (1993) Effect of stain, technician, number of spermatozoa evaluated and slide preparation on assessment of spermatozoal viability by light microscopy. *Journal of Equine Veterinary Science* 13(8), 449–455.

Hernandez-Jauregui, P.A., Sosa, A. and Gonzalez-Angulo, A. (1975) The fine structure of the glycocalyx of equine spermatozoa: a high resolution cytochemical study. *Journal of Reproduction and Fertility, Supplement* 23, 91–94.

Hiipakka, R.A. and Hammerstedt, R.H. (1978) 2-Deoxyglucose transport and phosphorylation by bovine sperm. *Biology of Reproduction* 19, 368–379.

Hillman, R.B., Olar, T.T. and Squires, E.L. (1980) Temperature of the artificial vagina and its effect on seminal quality and behavioural characteristics of stallions. *Journal of the American Veterinary Medical Association* 177(8), 720–722.

Hinrichs, K. and Hunt, P.R. (1990) Ultrasound as an aid to diagnosis of granulosa cell tumour in the mare. *Equine Veterinary Journal* 22, 99–103.

Hjinskaja, T. (1956) Vlijanie nekotoryh faktorov na ustoicivostj semeni zerebcov k zamorazivaniju pri temperature −70°. (The effect of various factors on stallion spermatozoa frozen to −70°.) *Konevodstvo* 26, 11, 32–35. *Animal Breeding Abstracts* 25, 574.

Hochereau-de Riviers, M.T. (1963) The constancy of the relative frequencies of the stages of the cycle of the seminiferous epithelium in the bull and rat. *Annals of Biology, Animal Biochemistry and Biophysics* 3, 93–102.

Hochereau-de Riviers, M.T. (1980) In: Jutisz, M. and McKerns, K.W. (eds) *Synthesis and Release of Adenohypophyseal Hormones.* Plenum Press, New York, p. 307.

Hochereau-de Riviers, M.T., Monet Kuntz, C. and Couret, M. (1987) Spermatogenesis and sertoli cell numbers and function in rams and bulls. *Journal of Reproduction and Fertility, Supplement* 34, 101–104.

Hochereau-de Riviers, M.T., Courtens, J.-L. and Couret, M. (1990) Spermatogenesis in mammals and birds. In: Laming, G.E. (ed.) *Marshall's Physiology of Reproduction, Volume 2, Reproduction in the Male,* 4th edn. Churchill Livingstone, London, pp. 106–182.

Hochi, S., Korosue, K., Choi, Y.H. and Oguri, N. (1996) *In vitro* capacitation of stallion spermatozoa assessed by the lysophosphatidylcholine-induced acrosome reaction and the penetration rate into *in-vitro* matured, zona free mare oocytes. *Journal of Equine Veterinary Science* 16(6), 244–248.

Holstein, A.F. and Roosen-Runge, E.C. (1981) *Atlas of Human Spermatogenesis.* Grosse Verlag, Berlin, 140 pp.

Holtan, D.W., Douglas, R.H. and Ginther, O.J. (1977) Estrus, ovulation and conception following synchronisation with progestagens, prostaglandin F2α and human CG in pony mares. *Journal of Animal Science* 44(3), 431–437.

Horse Race Betting Levy Board (1997) *Codes of Practice 1997–1999 on Contagious Equine Metritis, Klebsiella Pneumoniae, Pseudomonas Aeroginosa, Equine Viral Arteritis and Equine Herpes Virus 1.* Horse Race Betting Levy Board, London.

Hough, S.R. and Parks, J.E. (1994) Platelet-activating factor acetylhydrolase activity in seminal plasma from the bull, stallion, rabbit and rooster. *Biology of Reproduction* 50(4), 912–916.

Householder, D.D., Pickett, B.W., Voss, J.L. and Olar, T.T. (1981) Effect of extender, number of spermatozoa and hCG on equine fertility. *Equine Veterinary Science* 1, 9–13.

Housley, M.D. and Stanley, K.K. (1982) *Dynamics of Biological Membranes.* John Wiley and Sons, New York.

Huckins, C. (1971) The spermatogonal stem cell population in adult rats. 1. Their morphology, proliferation and maturation. *Anatomical Record* 169, 533–558.

Huckins, C. (1978) The morphology and kinetics of spermatogonal degeneration in normal adult rats: an analysis using a simplified classification of germinal epithelium. *Anatomical Record* 190, 905–911.

Huek, C. (1990) Untersuchungen zur Flussigkonservierung von Pferdesperma unter Verwendung, Verschiedner kuhl-ond Transportsystene-Laborstudie. Inaugural Dissertation, Hannover (cited in Katila, *et al.*, 1997).

Huffel, X.M. van, Varner, D.D., Hinrichs, K., Garcia, M.C., Stremienski, P.J. and Kenney, R.M. (1985) Photomicrographic evaluation of stallion spermatozoal motility characteristics. *American Journal of Veterinary Research* 46(6), 1272–1275.

Hughes, J.P. and Loy, R.G. (1969) Investigations on the effect of intrauterine inoculations of *Streptococcus Zooepidemicus* in the mare. *Proceedings of the 15th Annual Convention of the American Association of Equine Practitioners.* Houston, Texas, pp. 289–292.

Hughes, J.P. and Loy, R.G. (1970) Artificial insemination in the equine. A comparison of natural breeding and AI of mares using semen from six stallions. *Cornell Veterinary* 60, 463–475.

Hughes, J.P. and Loy, R.G. (1975) The relation of infection to infertility in the mare and stallion. *Equine Veterinary Journal* 7, 155–159.

Hughes, J.P. and Loy, R.G. (1978) Variations in ovulation response associated with the use of prostaglandins to manipulate the lifespan of the normal dioestrus corpus luteum or prolonged corpus luteum in the mare. *Proceedings of the 24th Annual Convention of the American Association of Equine Practitioners,* 173–175.

Hughes, J.P. and Stabenfeldt, G.H. (1977) The use of hormones in reproductive management of the mare. *Australian Veterinary Journal* 53, 258–261.

Hughes, J.P., Loy, R.G., Atwood, C., Astbury, A.C. and Burd, H.E. (1966) The occurrence of *Pseudomonas* in the reproductive tract of mares and its effect on fertility. *Cornell Veterinarian* 56, 595–600.

Hughes, J.P., Asbury, A.C., Loy, R.G. and Burd, H.E. (1967) The occurrence of *Pseudomonas* in the reproductive tract of stallions and its effect on fertility. *Cornell Veterinarian* 57, 53–69.

Hunter, R.H.F. (1982) *Reproduction in Farm Animals.* Longman Press, London, UK, 149 pp.

Hunter, R.H.F. (1990) Gamete lifespan in the mare's genital tract. *Equine Veterinary Journal* 22(6), 378–379.

Hurtgen, J.P. (1987) Stallion genital abnormalities. In: Robinson, N.E. (ed.) *Current Therapy in Equine Medicine, 2.* W.B. Saunders Co., Philadelphia, pp. 558–562.

Hurtgen, J.P. (1997) Commercial freezing of stallion semen. *Proceedings of the 19th Bain-Fallon Memorial Lectures, Equine Reproduction, Australian Equine Veterinary Association.* Manly, New South Wales, Australia, pp. 17–23.

Huszar, G., Vigue, L. and Corrales, M. (1990) Sperm creatine kinase activity in fertile and infertile oligospermic men. *Journal of Andrology* 11, 40–45.

Hyland, J.H. and Bristol, F. (1979) Synchronisation of oestrus and timed insemination of mares. *Journal of Reproduction and Fertility, Supplement* 27, 251–255.

Ijaz, A. and Ducharme, R. (1995) Effects of various extenders and taurine on survival of stallion sperm cooled to 5°C. *Theriogenology* 44(7), 1039–1050.

Inskeep, P.B. and Hammerstedt, R.H. (1985) Endogenous metabolism by sperm in response to altered cellular ATP requirements. *Journal of Cellular Physiology* 123, 180–190.

Ionata, L.M., Pickett, B.W., Heird, J.C., Lehner, P.N. and Squires, E.L. (1991a) Characteristics of stallions in novel and in routine semen collection sites, with or without restraint. *Theriogenology* 36(6), 913–921.

Ionata, L.M., Anderson, T.M., Pickett, B.W., Heird, J.C. and Squires, E.L. (1991b) Effect of supplementary sexual preparation on semen characteristics of stallions. *Theriogenology* 36(6), 923–937.

Irons, M.J. and Clermont, Y. (1982) Kinetics of fibrous sheath formation in the rat spermatid. *American Journal of Anatomy* 165, 121–130.

Irvine, C.H.G. (1993) Prostaglandins. In: McKinnon, A.O. and Voss, J.L. (eds) *Equine Reproduction*. Lea and Febiger, Philadelphia, pp. 319–325.

Irvine, C.H.G. and Alexander, S.L. (1991) Effect of sexual arousal on gonadotrophin releasing hormone, luteinising hormone and follicle stimulating hormone secretion in stallions. *Journal of Reproduction and Fertility, Supplement* 44, 135–143.

Irvine, C.H.G. and Alexander, S.L. (1993) GnRH. In: McKinnon, A.O. and Voss, J.L. (eds) *Equine Reproduction*. Lea and Febiger, Philadelphia, pp. 37–45.

Irvine, C.H.G., Alexander, S.L. and Turner, J.E. (1986) Seasonal variation in the feedback of sex steroid hormones on serum LH concentration in the male horse. *Journal of Reproduction and Fertility* 76, 221.

Ivanoff, E.I. (1922) On the use of artificial insemination for zootechnical purposes in Russia. *Journal of Agricultural Science* 12, 244.

Jacob, R.J., Cohen, D., Bouchey, D., Davis, T. and Borchelt, J. (1988) Molecular pathogenesis of equine coital exanthema: identification of a new equine herpesvirus isolated from lesions reminiscent of coital exanthema in a donkey. In: Powell, D.G. (ed.) *Equine Infectious Diseases V: Proceedings of the 5th International Conference of Equine Infectious Diseases*. University Press of Kentucky, Lexington, pp. 140–146.

Jainudeen, M.R. and Hafez, E.S.E. (1993) Reproductive failure in males. In: Hafez, E.S.E. (ed.) *Reproduction in Farm Animals*, 6th edn. Lea and Febiger, Philadelphia, pp. 287–297.

James, J. (1998) Electronmicroscopic evaluation of fresh and frozen equine spermatozoa. Msc Thesis, University of Wales, Aberystwyth.

Jasko, D.J. (1992) Evaluation of stallion semen. *The Veterinary Clinics of North America, Equine Practitioners* 8(1), 149–165.

Jasko, D.J., Little, T.V., Smith, K., Lein, D.H. and Foote, R.H. (1988) Objective analysis of stallion sperm motility. *Theriogenology* 30(6), 1159–1167.

Jasko, D.J., Smith, K., Little, T.V., Lein, D. and Foote, R.H. (1989) A spectrophotometric procedure for the determination of objective measurements of equine spermatozoan motility. *Theriogenology* 31(5), 945–954.

Jasko, D.J., Lein, D.H. and Foote, R.H. (1990a) Determination of the relationship between sperm morphologic classifications and fertility in stallions: 66 cases (1987–1988). *Journal of the American Veterinary Association* 197(3), 389–394

Jasko, D.J., Lein, D.H. and Foote, R.H. (1990b) A comparison of two computer automated semen analysis instruments for the evaluation of sperm motion characteristics in the stallion. *Journal of Andrology* 11, 453–459.

Jasko, D.J., Lein, D.H. and Foote, R.H. (1991a) Stallion spermatozoal morphology and its relationship to spermatozoal motility and fertility. *Proceedings of 37th Annual Convention of the American Association of Equine Practitioners.* San Francisco, pp. 211–221.

Jasko, D.J., Lein, D.H. and Foote, R.H. (1991b) The repeatability and effect of season on seminal characteristics and computer-aided sperm analysis in the stallion. *Theriogenology* 35(2), 317–327.

Jasko, D.J., Moran, D.M., Farlin, M.E. and Squires, E.L. (1991c) Effect of semen plasma dilution or removal on spermatozoa motion characteristics of cooled stallion semen. *Theriogenology* 35, 1059–1067.

Jasko, D.J., Hathaway, J.A., Schaltenbrand, V.L., Simper, W.D. and Squires, E.L. (1992a) Effect of seminal plasma and egg yolk on motion characteristics of cooled stallion spermatozoa. *Theriogenology* 37, 1241–1252.

Jasko, D., Little, T.V., Lein, D.H. and Foote, R.H. (1992b) Comparison of spermatozoal movement and semen characteristics with fertility in stallions: 64 cases (1987–1988). *Journal of the American Veterinary Association* 200(7), 979–985.

Jasko, D.J., Martin, J.M. and Squires, E.L. (1992c) Effect of insemination volume and concentration of spermatozoa on embryo recovery in mares. *Theriogenology* 37, 1233–1239.

Jasko, D.J., Bedford, S.J., Cook, N.L., Mumford, E.C., Squires, E. and Pickett, B.W. (1993a) Effect of antibiotics on motion characteristics of a cooled stallion spermatozoa. *Theriogenology* 40, 885–893.

Jasko, D.J. Moran, D.M., Farlin, M.E., Squires, E.L., Amann, R.P. and Pickett, B.W. (1993b) Pregnancy rates utilising fresh, cooled and frozen–thawed stallion semen. *Proceedings of the 38th Annual Convention of the American Association of Equine Practitioners.* Orlando, Florida, pp. 649–660.

Jaskowski, L. (1951) Observations on the use of AI for the control of infectious anaemia in the horse. *Medical Veterinarian* 7, 239–242.

Jernstrom, M. (1991) A personal view of the use of frozen stallion semen for ten years. *Swensk Veterinartidning* 43(13), 571–573.

Jeyendran, R.S., Van der Ven, H.H., Perez-Pelaez, M., Crabo, B.G. and Zanveld, L.J.D. (1984) Development of an assay to assess the functional integrity of the human sperm membrane and its relationship to other semen characteristics. *Journal of Reproduction and Fertility* 70, 219–228.

Jimenez, C.F. (1987) Effects of Equex STM and equilibration time on the pre-freeze and post thaw motility of equine epididymal spermatozoa. *Theriogenology* 28(6), 773–782.

Jochle, W. and Trigg, T.E. (1994) Control of ovulation in the mare with ovuplant a short-term release implant (STI) containing GnRH analogue deslorelin acetate. Studies from 1990–1994. A review. *Journal of Equine Veterinary Science* 14(12), 632–644.

Johnson, L. (1985) Increased daily production in the breeding season of stallions is explained by an elevated population of spermatogonia. *Biology of Reproduction* 32(5), 1181–1190.

Johnson, L. (1986a) A new approach to quantification of sertoli cells that avoids problems associated with irregular nuclear surface. *Anatomical Record* 214, 231–237.

Johnson, L. (1986b) Review article: spermatogenesis and aging in the human. *Journal of Andrology* 7, 331–354.

Johnson, A.L. (1986c) Pulsatile administration of gonadotrophin releasing hormone advances ovulation in cycling mare. *Biology of Reproduction* 35, 1123–1130.

Johnson, L. (1991a) Spermatogenesis. In: Cupps, D.T. (ed.) *Reproduction in Domestic Animals*, 3rd edn. Academic Press, New York, pp. 173–219.

Johnson, L. (1991b) Seasonal differences in equine spermatogenesis. *Biology of Reproduction* 44, 284–291.

Johnson, L.A. (1997) Flow cytometric separation of sperm in domestic animals. *Proceedings of ESHRE Campus 1997, Sex Ratio and Sex Selection.* Maastricht, The Netherlands, pp. 15–20.

Johnson, A.L. and Becker, S.E. (1987) Effects of physiologic and pharmacologic agents on serum prolactin concentrations in the non pregnant mare. *Journal of Animal Science* 65, 1292–1297.

Johnson, L. and Neaves, W.B. (1981) Age related changes in the leydig cell population, seminiferous tubules and sperm production in stallions. *Biology of Reproduction* 24, 703–712.

Johnson, L. and Tatum, M.E. (1988) Sequence of seasonal changes in numbers of sertoli cells, leidig cells and germ cells in adult stallions. *11th International Congress on Animal Reproduction and AI, University College Dublin, Volume 3,* brief communication. University of College Dublin, Dublin, pp. 373–374.

Johnson, L. and Tatum, M.E. (1989) Temporal appearance of seasonal changes in numbers of sertoli cells, leidig cells and germ cells in stallions. *Biology of Reproduction* 40, 994–999.

Johnson, L. and Thompson, D.L. Jr (1983) Age related and seasonal variation in sertoli cell population, daily spermatozoa production and serum concentrations of FSH, LH and testosterone in stallions. *Biology of Reproduction* 29, 777–789.

Johnson, L. and Thompson, D.L. Jr (1987) Effect of seasonal changes in leidig cell number on the volume of smooth endoplasmic reticulum in leidig cells and intertesticular testis content in stallions. *Journal of Reproduction and Fertility* 81, 227–232.

Johnson, L., Amann, R.P. and Pickett, B.W. (1978a) Scanning electron microscopy of the equine seminiferous tubules. *Fertility and Sterility* 29, 208–215.

Johnson, L., Amann, R.P. and Pickett, B.W. (1978b) Scanning electron microscopy of the epithelium and spermatozoa in the equine excurrent duct system. *Journal of Veterinary Research* 39, 1428–1434.

Johnson, L., Amann, R.P. and Pickett, B.W. (1980) Maturation of equine epididymal spermatozoa. *American Journal of Veterinary Research* 41, 1190–1196.

Johnson, L.A., Flook, J.P. and Hawk, H.W. (1989) Sex preselection in rabbits: live births from X and Y sperm separated by DNA and cell sorting. *Biology of Reproduction* 41, 199–203.

Johnson, L., Hardy, V.B. and Martin, M.T. (1990) Staging equine seminiferous tubules by Nomarski optics in unstained histologic sections and in tubules mounted *in toto* to reveal the spermatogenic wave. *Anatomical Record* 227(2), 167–174.

Johnson, L., Varner, D.D., Tatum, M.E. and Scrutchfield, W.L. (1991a) Season but not age affects sertoli cell number in adult stallions. *Biology of Reproduction* 45(3), 404–410.

Johnson, L., Varner, D.D. and Thompson, D.L. (1991b) Effect of age and season on the establishment of spermatogenesis in the horse. *Journal of Reproduction and Fertility, Supplement* 44, 87.

Johnson, L., Cater, G.K., Varner, D.D., Taylor, T.S., Blanchard, T.L. and Rembert, M.S. (1994) The relationship of daily sperm production with number of sertoli cells and testicular size in adult horses: role of primitive spermatogonia. *Journal of Reproduction and Fertility* 100(1), 315–321.

Johnson, L., Blanchard, T.L., Varner, D.D. and Scrutchfield, W.L. (1997) Factors affecting spermatogenesis in the stallion. *Theriogenology* 48(7), 1199–1216.

Johnson, L.A., Welch, G.R., Rens, W. and Dobrinsky, J.R. (1998) Enhanced cytometric sorting of mammalian X and Y sperm: high speed sorting and orientating nozzle for artificial insemination. *Theriogenology* 49(1), 361.

Johnstone, F. (1998) The effect of selected environmental variables on reproductive function in the horse. Msc thesis, University of Wales, Aberystwyth.

Jones, L.S. and Berndtson, W.E. (1986) A quantitative study of sertoli cell and germinal cell populations as related to sexual development and ageing in the stallion. *Biology of Reproduction* 35(1), 138–148.

Jones, R.L. and Martin, I.A.C. (1965) Deep freezing ram spermatozoa: the effects of milk, yolk citrate and synthetic diluents containing sugar. *Journal of Reproduction and Fertility* 10, 413–423.

Jones, R., Mann, T. and Sherins, R (1979) Reproductive breakdown of phospholipids in human spermatozoa, spermicidal properties of fatty acid peroxides and protective action of seminal plasma. *Fertility and Sterility* 31, 531–537.

Jones, R.L., Squires, E. and Slade, N.P. (1984) The effect of washing on the aerobic bacterial flora of the stallion's penis. *Proceedings of the American Association of Equine Practitioners* 30, 9–15.

Jones, W.R. (1980) Immunological factors in male and female infertility. In: Hearn, J.P. (ed.) *Immunological Aspects of Reproduction and Fertilisation Control*. MTP Press, Lancaster.

Julienne, P. (1987) Artificial insemination with fresh semen from French Trotter stallions and with frozen semen from French Saddle stallions at a stud farm. *13ème Journée de la Recherche Chevaline, 11 March 1987*. Centre de Études et de Recherche sur l'Économie et l'Organisation des Productions Animales, Paris, France, pp. 44–56.

Kainer, R.A. (1993) Reproductive organs of the mare. In: McKinnon, A.O. and Voss, J.L. (eds) *Equine Reproduction*. Lea and Febiger, Philadelphia, pp. 3–19.

Kamenev, N. (1955) Opyt po primeneniju molocnogo razbavitelja. (The use of milk diluent.) *Konevodstvo*, 25(2), 32–34. *Animal Breeding Abstracts*, 23, 1041.

Katila, T. (1995) Onset of duration of uterine inflammatory response of mares after insemination with fresh semen. *Biology of Reproduction, Monograph Equine Reproduction VI* 1, 515–517.

Katila, T. (1997) Procedures for handling fresh stallion semen. *Theriogenology* 48(7), 1217–1227.

Katila, T., Koskinen, E., Kuntsi, H. and Lindeburgh, H. (1988) Fertilisation after post ovulation insemination in mares. *Proceedings of the 11th International Congress in Animal Reproduction and AI*. University College Dublin, Ireland, 96.

Katila, T., Celebi, M. and Koskinen, E. (1996) Effect of timing of frozen semen insemination on pregnancy rate in mares. *Acta Veterinaria Scandinavica* 37(3), 361–365.

Katila, T., Combes, G.B., Varner, D.D. and Blanchard, T.L. (1997) Comparison of three containers used for the transport of cooled stallion semen. *Theriogenology* 48(7), 1085–1092.

Katz, D.F. (1991) Characteristics of sperm motility. *Annals of the New York Academy of Sciences* 637, 420–426.

Kayser, J.P.R. (1990) Effects of linear cooling rates on motion characteristics of stallion spermatozoa. MSc Thesis, Colorado State University.

Kayser, J.P., Amann, R.P., Shideler, R.K., Squires, E.L., Jasko, D.J. and Pickett, B.W. (1992) Effects of linear cooling rates on motion characteristics of stallion spermatozoa. *Theriogenology* 38, 601–614.

Kemp, B. and den Hartog, L.A. (1989) The influence of energy and protein intake on the reproductive performance of the breeding boar: a review. *Animal Reproduction Science* 20, 103–113.

Kemp, B., Grooten, H.J.G., den Hartog, L.A., Luiting, P. and Verstegen, M.W.A. (1988) The effect of high protein intake on sperm production in boars at two semen collection frequencies. *Animal Reproduction Science* 17, 65–71.

Kenney, R.M. (1975) Clinical fertility evaluation of the stallion. *Proceedings of the American Association of Equine Practitioners*, 336–355.

Kenney, R.M. and Cooper, W.L. (1974) Therapeutic use of a phantom for semen collection from a stallion. *Journal of the American Veterinary Medical Association* 165, 706–707.

Kenney, R.M. and Varner, D.D. (1986) Insemination artificielle chez les chevaux. (Artificial breeding of horses.) *Methodes Modernes de Reproduction dans l'Esece Chevaline*, Colloque International, Paris, pp. 2–37.

Kenney, R.M., Kingston, R.S., Rajamannan, A.H. and Rambers, C.F. Jr (1971) Stallion semen characteristics for predicting fertility. *Proceedings of the 17th Annual Convention of the American Association of Equine Practitioners*. Chicago, pp. 53–67.

Kenney, R.M., Bergman, R.V., Cooper, W.L. and Morse, G.W. (1975) Minimal contamination techniques for breeding mares: techniques and preliminary findings. *Proceedings of 21st Annual Convention of the American Association of Equine Practitioners*, pp. 327–335.

Kenney, R.M., Evenson, D.P., Garcia, M.C. and Love, C.C. (1995) Relationships between sperm chromatin structure, motility and morphology of ejaculated sperm and seasonal pregnancy rate. *Biology of Reproduction, Monograph Equine Reproduction VI*, 1, 647–653.

Kikuchi, N., Iguchi, I. and Hiramune, T. (1987) Capsule types of *Kebsiella pneumoniae* isolated from the genital tracts of mares with metritis, extra genital sites of healthy mares and the genital tracts of stallions. *Veterinary Microbiology* 15, 219–228.

Kimball, J.W. (1983) Cell division. In: *Biology*, 5th edn. Addison Wesley Publishing Company, pp. 172–188.

King, T.E. and Mann, T. (1959) Sorbitol metabolism in spermatozoa. *Proceedings of the Royal Society of Biology* 151, 226–243.

Kleeman, J.M., Moore, R.W. and Peterson, R.E. (1990) Inhibition of testicular steroidogenesis in 2,3,7,8-tetrachlorodibenzo-*p*-dioxin treated rats. Evidence that the key lesion occurs prior to or during pregnenolone formation. *Toxicology and Applied Pharmacology* 106, 112–125.

Klem, M.E., Kreider, J.L., Pruitt, J.B. and Potter, G.D. (1986) Motility and fertility of equine spermatozoa extended in bovine serum albumin and sucrose. *Theriogenology* 26(5), 569–576.

Klinefelter, G.R., Laskey, J.W. and Roberts, N.L. (1991) *In vitro/in vivo* effects of ethane dimethanesulfonate on leidig cells of adult rats. *Toxicology and Applied Pharmacology* 107, 460–471.

Kloppe, L.H., Varner, D.D., Elmore, R.G., Bretzlaff, K.N. and Shull, J.W. (1988) Effect of insemination timing on the fertilising capacity of frozen/thawed equine spermatozoa. *Theriogenology* 29(2), 429–439.

Klug, E. (1986) Present position of horse AI. *Tierzuchter* 38(9), 380–382.

Klug, E. (1992) Routine artificial applications in the Hannovarian sport horse breeding association. *Animal Reproduction Science* 28, 39–44.

Klug, E. and Schmid, D.O. (1982) Success of AI and its monitoring by blood typing. *Deutsche Tierärztliche Wochenschrift* 89(5), 181–184.

Klug, E. and Tekin, N. (1991) The Celle Central AI Centre and its new organisation. *Lalahan Hayvanclk Arastrmo Enstitusu Dergisi* 31(1–2), 123–129.

Klug, E., Treu, H., Hillmann, H. and Heinze, H. (1975) Results of insemination of mares with fresh and frozen stallion semen. *Journal of Reproduction and Fertility, Supplement* 23, 107–110.

Klug, E., Deegen, Liesk, R., Freytag, K., Martin, J.C., Gunzel, A.R. and Bader, H. (1979) The effect of vesiculectomy on seminal characteristics in the stallion. *Journal of Reproduction and Fertility, Supplement* 27, 61–66.

Kneissl, S. (1993) Freezing of stallion semen. Effect of semen collection technique, centrifugation, storage method and freezing method on sperm motility and membrane integrity. Thesis, Tierärztliche Hochschule Hannover, Germany, 143 pp.

Knickerbocker, J., Sawyer, H.R., Amann, R.P., Tekpetey, F.R. and Niswender, G.D. (1988) Evidence for the presence of oxytocin in the ovine epididymis. *Biology of Reproduction* 39, 391–397.

Ko, T.H. and Lee, S.E. (1993) Establishment of a biological assay for the fertilising ability of equine spermatozoa. *Korean Journal of Animal Science* 35(3), 175–179.

Koehler, J.K., Smith, D. and Karp, L.E. (1984) The attachment of acrosome-intact sperm to the surface of zona-free hamster oocytes. *Gamete Research* 9(2), 197–205.

Kosiniak, K. (1975) Characteristics of successive jets of ejaculated semen of stallions. *Journal of Reproduction and Fertility, Supplement* 23, 59–61.

Koskinen, E., Junnila, M., Katila, T. and Soini, H. (1989) A preliminary study on the use of betaine as a cryoprotective agent in deep freezing of stallion semen. *Journal of Veterinary Medicine,* Series A 36(2), 110–114.

Koskinen, E., Lindeberg, H., Kuntsi, H., Ruotsalainen, L. and Katila, T. (1990) Fertilisation of mares after post ovulation insemination. *Journal of Veterinary Medicine,* Series A 37(1), 77–80.

Koskinen, E., Martila, P. and Katila, T. (1997) Effect of 19-norandrostenololylaurate on semen characteristics of colts. *Acta Veterinaria Scandinavica* 38(1), 41–50.

Kotilainen, T., Huhtinen, M. and Katila, T. (1994) Sperm induced leukocytosis in the equine uterus. *Theriogenology* 41(3), 629–636.

Kreider, J.L., Tindall, W.C. and Potter, G.D. (1985) Inclusion of bovine serum albumin in semen extenders to enhance maintenance of stallion spermatozoa motility. *Theriogenology* 23, 399–408.

Kroger, B. (1991) Use of the swim-up method Verfahren als Beurteilungskriterium fur Hengstsamen. Thesis, Tierärztliche Hochschule Hannover, Germany.

Krogsrud, J. and Onstad, O. (1971) Equine coital exantheme – isolation of a virus and transmission experiments. *Acta Veterinaria Scandinavica* 12, 1–14.

Kroning, T. (1986) Electronmicroscopic studies on the differentiation of male germ cells in the horse. Thesis, Tierärztliche Fakultat der Ludwig-Maximillions-Universitat Munchon, Germany.

Kubiak, J.R., Crawford, B.H., Squires, E.L., Wrigley, R.H. and Ward, G.M. (1987) The influence of energy intake and percentage of body fat on the reproductive performance of non pregnant mares. *Theriogenology* 28, 587–598.

Kuhne, A. (1996) Effect of seminal plasma on the suitability for freezing of stallion semen. Thesis, Tierärztlicher Hochschule Hannover, Germany.

Kuhr, J. (1957) Charakteristika hrebciho semene a nektere formy jeho redeni. (The characters of stallion semen and methods of diluting it.) *Sborn. Csl. Akad. Zemed. Ved, Zivoc. Vyr,* 2(30), 557–574. *Animal Breeding Abstracts* 25, 1692.

Kuklin, A.D. (1983) Artificial insemination of horses. *Vetereinariya, Moscow, USSR* 7, 57–58.

Kumi-Diaka, J. and Badtram, G. (1994) Effect of storage on sperm membrane integrity and other functional characteristics of canine spermatozoa: *in vitro* bioassay for canine semen. *Theriogenology* 41, 1355–1366.

Kuznetsov, M.P. (1956) Artificial insemination of sheep in the USSR. *Proceedings of the 3rd International Congress of Animal Reproduction,* Cambridge, p. 64.

Ladd, P.W. (1985) The male genital system. In: Jubb, K.V.F., Kennedy, P.C. and Palmer, N. (eds) *Pathology of Domestic Animals,* 3rd edn, Vol. 3. Academic Press, New York, pp. 409–459.

Lake (1971) The male in reproduction. In: Bell, D.J. and Freeman, B.M. (eds) *Physiology and Biochemistry of the Domestic Fowl.* Academic Press, New York.

Lambert, W.V. and McKenzie, F.F. (1940) Artificial insemination in livestock breeding. *Circular US Department of Agriculture,* No. 567. *Animal Breeding Abstracts,* 10, 7.

Lampinen, K. and Kattila, T. (1994) Effects of lubricants on the motility of horse spermatozoa. *Suomen Eainlaakarilehti* 100(6), 380–383.

Landeck, A.G. (1997) Studies on the seasonal occurrence and correlations of testicular sex hormones of the stallion in blood and seminal plasma, their distribution in the ejaculate, and relationships with ejaculate quality. Justus-Liebig-Universität, Fachbereich Veterinarmedizin, Germany.

Lang, A.L., Vogelsang, M.M. and Potter, G.D. (1995) Circadian patterns of plasma LH and testosterone concentrations in stallions in summer and winter. *Proceedings of the 14th Equine Nutrition and Physiology Symposium.* Ontario, California, pp. 284–289.

Lange, J., Matheja, S., Klug, E., Aurich, C. and Aurich, J.E. (1997) Influence of training and competition on the endocrine regulation of testicular function and on semen parameters in the stallion. *Reproduction in Domestic Animals* 32(6), 297–302.

Langford, G.A., Marcus, G.J., Hackett, A.J., Ainsworth, L., Wolynetz, M.S. and Peters, H.E. (1979) A comparison of fresh and frozen semen in the insemination of confined sheep. *Canadian Journal of Animal Science* 59, 685–691.

Langkammer, S. (1994) Investigations on the keeping properties of fractionated and non fractionated fresh stallion semen. Thesis, Tierärztliche Hochschule Hannover, Germany.

Langlais, J. and Roberts, K.D. (1985) A molecular membrane model of sperm capacitation and the acrosome reaction of mammalian spermatozoa. *Gamete Research* 12, 183–224.

Laplaud, M. and Cassou, R. (1948) Recherches sur l'etectro-ejaculate chez le taureau et le verrant. (Research on electro ejaculation in the bull and the ram.) *Comptes Rendus des Séances de la Société Biologie* 142, 726.

Laplaud, M., Bruneel, R. and Galland, H. (1950) Nouveau modele de vagin artifficiel pour chevaux et baudets – technique de la collete. (New model of an artificial vagina for horses and asses – technique for collection.) *Comptes Rendus de l'Academie d'Agriculture de France* 36, 351–354. *Animal Breeding Abstracts*, 19, 48.

Lapwood, K.R. and Martin, I.A.C. (1966) The use of monosaccharides, disaccharides and trisaccharides in synthetic diluents for storage of ram spermatozoa at 37°C and 5°C. *Australian Journal of Biological Science* 19, 655–671.

Larsson, K. (1978) Current research on the deep freezing of boar semen. *World Review of Animal Production* 14(4), 59–64.

Lawson, K.A. (1996) Longevity of stallion semen in various extenders when chilled to 4°C. MSc Thesis, The University of Wales, Aberystwyth.

Lawson, K.A. and Davies Morel, M.C.G. (1996) The use of mare's milk as a seminal extender for chilled stallion semen. *Warwick Horse Conference February 1996 New Developments in Equine Studies*.

Le Blanc, M.M. (1993) Endoscopy. In: McKinnon, A.O. and Voss, J.L. (eds) *Equine Reproduction*. Lea and Febiger, Philadelphia, pp. 255–257.

Le Blanc, M.M. (1995) Ultrasound of the reproductive tract In: Kobluk, C.N., Ames, T.R. and Goer, R.J. (eds) *The Horse – Diseases and Clinical Management*, Vol. 2. Saunders and Company, Philadelphia, pp. 926–935.

Le Blanc, M., Ward, L., Tran, T. and Widders, P. (1991). Identification and opsonic activity of immunoglobulins recognising *Streptococcus zooepidemicus* antigens in uterine fluids of mares. *Journal Reproduction and Fertility, Supplement* 44, 289–296.

Le Bland, C.P. and Clermont, Y. (1952) Spermiogenesis of rat, mouse, hamster and guinea pig as revealed by the periodic acid-fuchs in sulfurous acid technique. *American Journal of Anatomy* 90, 167–215.

Leatham, J.H. (1970) Nutrition. In: Johnson, A.D., Gomes, W.R. and Van Denmark, N.L. (eds) *The Testis*. Academic Press, New York, pp. 169–205.

Leendertse, I.P., Asbury, A.C., Boening, K.J. and Saldern, F.C. von (1990) Successful management of persistant urination during ejaculation in a Thoroughbred stallion. *Equine Veterinary Education* 2(2), 62–64.

Leone, E. (1954) Ergothionine in the equine ampullar secretions. *Nature* (London) 174, 404–407.

Leopold, S. (1994) Factors affecting Ca^{2+} flux in equine spermatozoa. Thesis, Tierärztliche Hochschule Hannover, Germany, 111 pp.

Ley, W.B., Lessard, P. and Pyle, H. (1991) Sperm chromatin structure assay of stallion spermatozoa: methods of sperm preservation. *Proceedings of the Society of Theriogenology 1991*. Society of Theriogenology, San Diego, California.

Lipczynski, A. and Deskur, S. (1958) Some observations on a season of AI of mares. *Ptodnosc I nieplodnosc zwierzat domowych. Zesz. Probl. Postep. Nauk. Roln. Polsk. Akad. Nauk,* 11, 175–178. *Animal Breeding Abstracts,* 26, 1773.

Lippe, A. (1986) Studies on the feasibility of using the hamster ova penetration test for estimating the fertilizing ability of stallion spermatozoa. Thesis, Tierärtzliche Hochschule Hannover, Germany.

Little, T.V. (1998) Accessory sex glands and internal reproductive tract evaluation. In: Rantanen, N.W. and McKinnon, A.O. (eds) *Equine Diagnostic Ultrasonography.* Williams and Wilkins, Baltimore, pp. 271–288.

Little, T.V. and Woods, G.L. (1987) Ultrasonography of accessory sex glands in the stallion. *Journal of Reproduction and Fertility, Supplement* 35, 87–94.

Lofstedt, R. (1997) Standard approach to the use of progesterone and oestradiol. cited by Revell, S., Wales, personal communication.

Loir, M. (1972) Metabolism of ribonucleic acid and proteins in spermatocytes and spermatids of the ram. *Annals of Biology, Animal Biochemistry and Biophysics* 12, 203–219.

Long, P.L., Pickett, B.W., Sawyer, H.R., Shidler, R.K., Amann, R.P. and Squires, E.L. (1993) Relationship between morphologic and seminal characteristics of equine spermatozoa. *Journal of Equine Veterinary Science* 10, 298–300.

Loomis, P.R. (1993) Factors affecting the success of AI with cooled, transported semen. *Proceedings of the 38th Annual Convention of the American Association of Equine Practitioners* 2, pp. 629–647.

Loomis, P.R., Amann, R.P., Squires, E.L. and Pickett, B.W. (1983) Fertility of unfrozen and frozen stallion spermatozoa extended in EDTA–lactose–egg yolk and packaged in straws. *Journal of Animal Science* 56, 687–693.

Lopez, A.M.S. and Souza, W. (1991) Distribution of filipin–sterol complexes in the plasma membrane of stallion spermatozoa during the epididymal maturation process. *Molecular Reproduction and Development* 28(2), 158–168.

Lorton, S.P., Sullivan, J.J., Bean, B., Kaprotl, M., Kellgren, M. and Marshall, C. (1988a) A new antibiotic combination for frozen bovine semen. 2. Evaluation of seminal quality. *Theriogenology* 29, 593–607.

Lorton, S.P., Sullivan, J.J., Bean, B., Kaprotl, M., Kellgren, M. and Marshall, C. (1988b) A new antibiotic combination for frozen bovine semen. 3. Evaluation of fertility. *Theriogenology* 29, 609–614.

Love, C.C. (1992) Semen collection techniques. *Veterinary Clinics of North America, Equine Practice* 8(1), 111–128.

Love, C.C. and Varner, D.D. (1998) Ultrasonography of the scrotal contents and penis of stallion. In: Rantanen, N.W. and McKinnon, A.O. (eds) *Equine Diagnostic Ultrasonography.* Williams and Wilkins, Baltimore, pp. 253–270.

Love, C.C., Loch, W.L., Bristol, F., Garcia, M.C. and Kenney, R.M. (1989) Comparison of pregnancy rates achieved with frozen semen using two packaging methods. *Theriogenology* 31, 613–622.

Love, C.C., Garcia, M.C., Riera, F.R. and Kenney, R.M. (1991) Evaluation of measures taken by ultrasonography and calipers to estimate testicular volume and predict daily sperm output in the stallion. *Journal of Reproduction and Fertility, Supplement* 44, 99–105.

Love, C.C., McDonnell, S.M. and Kenney, R.M. (1992) Manually assisted ejaculation in a stallion with erectile dysfunction subsequent to paraphimosis. *Journal of the American Veterinary Medical Association* 200(9), 1357–1359.

Loy, R.G. and Swann, S.M. (1966) Effects of exogenous progestagens on reproductive phenomena in mares. *Journal of Animal Science* 25, 821–825.

Loy, R.G., Buell, J.R., Stevenson, W. and Hamm, D. (1979) Sources of variation in response intervals after prostaglandin treatment in mares with functional corpus lutea. *Journal of Reproduction and Fertility, Supplement* 27, 229–235.

Loy, R.G., Pernstein, R., O'Conna, D. and Douglas, R.H. (1981) Control of ovulation in cycling mares with ovarian steroids and prostaglandin. *Theriogenology* 15, 191–197.

Lyngset, O., Aamdal, J. and Velle, W. (1965) Artificial insemination in the goat with deep frozen and liquid semen after hormonal synchronisation of oestrus. *Nordisk Veterinär Medicin* 17, 178–181.

Mackintosh, M.E. (1981) Bacteriological techniques in the diagnosis of equine genital infection. *Veterinary Record* 108, 52–55.

Macleod, J. and McGee, W.R. (1950) The semen of the Thoroughbred. *Cornell Veterinarian* 40, 233–248.

MacMillan, K.L. (1979) Factors influencing conception rates to artificial breeding in New Zealand dairy herds: a review. *Proceedings of the New Zealand Society of Animal Production* 39, 129–137.

Madsen, M. and Christensen, P. (1995) Bacterial flora of semen collected from Danish warmblood stallions by artificial vagina. *Acta Veterinaria Scandinavica* 36(1), 1–7.

Magistrini, M. and Vidament, M. (1992) Artificial insemination in horses. *Recueil de Medicine Vétérinaire Special: Reproduction des Equides* 168(11–12), 959–967.

Magistrini, M., Chanteloube, P. and Palmer, E. (1987) Influence of season and frequency of ejaculation on production of stallion semen for freezing. *Journal of Reproduction and Fertility, Supplement* 35, 127–133.

Magistrini, M., Couty, I. and Palmer, E. (1992a) Interactions between sperm packaging, gas environment, temperature and diluent on fresh stallion sperm survival. *Acta Veterinaria Scandinavica, Supplement* 88, 97–110.

Magistrini, M., Couty, I. and Palmer, E. (1992b) Factors influencing stallion sperm survival. *Proceedings of the 12th International Congress on Animal Reproduction* 4, 1888–1890.

Magistrini, M., Seguin, F., Beau, P., Akoka, S., Le Pape, A. and Palmer, E. (1995) ^{1}H Nuclear magnetic resonance analysis of stallion genital tract fluids and seminal plasma: contribution of the accessory sex glands to the ejaculate. *Biology of Reproduction, Monograph Equine Reproduction VI* 1, 599–607.

Magistrini, M., Vidament, M., Clement, F. and Palmer, E. (1996) Fertility prediction in stallions. *Animal Reproduction Science* 42, 181–188.

Magistrini, M., Guitton, E., Levern, Y., Nicolle, J.C., Vidament, M., Kerboeuf, D. and Palmer, E. (1997) New staining methods for sperm evaluation estimated by microscopy and flow cytometry. *Theriogenology* 48, 1229–1235.

Makler, A. (1978) The thickness of microscopically examined seminal sample and its relationship to sperm motility estimation. *International Journal of Andrology* 1, 213–219.

Malmgren, L. (1992) Sperm morphology in stallions in relation to fertility. *Acta Veterinaria Scandinavica, Supplement* 88, 39–47.

Malmgren, L. (1997) Assessing the quality of raw semen: a review. *Theriogenology* 48, 523–530.

Malmgren, L., Kamp, B., Wockener, A., Boyle, M. and Colenbrander, B. (1994) Motility, velocity and acrosome integrity of equine spermatozoa stored under different conditions. *Reproduction in Domestic Animals* 29(7), 469–476.

Mann, T. (1954) *The Biochemistry of Sperm*. Methuen and Company, London, 32 pp.

Mann, T. (1964a) Metabolism of spermatozoa. Fructolysis, respiration and sperm energetics. In: Mann, T. (ed.) *The Biochemistry of Semen and the Male Reproductive Tract*. Barns and Noble, New York, pp. 265–307.

Mann, T. (1964b) *Biochemistry of Semen and the Male Reproductive Tract*. Methuen and Company, London, 494 pp.

Mann, T. (1975) Biochemistry of stallion semen. *Journal of Reproduction and Fertility, Supplement* 23, 47–52.

Mann, T. and Lutwak-Mann, C. (1981) *Male Reproductive Function and Semen.* Springer-Verlag, New York, 193 pp.

Mann, T., Leone, E. and Polge, C. (1956) The composition of stallion's semen. *Journal of Endocrinology* 13, 279–290.

Mann, T., Minotakis, C.S. and Polge, C. (1963) Semen composition and metabolism in the stallion and jackass. *Journal of Reproduction and Fertility* 5, 109–122.

Martin, G.B. and Walken-Brown, S.W. (1995) Nutritional influences on reproduction in mature male sheep and goats. *Journal of Reproduction and Fertility, Supplement* 49, 437–439.

Martin, J.C., Klug, E. and Gunzel, A.R. (1979) Centrifugation of stallion semen and its storage in large volume straws. *Journal of Reproduction and Fertility* 27, 47–51.

Marusi, A. and Ferroni, O. (1993) Influence of an artificial photoperiod on sexual maturity and reproductive function of stallions. *Ippologia* 4(1), 57–60.

Massanyi, L. (1988) The ultrstructure of stallion spermatozoa. *Acta Zootechnica Universitatis Agriculturae Nitra* 44, 23–33.

Mather, E.C., Refsal, K.R., Gustafisson, B.K., Seguin, B.E. and Whitmore, H.C. (1979) The use of fiber-optic techniques in clinical diagnosis and visual assessment of experimental intra-uterine therapy in mares. *Journal of Reproduction and Fertility, Supplement* 25, 293–297.

Matousek, J. (1968) Seminal vesicle fluid substance of bulls sensitizing spermatozoa to cold shock. *Proceedings of the 6th International Congress on Animal Reproduction and Artificial Insemination.* Paris, pp. 553–555.

Matveev, L.V. (1986) The effect of catalase and hydrogen peroxide on semen quality of stallions. *Puti uskoreniya nauch-tekh progressa v konevod,* Moscow, pp. 126–132.

Mayer, D.T., Squires, C.D., Bogart, R. and Oloufa, M.M. (1951) The technique for characterising mammalian spermatozoa as dead or living by differential staining. *Journal of Animal Science* 10, 226–235.

Mayhew, I.G. (1990) Neurological aspects of urospermia in the horse. *Equine Veterinary Education* 2(2), 68–69.

Mazur, P. (1984) Freezing of living cells. Mechanisms and implications. *American Journal of Physiology* 247, C125–C142.

Mazur, P. (1985) Basic concepts in freezing cells. In: Johnson, L.A. and Larsson, K (eds) *Proceedings of the International Conference on Deep Freezing Boar Semen.* Swedish University of Agricultural Science, Uppsala, pp. 91–111.

Mazur, P. and Cole, K.W. (1989) Roles of unfrozen fraction, salt concentration and changes in cell volumes in the survival of frozen human erythrocytes. *Cryobiology* 26, 1–29.

Mazurova, J. and Mazura, F. (1988) Haemolytic streptococci in the genital organs of horses. *Sbornik Vedeakych Praci Ustredniho Statniho Veterinarniho Ustavuv Praze* 17, 51–62.

Mazzari, G., du Mensil du Buisson, F. and Ortavant, R. (1968) Action of temperature on spermatogenesis, sperm production and fertility of the boar. *Proceedings of the VI Congress on Reproduction in Animals and AI,* Vol. 1, pp. 305.

McCollum, W.H., Timoney, P.J., Roberts, A.W., Willard, J.E. and Carswell, G.D. (1988) Response of vaccinated and non-vaccinated mares to artificial insemination with semen from stallions persistently infected with equine arteritis virus. In: Powell, D.G. (ed.) *Equine Infectious Diseases V: Proceedings of the Fifth International Conference.* Univerity Press of Kentucky, Lexington, Kentucky, pp. 13–18.

McDonnell, S.M. (1992) Ejaculation: physiology and dysfunction, stallion management. *Veterinary Clinics of North America, Equine Practice* 8, 57–70.

McDonnell, S.M. and Love, C.C. (1990) Manual stimulated collection of semen from stallions. Training time, sexual behaviour and semen. *Theriogenology* 33, 1201–1210.

McDonnell, S.M. and Love, C.C. (1991) Xylazine induced ex-copulatory ejaculation in stallions. *Theriogenology* 36, 73–76.

McDonnell, S.M. and Odian, M.J. (1994) Impramine and xylazine induced ex copula ejaculation in stallions. *Theriogenology* 41(5), 1005–1010.

McDonnell, S.M. and Turner, R.M.O. (1994) Post-thaw motility and longevity of motility of impramine-induced ejaculate of pony stallions. *Theriogenology* 42(3), 475–481.

McDonnell, S.M., Garcia, M. and Kenney, R.M. (1987) Imipramine-induced erection, masturbation and ejaculation in male horses. *Pharmacology, Biochemistry and Behaviour* 27, 187–191.

McDonnell, S.M., Pozor, M.A., Beech, J. and Sweeney, R.W. (1991) Use of manual stimulation for the collection of semen from an atactic stallion unable to mount. *Journal of the American Veterinary Medical Association* 199(6), 753–754.

McDonnell, S.M., Love, C.C., Pozor, M.A. and Diehl, N.K. (1992) Phenylbutazone treatment in breeding stallions: preliminary evidence for no effect on semen or testicular size. *Theriogenology* 37(6), 1225–1232.

McEntee, K. (1970) The male genital system. In: Jubb, K.V.F. and Kennedy, P.C. (eds) *Pathology of Domestic Animals*, 2nd edn. Academic Press, New York, pp. 450–454.

McKinnon, A.O. and Carnevale, E.M. (1993) Ultrasonography. In: McKinnon, A.O. and Voss, J.L (eds) *Equine Reproduction*. Lea and Febiger, Philadelphia, pp. 221–225.

McKinnon, A.O. and Voss, J.L. (1993) Breeding the problem mare. In: McKinnon, A.O. and Voss, J.L (eds) *Equine Reproduction*. Lea and Febiger, Philadelphia, pp. 369–378.

McKinnon, A.O., Squires, E.L., Carnevale, E.M., Harrison, L.A., Frantz, D.D., McChesney, A.E. and Shideler, R.K. (1987) Diagnostic ultrasonography of uterine pathology in the mare. *Proceedings of the American Association of Equine Practitioners*, 605–622.

McKinnon, A.O., Voss, J.L., Trotter, G.W., Pickett, B.W., Shideler, R.K. and Squires, E.L. (1988) Hemospermia of the stallion. *Equine Practice* 10(9), 17–23.

Meinert, C., Silva, J.F.S., Kroetz, I., Klug, E., Trigg, T.E., Hoppen, H.O. and Jochle, W. (1993) Advancing the time of ovulation in the mare with a short term implant releasing the GnRH analogue deslorelin. *Equine Veterinary Journal* 25, 65–68.

Meistrich, M.L. (1989) Interspecies comparison and qualitative extrapolation of toxicity to the human male reproductive system. In: Working, P.K. (ed.) *Toxicity of the Male and Female Reproductive Systems*. Hemisphere Publishing, New York, pp. 303–321.

Menkveld, R., Rhemrev, J.P.T., Franken, D.R., Vermenden, J.P.W. and Kruger, F.F. (1996) Acrosomal morphology as a novel criterion for male-fertility diagnosis, relation with acrosin activity, morphology (strict criteria) and fertilisation *in vitro*. *Fertility and Sterility* 65(3), 637–644.

Merilan, C.P. and Loch, W.E. (1987) The effect of artificial vagina liners on liveability of stallion spermatozoa. *Equine Veterinary Science* 7(4), 226–228.

Merkies, K. and Buhr, M.M. (1998) Epididymal maturation affects calcium regulation in equine spermatozoa exposed to heparin and glucose. *Theriogenology*, 49(3), 683–695.

Merkt, H. (1984) Production and use of stallion semen for AI of the mare. In: *Le Cheval, Reproducion, Selection, Alimentation, Exploitation*. Institut National de la Recherche Agronomique, Paris, pp. 129–131.

Merkt, H., Klug, E., Bohm, K.H. and Weiss, R. (1975a) Recent observations concerning *Klebsiella* infections in stallions. *Journal of Reproduction and Fertility, Supplement* 23, 143–145.

Merkt, H., Klug, E., Krause, D. and Bader, H. (1975b) Results of long-term storage of stallion semen frozen by pellet method. *Journal of Reproduction and Fertility, Supplement* 23, 105–106.

Mesnil du Buisson, F. du (1994) Artificial insemination of domestic animals (except cattle) in France and its development. *Comptes Rendus de l'Academie d'Agriculture de France* 80(3), 89–106.

Metz, K.W., Berger, T. and Clegg, E.D. (1990) Absorbtion of seminal plasma proteins by boar spermatozoa. *Theriogenology* 34, 691–700.

Meyers, P.J. (1997) Control and synchronisation of oestrus cycle and ovulation. In: Youngquist, R.S. (ed.) *Current Therapy in Large Animal Theriogenology.* W.B. Saunders, Philadelphia, pp. 96–102.

Meyers, S.A., Liu, I.K.M., Overstreet, J.W. and Drobnis, E.Z. (1995a) Induction of acrosome reactions in stallion sperm by equine zona pellucida, porcine zona pellucida and progesterone. *Biology of Reproduction, Monograph Equine Reproduction VI* 1, 739–744.

Meyers, S.A., Overstreet, J.W., Lui, I.K.M. and Drobnis, E.Z. (1995b) Capacitation *in vitro* of stallion spermatozoa: comparison of progesterone induced acrosome reaction in fertile and sub-fertile males. *Journal of Andrology* 6(1), 47–54.

Meyers, S.A., Liu, I.K.M., Overstreet, J.W., Vadas, S. and Drobnis, E.Z. (1996) Zona pellucida binding and zona induced acrosome reactions in horse spermatozoa: comparisons between fertile and sub fertile stallions. *Theriogenology* 46(7), 1277–1288.

Mihailov, N.I. (1956) Primenenie moloka v kacestve razbavitelja semini seljskohozjaistvennyh zivotnyh. (Milk as a diluent for livestock semen.) *Zivotnovodstvo,* 4, 74. *Animal Breeding Abstracts* 24, 953.

Milovanov, V.K. (1934) *Iskusstvennoe osemenenie s-h zivotnyh. Prakticeskoe rukovodstvo. (Artificial Inseminatiom of Livestock. Practical Handbook.)* Seljhozgiz, Moscow-Leningrad.

Milovanov, V.K. (1938) *Isskusstvenoye Ossemenebie selsko-khoziasvennykh jivotnykh. (The Artificial Insemination of Farm Animals.)* Seljhozgiz, Moscow.

Mintscheff, P. and Prachoff, R. (1960) Attempts to improve the fertility of mares by starvation during oestrus. *Zuchthygiene. FortPH StorBesam. Haustiere* 4, 40–48.

Mohan, G. and Sahni, K.L. (1991) Comparative efficiency of certain extenders containing various levels of yolk for preservation of buffalo semen at 5°C. *Indian Journal of Animal Science* 61(7), 725–727.

Moore, R.Y. (1978) The innervation of the pineal gland. *Proceedings of Reproductive Biology* 4, 1–29.

Moore, H.D.M., Hall, G.A. and Hibbitt, K.G. (1976) Seminal plasma proteins and the reaction of spermatozoa from intact boars and from boars without seminal vesicles to cooling. *Journal of Reproduction and Fertility* 47, 39–45.

Moran, D.M., Jasko, D.J., Squires, E.L. and Amann, R.P. (1992) Determination of temperature and cooling rate which induce cold shock in stallion spermatozoa. *Theriogenology* 38, 999–1012.

Morris, R.P., Rich, G.A., Ralston, S.L., Squires, E.L. and Pickett, B.W. (1987) Follicular activity in transitional mares as affected by body condition and dietary energy. *Proceedings of the 10th Equine Nutrition and Physiology Symposium,* pp. 93–99.

Morse, M.J. and Whitmore, W.F. (1986) Neoplasms of the testis. In: Walsh, P.C., Gittes, R.F., Perlmutter, A.D. and Stomey, T.A. (eds) *Cambell's Urology,* Vol. 2, 5th edn. W.B. Saunders, Philadelphia, pp. 1535–1575.

Mortimer, D., Goel, N. and Shu, M.A. (1988) Evaluation of Cell Soft automated semen analysis system in routine laboratory setting. *Fertility and Sterility* 50, 960–968.

Moses, D.F., Delasheras, M.A., Valcarcel, A., Perez, L. and Baldassarre, H. (1995) Use of computerized analyser for the evaluation of frozen–thawed ram spermatozoa. *Andrologia* 27(1), 25–29.

Muller, Z. (1982) Fertility of frozen equine semen. *Journal of Reproduction and Fertility, Supplement* 32, 47–51.

Muller, Z. (1987) Practicalities of insemination of mares with deep frozen semen. *Journal of Reproduction and Fertility, Supplement* 35, 121–125.

Muller, Z. and Cekrit, P. (1991) Artificial insemination of mares with frozen semen in Czechoslovakia. *Mat. Symp. Rozrod. Koni. Krakow,* pp. 47–50.

Mumford, E.L., Squires, E.L., Jochle, E., Harrison, L.A., Nett, T.M. and Trigg, T.E. (1995) Use of deslorelin short term implants to induce ovulation in cycling mares during 3 consecutive estrous cycles. *Animal Reproduction Science* 39, 129–140.

Murray, P.J., Rowe, J.B., Pethick, D.W. and Adams, N.R. (1990) The effect of nutrition on testicular growth in Merino ram. *Australian Journal of Agricultural Research* 41, 185–195.

Myles, D.G. and Primakoff, P. (1991) Spermatozoa proteins that serve as receptors for the zona pellucida and their past testicular modification. *Annals of the New York Academy of Sciences* 637, 486–493.

Naden, J., Amann, R.P. and Squires, E.L. (1990) Testicular growth, hormone concentrations, seminal characteristics and sexual behaviour in stallions. *Journal of Reproduction and Fertility* 88, 167–176.

Nagase, H. (1967) Studies on the freezing of stallion semen. II. Factors affecting survival rates of stallion spermatozoa after freezing and thawing and results of fertility trial. *Japanese Journal of Animal Reproduction* 12, 52–57. *Animal Breeding Abstracts* 35, 195.

Nagase, H. and Niwa, T. (1964) Deep freezing bull semen in concentrated pellet form. I. Factors affecting survival of spermatozoa. *Proceedings of the 5th International Congress on Animal Reproduction and AI.* Trento, 4, 400–415.

Nagase, H., Soejima, A., Nowa, T., Oshida, H., Sagara, Y., Ishizaki, N. and Hoshi, S. (1966) Studies on the freezing storage of stallion semen. Fertility results of semen in concentrated pellet form. *Japanese Journal of Animal Reproduction* 12, 48–51.

Nash, J.G. Jr, Voss, J.L. and Squires, E.L. (1980) Urination during ejaculation in a stallion. *Journal of the American Veterinary Medical Association* 176, 224–227.

Naumenkov, A. and Romankova, N. (1979) Freezing stallion semen. *Konevodstvoi Konnyi Sport* 9, 13. *Animal Breeding Abstracts* 47, 4074.

Naumenkov, A. and Romankova, N. (1981) An improved semen diluent. *Konevodstvoi Konnyi Sport* 4, 34. *Animal Breeding Abstracts* 49, 6207.

Naumenkov, A. and Romankova, N. (1983) Improving diluent composition and handling for stallion sperm. *Nauchnye Trudy. Vsesoyuznyi Nauchno Issledovatel'skii Institut Konevodstva,* 38–47. *Animal Breeding Abstracts* 51, 6370.

Neely, D.P., Kindahl, H., Stabenfeldt, G.H., Edquist, L.-E. and Hughes, J.P. (1979) Prostaglandin release patterns in the mare: physiological, pathophysiological and therapeutic responses. *Journal of Reproduction and Fertility, Supplement* 27, 181–189.

Nessau, S. (1994) Structural and functional characterisation of the main protein components in stallion seminal plasma. Thesis, Tierärztliche Hochschule Hannover, Germany.

Nett, T.M. (1993) Reproductive peptide and protein hormones. In: McKinnon, A.O. and Voss, J.L. (eds) *Equine Reproduction.* Lea and Febiger, Philadelphia, pp. 109–113.

Neu, S.M., Timoney, P.J. and Lowry, S.R. (1992) Changes in semen quality in the stallion following experimental infection with equine arteritis virus. *Theriogenology* 37(2), 407–431.

Newcombe, J.R. and Wilson, M.C. (1997) The use of progesterone releasing intravaginal devices to induce estrus and ovulation in anoestrus standardbred mares in Australia. *Equine Practice* 19(6), 13–21.

Nielsen, F. (1938) En Kunstig Vagina til Hengst og Tyr. (An artificial vagina for use in horses and cattle.) *Maanedsskr. Dyrlaeg.* 49, 573.

Nishikawa, Y. (1959a) Semen properties and artificial insemination in horses. In: *Studies on Reproduction in Horses.* Koei, Kyoto, Japan, 208 pp.

Nishikawa, Y. (1975) Studies on the preservation of raw and frozen horse semen. *Journal of Reproduction and Fertility, Supplement* 23, 99–104.

Nishikawa, Y. and Wade, Y. (1951) On artificial insemination in the horse. V. On the properties of semen and the factors affecting them. *Bulletin of the National Institute of Agricultural Science (Chiba)* 1, 13–28. *Animal Breeding Abstracts* 20, 60.

Nishikawa, Y., Wade, Y. and Onuma, H. (1951) Studies on artificial insemination in the horse. VI. Morphological studies on horse spermatozoa. *Bulletin of the National Institute of Agricultural Science (Chiba)* 1, 29–36. *Animal Breeding Abstracts* 20, 62.

Nishikawa, Y., Wade, Y. and Shinomiya, S. (1968) Studies on the deep freezing of horse spermatozoa. *Proceedings of the 7th International Congress on Animal Reproduction and Artificial Insemination.* Munich, 2, 1589–1591.

Nishikawa, Y., Iritani, A. and Shinomiya, S. (1972a) Studies on the protective effects of egg yolk and glycerol on the freezability of horse sperm. *Proceedings of the International Congress on Animal Reproduction and Artificial Insemination* 2, 1545–1549.

Nishikawa, Y., Wade, Y. and Shinomiya, S. (1972b) Our experimental results and methods of deep freezing of horse spermatozoa. *Proceedings of the International Congress on Animal Reproduction and Artificial Insemination* 2, 1539–1543.

Nishikawa, Y., Wade, Y. and Shinomiya, S. (1976) Results of conception tests of frozen horse semen during the past ten years. *Proceedings of the 15th International Congress on Animal Reproduction and Artificial Insemination* 4, 1034–1037.

Oakberg, E.F. (1960) Irradiation damage to animals and its effect on their reproductive capacity. *Journal of Dairy Science, Supplement* 43, 54–64.

Oba, E., Bicudo, S.Od., Pimentel, S.L., Lopes, R.S., Simonetti, F. and Hunziker, R.A. (1991) Quantitative and qualitative evaluation of stallion semen. In: *Anais, IX Congresso Brasileiro de Reproducao Animal.* Belo Horizonte, Brazil, 22 a 26 de Junho de 1991. Vol. II. Colegio Brasileiro de Reproducao Animal, Belo Horizonte, Brazil, 442 pp.

Oba, E., Bicudo, S.D., Pimentel, S.L., Lopes, R.S., Simonetti, F. and Hunziker, R.A. (1993) Quantitative and qualitative evaluation of stallion semen. *Revista Brasileira de Reproducao Animal* 17(1–2), 57–74.

Oetjen, M. (1988) Use of different acrosome stains for evaluating the quality of fresh and frozen semen. Thesis, Tierärztliche Hochschule Hannover, Germany.

Oettle, E.E. (1986) Using a new acrosome stain to evaluate sperm morphology. *Veterinary Medicine* 1986, 263–266.

Oh, R. and Tamaoki, B. (1970) Steroidogenesis in equine testis. *Acta Endocrinologica* 64, 1–16.

Oh, W.Y., Park, N.G., Kim, Y.H., Lee, S.S., Kim, H.S., Kim, J.K. and Shin, W.J. (1994) Studies on the common semen characteristics in highly fertile Cheju horses. *RDA Journal of Agricultural Science, Livestock* 36(2), 552–557.

Olbrycht, T. (1935) Artificial insemination in mares. *Przeglad Wet.* 12, 1–28.

Ortavant, R. (1959) The development and duration of the spermatogenic cycle in the ram. Part I. Description and histological study of the spermiogenic cycle. *Annals of Zootechnology* 8, 183–271.

Ortavant, R., Courot, M. and Hochereau-de Riviers, M.T. (1977) Spermatogenesis in domestic mammals. In: Cole, H.H. and Cupps, P.T. (eds) *Reproduction in Domestic Animals*, 3rd edn. Academic Press, New York, pp. 203–227.

Oshida, H. (1968) Studies on the freezing of stallion semen. III. Pellet frozen semen preserved in liquid nitrogen. *Animal Breeding Abstracts* 36, 387.

Oshida, H., Mikawa, T., Horiuchi, S., Takahashi, H., Tomizuka, T. and Nagase, H. (1967) Studies on the freezing of stallion semen. III. Pellet frozen semen preserved in liquid nitrogen. *Japanese Journal of Animal Reproduction* 13, 136–140.

Oshida, H., Tomizuka, T., Masaki, J., Hanada, A. and Nagase, H. (1972) Some observations on freezing stallion semen. *Proceedings of the 7th International Congress of Animal Reproduction and AI.* Munich, 2, 1535–1537.

Overstreet, J.W., Yanagimachi, R., Katz, D.F., Hayashi, K. and Hanson, F.W. (1980) Penetration of human spermatozoa into the human zona pellucida and the zona-free hamster egg: a study of fertile donors and infertile patients. *Fertility and Sterility* 33, 534–542.

Pace, M.M. and Sullivan, J.J. (1975) Effect of timing of insemination, numbers of spermatozoa and extender components on the pregnancy rates in mares inseminated with frozen stallion semen. *Journal of Reproduction and Fertility, Supplement* 23, 115–121.

Padilla, A.W. and Foote, R.H. (1991) Extender and centrifugation effects on the motility patterns of slow cooled spermatozoa. *Journal of Animal Science* 60, 3308–3313.

Padilla, A.W., Tobback, C. and Foote, R.H. (1991) Penetration of frozen–thawed, zona-free hamster oocytes by fresh and slow-cooled stallion spermatozoa. *Journal of Reproduction and Fertility, Supplement* 44, 207–212.

Palacios Angola, A. and Zarco Quintero, L. (1996) Effect of replacing egg yolk in freezing extenders by bovine serum albumin, equine serum or bovine serum on post thawing sperm viability in stallion semen. *Veterinaria Mexico* 27(3), 221–227.

Palacios Angola, A., Valencia Mandez, J. and Zarco Quintero, L. (1992) Effect of size of straw and sperm concentration on acrosome damage and post-thawing motility in stallion semen. *Veterinaria Mexico* 23(4), 315–318.

Palmer, E. (1976) Different techniques for synchronisation of ovulation in the mare. *Proceedings of the 8th International Congress on Animal Reproduction and Artificial Insemination.* Krakow, 1, 191.

Palmer, E. (1979) Reproductive management of mares without detection of oestrus. *Journal of Reproduction and Fertility, Supplement* 27, 263–270.

Palmer, E. (1984) Factors affecting stallion semen survival and fertility. *Proceedings of the 10th International Congress on Animal Reproduction and Artificial Insemination* 3, 377–379.

Palmer, E. (1991) Automated-analysis of stallion sperm motility after thawing. *Contraception, Fertilité, Sexualité* 19(10), 855–863.

Palmer, E. and Jousset, B. (1975) Synchronisation of oestrus in mares with prostaglandin analogue and hCG. *Journal of Reproduction and Fertility, Supplement* 23, 269–274.

Palmer, E., Domerg, D., Fauquenot, A. and Sainte-Marie, T. de (1984) Artificial insemination of mares: results of five years of research and practical experience. *Le Cheval. Reproduction, Selection, Alimentation, Exploitation.* Institut National de la Recherche Agronomique, Paris, pp. 133–147.

Palmer, E., Draincourt, M.A. and Chevalier, F. (1985) Breeding without oestrus detection. Synchronisation of oestrus. *Bulletin des Groupements Techniques Vétérinaires* 1, 41–50.

Pantke, P., Hyland, J.H., Galloway, D.B., Liu, D.Y and Baker, H.W.G. (1992) Development of a zona pellucida sperm binding assay for the assessment of stallion fertility. *Australian Equine Veterinarian* 10(2), 91.

Pantke, P., Hyland, J.H., Galloway, D.B., Liu, D.Y and Baker, H.W.G. (1995) Development of a zona pellucida sperm binding assay for the assessment of stallion fertility. *Biology of Reproduction, Monograph Equine Reproduction VI* 1, 681–687.

Papa, F.O. (1989) Advances in stallion semen technology. *Anais VIII Congresso Brasileiro de Reproduao Animal, Belo Horizonte, Brazil, 11–14 July 1989.* Colegio Brasileiro de Reproducao Animal, Belo Horizonte, Brazil, pp. 318–339.

Papa, F.O. (1991) Artificial insemination within frozen stallion semen. *Anais IX Congresso Brasileiro de Reproduao Animal, Belo Horizonte, Brazil, 22–26 July 1991.* Vol. II. Colegio Brasileiro de Reproducao Animal, Belo Horizonte, Brazil, pp. 80–90.

Papa, F.O., Alvarenga, M.A., Lopes, M.D. and Campos Filho, E.P. (1990) Infertility of autoimmune origins in a stallion. *Equine Veterinary Journal* 22, 145–146.

Papa, F.O., Campus Filho, E.P., Alvarenga, M.A., Bicudo, S.D. and Meria, C. (1991) Effect of storage container on acrosome integrity and thermoresistance of frozen stallion semen. *Anais IX Congresso Brasileiro de Reproduao Animal, Belo Horizonte, Brazil, 22–26 July 1991.* Vol. II. Colegio Brasileiro de Reproducao Animal, Belo Horizonte, Brazil, pp. 448–458.

Papaioannou, K.Z., Murphy, R.P., Monks, R.S., Hynes, N., Ryan, M.P., Boland, M.P. and Roche, J.F. (1997) Assessment of viability and mitochondrial function of equine spermatozoa using double staining and flow cytometry. *Theriogenology* 48, 299–312.

Park, N.K., Oh, W.Y., Lee, S.S., Oh, C.A., Kang, S.W., Ko, S.B., Kang, M.S. and Kim, H.S. (1995) Studies on semen freezing in Cheju native stallions. *RDA Journal of Agricultural Science, Livestock* 37(1), 459–463.

Parker, G.U. and Thwaites, C.J. (1972) The effects of under nutrition on libido and semen quality in adult Merino rams. *Australian Journal of Agricultural Research* 23, 109–115.

Parks, J.E. and Hammerstedt, R.H. (1985) Developmental changes occurring in the phospholipids of rams epididymal spermatozoa plasma membrane. *Biology of Reproduction* 32, 653–668.

Parks, J.E. and Lynch, D.V. (1992) Lipid composition and thermotropic phase behaviour of boar, bull, stallion and rooster sperm membrane. *Cryobiology* 29(2), 255–266.

Parks, J.E., Arion, J.W. and Foote, R.H. (1987) Lipids of plasma membrane and outer acrosomal membrane from bovine spermatozoa. *Biology of Reproduction* 37, 1249–1258.

Parlevliet, J.M. (1997) Clinical aspects of stallion infertility. Thesis, University of Utrecht, The Netherlands.

Parlevliet, J.M., Kemp, B. and Colenbrander, E. (1994) Reproductive characteristics and semen quality in Main Dutch Warmblood stallions. *Journal of Reproduction and Fertility* 101(1), 183–187.

Parlevliet, J.M., Bleumink-Pluym, N.M.C., Houwers, D.J., Remmen, J.L.A.M., Sluijter, F.J.H. and Colenbrander, B. (1997) Epidemiologic aspects of *Taylorella equigenitalis*. *Theriogenology* 47, 1169–1177.

Parsutin, G.V. (1939) Metody sobiranija spermy u losadel. (Methods of collecting sperm fluid of stallions.) *Voprosy plodovitosti I rabotosposobnosti losadi.* Seljhozgiz, Moscow. pp. 141–148. *Animal Breeding Abstracts* 8, 223.

Parsutin, G.V. and Rumjanceva, E. Ju. (1953) Projavlenie polovyh refleksov u molodyh zerebcov. (The occurrence of sexual reflexes in young stallions.) *Konevodstvo* 23(7), 12–17. *Animal Breeding Abstracts* 21, 1594.

Pascoe, P.R. (1979) Observations of the length of declination of the vulva and its relation to fertility in the mare. *Journal of Reproduction and Fertility, Supplement* 27, 299–305.

Pascoe, P.R. and Bagust, T.J. (1975) Coital exanthema in stallions. *Journal of Reproduction and Fertility, Supplement* 23, 147–150.

Pascoe, P.R., Spradbrow, P.B. and Bagust, T.J. (1968) Equine coital exanthema. *Australian Veterinary Journal* 44, 485.

Pascoe, P.R., Spradbrow, P.B. and Bagust, T.J. (1969) An equine genital infection resembling coital exanthema associated with a virus. *Australian Veterinary Journal* 45, 166.

Pearson, H. and Weaver, B.M.Q. (1978) Priapism after sedation, neuroleptanalgesia and anaesthesia in the horse. *Equine Veterinary Journal* 10, 85–90.

Pecnikov, P.P. (1955) Fiziko-himiceskie svoistva semeni zerebca. (Physico-chemical properties of stallion semen.) *Voprosy fiziologii razmnozenija losadel.* Trud. Vesojuz. Nauc-issled. Inst. Konev. Seljhozgiz, Moscow. pp. 74–134. *Animal Breeding Abstracts* 24, 1493.

Pegg, D.T. and Diaper, M.D. (1989) The unfrozen fraction hypothesis of freezing injury to human erythrocytes: a clinical examination of the evidence. *Cryobiology* 26, 30–43.

Peippo, J., Huhtinen, M. and Kotilainen, T. (1995) Sex diagnosis of equine preimplantation embryos using a polymerase chain reaction. *Theriogenology* 44(5), 619–627.

Perkins, N. (1996) Equine reproductive ultrasonography. *Veterinary Continuing Education* No. 172, Massey University, pp. 129–146.

Perry, E.J. (1945) Historical. In: Perry, E.J. (ed.) *The Artificial Insemination of Farm Animals.* Rutgers University Press, New Brunswick, pp. 3–8.

Perry, E.J. (1968) *The Artificial Insemination of Farm Animals*, 4th edn. Rutgers University Press, New Brunswick.

Pesic, M.C., Stanic, M., Jevremovic, M., Czarnecki, J. and Arsenjevic, S. (1993) Influence of biofactors in sperm washing medium on sperm progressive motility. *International Journal of Thymology* 1(2), 106–110.

Peterson, F.B., McFeely, R.A. and David, J.S.E. (1969) Studies on the pathogenesis of endometritis in the mare. *Proceedings of the 15th Annual Convention of the American Association of Equine Practitioners*, 279–284.

Peterson, R.N. (1982) The sperm tail and mid piece. In: Zaneveld, L.J.D. and Chatterton, R.T. (eds) *Biochemistry of Mammalian Reproduction*. Wiley and Sons, New York, pp. 153–173.

Peterson, R.N., Bundman, D. and Freund, M. (1977) Binding of cytochalasin B to hexose transport sites in human spermatozoa and inhibition of binding by purines. *Biology of Reproduction* 17, 198–206.

Phillips, P.H. (1939) Preservation of bull semen. *Journal of Biological Chemistry* 130, 415–418.

Phillips, W.D. and Chilton, T.J. (1991) Chromosomes and cell division. In: *'A' Level Biology*. Oxford University Press, pp. 320–327.

Philpott, M. (1993) The danger of disease transmission by artificial insemination and embryo transfer. *British Veterinary Journal* 149, 339–369.

Piao, S. and Wang, Y. (1988) A study on the technique of freezing concentrated semen of horses (donkeys) and the effect of insemination. *Proceedings of the International Congress for Animal Reproduction and Artificial Insemination*, 3, pp. 286a–286c.

Pickering, B.T., Birkett, S.D., Guldenaar, S.E.F., Nicholson, H.D., Warley, R.T.S. and Yavachev, L. (1989) Oxytocin in the testis: what, where and why? *Annals of the New York Academy of Sciences* 564, 198–209.

Pickett, B.W. (1993a) Collection and evaluation of stallion semen for AI. In: McKinonn, A.O. and Voss, J.L. (eds) *Equine Reproduction*. Lea and Febiger, Philadelphia, pp. 705–714.

Pickett, B.W. (1993b) Seminal extenders and cooled semen. In: McKinonn, A.O. and Voss, J.L. (eds) *Equine Reproduction*. Lea and Febiger, Philadelphia, pp. 746–754.

Pickett, B.W. (1993c) Cryopreservation of semen. In: McKinonn, A.O. and Voss, J.L. (eds) *Equine Reproduction*. Lea and Febiger, Philadelphia, pp. 769–789.

Pickett, B.W. and Amann, R.P. (1987) Extension and storage of stallion spermatozoa. A review. *Equine Veterinary Science* 7, 289–302.

Pickett, B.W. and Amann, R.P. (1993) Cryopreservation of semen. In: McKinnon, A.O. and Voss, E.L. (eds) *Equine Reproduction*. Lea and Febiger, Philadelphia, pp. 768–789.

Pickett, B.W. and Back, D.G. (1973) Procedures for preparation, collection, evaluation and insemination of stallion semen. *Bulletin of the Colorado State University, Agricultural Experimental Station, Animal Reproduction Laboratory General Services* 935.

Pickett, B.W. and Shiner, K.A. (1994). Recent developments in AI in horse. *Livestock Production Science* 40, 31–36.

Pickett, B.W. and Voss, J.L. (1972) Reproductive management of stallions. *Proceedings of the 18th Annual Convention of the American Association of Equine Practitioners*, San Francisco, California, 18, 501–531.

Pickett, B.W. and Voss, E.L. (1975) The effect of semen extenders and sperm number on mare fertility. *Journal of Reproduction and Fertility, Supplement* 23, 95–98.

Pickett, B.W. and Voss, J.L. (1976) Reproductive management of the broodmare. *Animal Reproduction Laboratory General Series, Bulletin No. 961*, Colorado State University, Fort Collins, pp. 18–21.

Pickett, B.W., Faulkner, J.L. and Sutherland, T.N. (1970) Effect of month and stallion on seminal characteristics and sexual behaviour. *Journal of Animal Science* 31, 713.

Pickett, B.W., Back, D.G., Burquash, L.D. and Voss, J.L. (1974a) The effect of extenders, spermatozoal numbers and rectal palpation on equine fertility. *5th NAAB Technical Conference on Artificial Insemination and Reproduction,* pp. 47–58.

Pickett, B.W., Gebauer, M.R., Seidel, G.E. and Voss, J.L. (1974b) Reproductive physiology of the stallion: spermatozoal losses in the collection equipment and gel. *Journal of the American Veterinary Medical Association* 165, 708–710.

Pickett, B.W., Burwash, L.D., Voss, J.L. and Back, D.G. (1975a) Effect of seminal extenders on equine fertility. *Journal of Animal Science* 40, 1136–1143.

Pickett, B.W., Faulkner, J.L. and Voss, J.L. (1975b) Effect of season on semen characteristics of stallion semen. *Journal of Reproduction and Fertility, Supplement* 23, 25.

Pickett, B.W., Sullivan, J.J. and Seidel, G.E. Jr (1975c) Reproductive physiology of the stallion. V. Effect of frequency of ejaculation on seminal characteristics and spermatozoal output. *Journal of Animal Science* 40, 917–923.

Pickett, B.W., Sullivan, J.J. and Seidel, G.E. Jr (1975d) Effect of centrifugation and seminal plasma on motility and fertility of stallion and bull spermatozoa. *Fertility and Sterility* 26, 167–174.

Pickett, B.W., Faulkner, L.C., Seidel, G.E. Jr, Berndtson, W.E. and Voss, J.L. (1976) Reproductive physiology of the stallion. VI. Seminal and behavioural characteristics. *Journal of Animal Science* 43, 617–625.

Pickett, B.W., Voss, J.L. and Squires, E.L. (1977) Impotence and abnormal sexual behaviour in the stallion. *Theriogenology* 8, 329–347.

Pickett, B.W., Anderson, E.W., Roberts, A.D. and Voss, J.L. (1980) Freezability of first and second ejaculates of stallion semen. *Proceedings of the International Congress on Animal Reproduction and Artificial Insemination* 5, 339–347.

Pickett, B.W., Voss, J.L., Squires, E.L. and Amann, R.P. (1981) Management of the stallion for maximum reproductive efficiency. *Animal Reproduction Laboratory General Series Bulletin No. 1005,* Colorado State University, Fort Collins.

Pickett, B.W., Squires, E.L. and Voss, J.L. (1982) Techniques for training a stallion to use a phantom for seminal collection. *Journal of Equine Veterinary Science* 2, 66–67.

Pickett, B.W., Neil, J.R. and Squires, E.L. (1985) The effect of ejaculation frequency on stallion sperm output. *Proceedings of the 9th Equine and Physiology Society Symposium,* 290–295.

Pickett, B.W., Squires, E.L. and McKinnon, A.O. (1987) Procedures for collection, evaluation and utilisation of stallion semen for AI. *Animal Reproduction Laboratory Bulletin No. 03,* Colorado State University, Fort Collins.

Pickett, B.W., Voss, J.L., Bowen, R.A., Squires, E.L. and McKinnon, A.O. (1988a) Comparison of seminal characteristics of stallions that passed or failed seminal evaluation. In: *11th International Congress on Animal Production and Artificial Insemination,* June 26–30, Vol. 3, Paper 380. University College Dublin, Dublin, Republic of Ireland.

Pickett, B.W., Voss, J.L., Bowen, R.A., Squires, E.L. and McKinnon, A.O. (1988b) Seminal characteristics and total scrotal width (TSW) of normal and abnormal stallions. *Proceedings of the 33rd Annual Convention of the American Association of Equine Practitioners,* 485–518.

Pickett, B.W., Amann, R.P., McKinnon, A.O., Squires, E.L. and Voss, J.L. (1989) Management of the stallion for maximum reproductive efficiency. II. *Animal Reproduction Laboratory Bulletin No. 05,* Colorado State University, Fort Collins, pp. 121–125.

Pinkel, D., Garner, D.L., Gledhill, B.L., Lake, S., Stephenson, D. and Johnson, L.A. (1985) Flow cytometric determination of the proportions of X- and Y-chromosome-bearing sperm in samples purportedly separated from bull sperm. *Journal of Animal Science* 60, 1303–1307.

Pitra, C., Schafer, W. and Jewgenow, K. (1985) Quantitative measurement of the fertilising capacity of deep-frozen stallion semen by means of the hamster egg penetration test. *Monatschefte für Veterinarmedizin* 40(7), 235–237.

Polakoski, K.L. and Kopta, M. (1982) Seminal plasma. In: Zaneveld, L.J.D. and Catterton, R.T. (eds) *Biochemistry of Mammalian Reproduction*. John Wiley and Sons, New York.

Polge, C. and Rowson, L.E.A. (1952) Fertilizing capacity of bull spermatozoa after freezing at −79 degrees C. *Nature* 169, 626.

Polge, C., Smith, A.U. and Parkes, A.S. (1949) Revival of spermatozoa after vitrification and dehydration at low temperature. *Nature* 164, 666.

Polozoff, V. (1928) Die spermatozoenproduktion beim pferede. I. Mitteilung. *Pflugers Archiv. Ges. Physiol.* 218, 374–385.

Pool, K.C., Charneco, R. and Arns, M.J. (1993) The influence of seminal plasma from fractionated ejaculation on the cold storage of equine spermatozoa. *Proceedings of the 13th Conference of the Equine Nutrition and Physiology Society*, 395–396.

Pouret, E.J.M. (1982) Surgical techniques for the correction of pneumo and uro vagina. *Equine Veterinary Journal* 14, 249–250.

Pozor, M.A., McDonnell, S.M., Kenney, R.M. and Tischner, M. (1991) GnRH facilitates copulatory behaviour in geldings treated with testosterone. *Journal of Reproduction and Fertility* 44, 666–667.

Province, C.A. (1984) Cooling and storage of canine and equine spermatozoa. MSc Thesis, Colorado State University, Fort Collins.

Province, C.A., Amann, R.P., Pickett, B.W. and Squires, E.L. (1984) Extenders for preservation of canine and equine spermatozoa at 5°C. *Theriogenology* 22, 409–415.

Province, C.A., Squires, E.L., Pickett, B.W. and Amann, R.P. (1985) Cooling rates, storage temperature and fertility of extended equine spermatozoa. *Theriogenology* 23, 925–934.

Pruitt, J.A., Kreider, J.L., Potter, G.D., Bowen, J.M. and Evans, J.W. (1988) Fertility of stallion spermatozoa isolated on albumin gradients. *Journal of Equine Veterinary Science* 8(2), 153–155.

Pruitt, J.A., Arns, M.J. and Pool, K.C. (1993) Seminal plasma influences recovery of equine spermatozoa following *in vitro* culture (37°C) and cold storage (5°C). *Theriogenology* 39, 291.

Pursel, V.G. and Johnson, L.A. (1975a) Freezing of boar spermatozoa: fertilising capacity with concentrated semen and a new thawing procedure. *Journal of Animal Science* 40, 99–107.

Pursel, V.G. and Johnson, L.A. (1975b) Effect of extender components on fertilising capacity of frozen boar spermatozoa. *Journal of Animal Science* 41, 374.

Purswell, B.J., Dave, D.L., Caudle, A.B., Williams, D.J. and Brown, J. (1983) Spermatagglutins in serum and seminal fluid of bulls and their relationship to fertility classification. *Theriogenology* 20, 375–338.

Quinn, P.J. (1989) Principles of membrane stability and phase behaviour under extreme conditions. *Journal of Bioenergetics and Biomembranes* 21, 3–19.

Rabb, M.H., Thompson, D.L., Barry, B.E., Colburn, D.R., Garza, F. and Hehnke, K.E. (1989) Effects of sexual stimulation, with and without ejaculation, on serum concentrations of LH, FSH, testosterone, cortisol and prolactin in stallions. *Journal of Animal Science* 67, 2724–2729.

Raeside, J.I. (1969) The isolation of oestrone sulphate and oestradiol 17β sulphate from stallion testis. *Canadian Journal of Biochemistry* 47, 811–816.

Rajamannan, A.H.J., Zemjanis, R. and Ellery, J. (1968) Freezing and fertility studies with stallion semen. *Proceedings of the International Congress on Animal Reproduction and Artificial Insemination* 2, 1601–1604.

Ralston, S.L., Rich, G.A., Jackson, S. and Squires, E.L. (1986) The effect of vitamin A supplementation on sexual characteristics and vitamin A absorption in stallions. *Journal of Equine Veterinary Science* 6(4), 203–207.

Rasbech, N.O. (1959) Artificiel insemination I hesteavlen. (Artificial insemination in horses.) *Medlemsbl. Danske Dyrlaegeforen* 42(8), 9.

Rasbech, N.O. (1975) Ejaculation disorders of the stallion. *Journal of Reproduction and Fertility, Supplement* 23, 123–128.

Rath, D. and Brass, K. (1988) Ultrasound investigation of male genital organs. *Proceedings of 11th International Congress on Animal Reproduction and Artificial Insemination,* University College Dublin, Ireland, Vol. 23, Paper 382.

Rath, D., Leidig, C., Klug, E. and Krebs, H.C. (1987) Influence of chlorohexidine on seminal patterns in stallions. *Journal of Reproduction and Fertility* 35, 109–112.

Rauterberg, H. (1994) Use of glasswool Sephadex filtration for the collection of fresh semen from horses. Laboratory studies and field trials. Thesis, Tierärztliche Hochschule Hannover, Germany, 105 pp.

Reichart, M., Lederman, H., Hareven, D., Keden, P. and Bartoov, B. (1993) Human sperm acrosin activity with relation to semen parameters and acrosomal ultrastructure. *Andrologia* 25(2), 59–66.

Reifenrath, H. (1994) Use of L4 leucocyte absorption membrane filtration in AI in horses, using fresh or frozen semen. Thesis, Tierärztliche Hochschule Hannover, Germany.

Reifenrath, H., Sieme, H. and Klug, E. (1996) The use of L4 membrane filter in the filtration of horse semen. *Pferdeheilkunde* 12(5), 773–777.

Reis, C. (1991) An investigation on the storage of fresh stallion semen. Thesis, Lutwig-Maximilians-Universität Munchen, Germany, 64 pp.

Reiter, R.J. (1981) Pineal control of reproduction. In: Vidrio, E.A. and Galina, M.A. (eds) *Eleventh International Congress of Anatomy: Advances in the Morphology of Cells and Tissues.* Alan R. Liss, New York, pp. 349–355.

Revell, S.G. (1997) A sport horse for the future. *Proceedings of the British Society for Animal Science Equine Conference*, Cambridge, July 1997, (in press).

Revell, S.G. and Glossop, C.E. (1985) A long-life ambient temperature diluent for boar semen. *Animal Production* 8(3), 579–584.

Revell, S.G. and Mrode, R.A. (1994) An osmotic resistance test for bovine semen. *Animal Reproduction Science* 36, 77–86.

Revell, S.G., Pettit, M.T. and Ford, T.C. (1997) Use of centrifugation over idodixanol to reduce damage when processing stallion sperm for freezing. *Proceedings of the Joint Meeting of the Society for the Study of Fertility, British Fertility Society and British Andrology Society*, Abstract Series, No. 19, July 1997, Abstract No. 92, p. 38.

Rhynes, W.E. and Ewing, L.L. (1973) Testicular endocrine function in Hereford bulls exposed to high ambient temperature. *Endocrinology* 92, 509–515.

Rich, G.A., McGlothlin, D.E., Lewis, L.D., Squires, E.L. and Pickett, B.W. (1984) Effect of vitamin E supplement on stallion seminal characteristics and sexual behaviour. *Proceedings of the 10th International Congress on Animal Reproduction and Artificial Insemination,* Vol. 2, Paper No. 163. University of Illinois at Urbana-Champaign, Illinois.

Richardson, G.F and Wenkhoff, M.S. (1976) Semen collection from a stallion using a dummy mount. *Canadian Veterinary Journal* 17, 177–180.

Ricketts, S.W. (1975) The technique and clinical application of endometrial biopsy in the mare. *Equine Veterinary Journal* 7, 102–108.

Ricketts, S.W. (1981) Bacteriological examinations of the mare's cervix. Techniques and interpretation of results. *Veterinary Record* 108, 46–51.

Ricketts, S.W. (1993) Evaluation of stallion semen. *Equine Veterinary Education* 5(5), 232–237.

Ricketts, S.W. and Mackintosh, M.E. (1987) Role of anaerobic bacteria in equine endometritis. *Journal of Reproduction and Fertility, Supplement* 35, 343–351.

Ricketts, S.W., Young, A. and Medici, E.B. (1993) Uterine and clitoral cultures. In: McKinnon, A.O. and Voss, E.L. (eds) *Equine Reproduction.* Lea and Febiger, Philadelphia, pp. 234–245.

Rikmenspoel, M. (1964) Movements and active movements of bull sperm flagella as a function of temperature and viscosity. *Journal of Experimental Biology* 108, 205–230.

Riley, F.J. and Masters, W.H. (1956) Problems of male fertility. III. Bacteriology of human semen. *Fertility and Sterility* 7, 128–132.

Robbelen, I. (1993) A comparison of different methods of freezing stallion semen, effects of different diluents, equilibration times and freezing rates on sperm motility and membrane integrity. Thesis, Tierärztliche Hochschule Hannover, Germany, 135 pp.

Roberts, D.H. and Lucas, M.H. (1987) Equine infective anaemia. *Veterinary Annual* 27, 147–150.

Roberts, S.J. (1971a) Infertility in male animals. In: Roberts, S.J. (ed.) *Veterinary Obstetrics and Genital Diseases (Therigenology)*, 2nd edn. Comstock, Ithaca, New York, pp. 662–739.

Roberts, S.J. (1971b) AI in horses. In: Roberts, S.J. (ed.) *Veterinary Obstetrics and Genital Diseases (Theriogenology)*, 2nd edn. Comstock, Ithaca, New York, pp. 740–743.

Roberts, S.J. (1986a) Infertility in male animals (andrology). In: Roberts, S.J. (ed.) *Veterinary Obstetrics and Genital Diseases (Theriogenology)*, 3rd edn. Edwards Brothers, North Pomfret, Vermont, pp. 752–893.

Roberts, S.J. (1986b) Infertility in the mare. In: Roberts, S.J. (ed.) *Veterinary Obstetrics and Genital Diseases*, 3rd edn. Edwards Brothers, North Pomfret, Vermont, pp. 599–600.

Robertson, J.T. (1987) Small intestinal strangulation and obstruction. In: Robinson, E. (ed.) *Current Therapy in Equine Medicine*. W.B. Saunders, Philadelphia, pp. 47–51.

Roger, J.F. and Hughes, J.P. (1991) Prolonged pulsatile administration of gonadotrophin releasing hormone (GnRH) to fertile stallions. *Journal of Reproduction and Fertility, Supplement* 44, 155–168.

Roger, J.F., Kiefer, B.L., Evans, J.W., Neely, D.P. and Pacheco, C.A. (1979) The development of antibodies to human CG following its repeated injection in the cycling mare. *Journal of Reproduction and Fertility, Supplement* 27, 173–179.

Rogers, B.J. and Brentwood, B.J. (1982) Capacitation, acrosome reaction and fertilisation. In: Zaneveld, L.J.D. and Chatterton, R.T. (eds) *Biochemistry of Mammalian Reproduction*. John Wiley and Sons, New York, pp. 203–230.

Rogers, B.J. and Parker, R.A. (1991) Relationship between the human sperm hypo-osmotic swelling test and sperm penetration assay. *Journal of Andrology* 12, 152–158.

Roosen-Runge, E.C. and Giesel, L.O. (1950) Quantitative studies on spermatogenesis in the albino rat. *American Journal of Anatomy* 87, 1–30.

Roser, J.F. (1995) Endocrine profiles in fertile, subfertile and infertile stallions. Testicular response to human chorionic gonadotrophin in infertile stallions. *Biology of Reproduction Monograph, Equine Reproduction VI*, 1, 661–669.

Roser, J.F. (1997) Endocrine basis for testicular function in the stallion. *Theriogenology* 48(5), 883–892.

Roser, J.F. and Hughes, J.P. (1992) Prolonged pulsatile administration of gonadotrophin releasing hormone (GnRH) to fertile stallions. *Journal of Reproduction and Fertility, Supplement* 44, 155–168.

Rossdale, P.D. and Ricketts, S.W. (1980) The stallion. In: *Equine Stud Farm Medicine*, 2nd edn. Ballière Tindall, London, pp. 120–164.

Rossdale, P.D., Hunt, M.D.N., Peace, C.K., Hope, R., Ricketts, S.W. and Wingfield-Digby, N.Y. (1979) CEM: the case for AI. *Veterinary Record* 104, 536.

Rousset, H., Chanteloube, P., Magistrini, M. and Palmer, E. (1987) Assessment of fertility and semen evaluation of stallions. *Journal of Reproduction and Fertility, Supplement* 35, 25–31.

Rowley, M.S., Squires, E.L. and Pickett, B.W. (1990) Effect of insemination volume on embryo recovery in mares. *Equine Veterinary Science* 10, 298–300.

Rowson, L.E. and Murdoch, M.I. (1954) Electrical ejaculation in the bull. *Veterinary Record* 66, 326.

Rudak, E., Jacobs, P. and Yanagimachi, R. (1978) Direct analysis of chromosome constitution of human spermatozoa. *Nature (London)* 174, 911–913.

Rutten, D.R., Chaffaux, S., Valon, M., Deletang, F. and De Haas, V. (1986) Progesterone therapy in mares with abnormal oestrus cycles. *Veterinary Record* 119, 569–571.

Sack, W.O. (1991) Isolated male organs. In: *Rooney's Guide to the Dissection of the Horse*, 6th edn. Veterinary Text Books, Ithaca, New York, pp. 75–78.

Salisbury, G.W., Van Denmark, N.L. and Lodge, J.R. (1978) Part 2: The storage and the planting. In: *Physiology of Reproduction and AI of Cattle*, 2nd edn. W.H. Freeman, San Francisco, pp. 187–578.

Samper, J.C. (1995a) Diseases of the male system. In: Kobluk, C.N., Ames, T.R. and Goer, R.J. (eds) *The Horse, Diseases and Clinical Management*, Vol. 2. W.B. Saunders, Philadelphia, pp. 937–972.

Samper, J.C. (1995b) Stallion semen cryopreservation: male factors affecting pregnancy rates. *Proceedings of the Society for Theriogenology*. San Antonio Texas, pp. 160–165.

Samper, J.C. (1997) Reproductive anatomy and physiology of breeding stallion. In: Youngquist, R.S. (ed.) *Current Therapy in Large Animal Theriogenology*. W.B. Saunders, Philadelphia, pp. 3–12.

Samper, J.C. and Crabo, B.G. (1988) Filtration of capacitated spermatozoa through filters containing glass wool and/or Sephadex. In: *11th International Congress on Animal Production and Artificial Insemination, June 26–30*, Vol. 3, Paper 294. University College Dublin, Republic of Ireland.

Samper, J.C. and Crabo, B.G. (1993) Assay of capacitated, freeze-damaged and extended stallion spermatozoa by filtration. *Theriogenology* 39(6), 1209–1220.

Samper, J.C. and Gartley, C.J. (1991) New perspectives in stallion semen evaluation. *Proceedings of the Society of Theriogenology* 1991, pp. 162–170.

Samper, J.C., Loseth, K.J. and Crabo, B.G. (1988) Evaluation of horses spermatozoa with Sephadex filtration using three extenders and three dilutions. In: *Proceedings of the 11th International Congress on Animal Production and Artificial Insemination, June 26–30*, Vol. 3, Brief communications, 294. University College Dublin, Republic of Ireland.

Samper, J.C., Behnke, E.J., Byers, A.P., Hunter, A.G. and Crabo, B.G. (1989) *In vitro* capacitation of stallion spermatozoa in calcium-free Tyrode's medium and penetration of zona-free hamster eggs. *Theriogenology* 31(4), 875–884.

Samper, J.C., Hellander, J.C. and Crabo, B.G. (1991) Relationship between the fertility of fresh and frozen stallion semen and semen quality. *Journal of Reproduction and Fertility, Supplement* 44, 107–114.

Sanchez, R.A., Frey, G.W. von and Reyers, S.M. de los (1995) Effect of semen diluents and seminal plasma on the preservation of refrigerated stallion semen. *Veterinaria Argentina* 12(113), 172–178.

Satir, P. (1974) How cilia move. *Scientific American* 231(10), 44–52.

Satir, P. (1984) The generation of cilary motion. *Journal of Protozoology* 31, 8–12.

Savage, N.C. and Liptrap, R.M. (1987) Induction of ovulation in cyclic mares by administration of a synthetic prostaglandin, fenprostalene, during estrus. *Journal of Reproduction and Fertility, Supplement* 35, 239–243.

Savard, K. and Goldziecher, J.W. (1960) Biosynthesis of steroids in stallion testis. *Endocrinology* 66, 617–625.

Schafer, W. and Baum, W. (1964) Tiefgefrierung von Pferdsperma bei −79°C unter Verwendung von CO_2-Eis. *Fortplf. Besam. Aufzucht Haust* 1, 105–111.

Schanbacher, B.D. and Pratt, B.R. (1985) Responses of cryptorchid stallions to vaccination against luteinising hormone releasing hormone. *Veterinary Record* 116, 74–75.

Schell, F.G. (1948) Equine insemination pros and cons. *North American Veterinarian* 29, 413–417. *Animal Breeding Abstracts* 17, 437 (1948).

Scherbarth, R., Pozvari, M., Heilkenbrinker, T. and Mumme, J. (1994) Genital microbial flora of the stallion – microbiological examination of presecretion samples between 1972 and 1991. *Deutsche Tierärztliche Wochenschrift* 101(1), 18–22.

Schmidt, K. (1950) Die praktische durchfuhrung der kunstlichen besamung beim pferd. *Mh. Vet. Med.* 5(1–5), 29–34. *Animal Breeding Abstracts* 21, 79.

Schneider, R.K., Milne, D.W. and Kohn, C.W. (1982) Acquired inguinal hernia in the horse: a review of 27 cases. *Journal of the American Veterinary Medical Association* 180, 317–332.

Schrop, H. (1992) Effect of membrane protectant lecithin and phenylmethane sulphonylfluoride on sperm motility and acrosome integrity in thawed semen, evaluated by immunofluorescence or Spermac staining. Thesis, Tierärztliche Hochschule Hannover, Germany, 110 pp.

Schumacher, J. and Riddell, M.G. (1986) Collection of stallion semen without a mount. *Theriogenology* 26(2), 245–250.

Schumacher, J. and Varner, D.D. (1993) Neoplasia of the stallion's reproductive tract. In: McKinnon, A.O. and Voss, J.L. (eds) *Equine Reproduction*. Lea and Febiger, Philadelphia, pp. 871–878.

Schumacher, J., Varner, D.D., Schmitz, D.G. and Blanchard, T.L. (1995) Urethral defects in geldings with hematuria and stallions with haemospermia. *Veterinary Surgery* 24(3), 250–254.

Seamens, M.C., Roser, J.F., Linford, R.L., Lui, I.K.M. and Hughes, J.P. (1991) Gonadotrophin and steroid concentrations in jugular and testicular venous plasma in stallions before and after GnRH injection. *Journal of Reproduction and Fertility, Supplement* 44, 57–67.

Seidel, G.E. Jr, Herickhoff, L.A., Schenk, J.L., Doyle, S.P. and Green, R.D. (1998) Artificial insemination of heifers with cooled, unfrozen sexed semen. *Theriogenology* 49(1), 365.

Sembrat, R.F. (1975) The acute abdomen of the horse: epidemiologic considerations. *Veterinary Surgery* 4, 34–38.

Serban, M., Georgescu, G., Lozinschi, A. and Suteau, M. (1982) Comparative studies on the activity of certain enzymes in the seminal plasma of stallions of different breeds. *Lucrari Stiintifice Institutul Agronomic N. Balcescu D (Zootechnie)* 25, 63–68.

Serebrovsky, A.S. and Sokolovskaya, I.I. (1934) Electroejakuljacia u Pitic. (Electroejaculation in chickens.) *Prob. Zhivotn. Moscow* 5, 57. *Animal Breeding Abstracts* 3, 73 (1935).

Sertich, P.L. (1998) Ultrasonography of the genital tract of the mare. In: Reef, V.B. (ed.) *Equine Diagnostic Ultrasound*. W.B. Saunders, Philadelphia, pp. 405–424.

Setchell, B.P. (1982) Spermatogenesis and spermatozoa. In: Austin, C.R. and Short, R.V. (eds) *Reproduction in Mammals, 1. Germ Cells and Fertilisation*, 2nd edn. Cambridge University Press, New York, pp. 63–101.

Setchell, B.P. (1991) Male reproductive organs and semen. In: Cupps, P.T. (ed.) *Reproduction in Domestic Animals*, 4th edn. Academic Press, London, pp. 221–249.

Setchell, B.P. and Brooks, D.E. (1988) In: Knobil, E. and Neill, J.D. (eds) *Physiology of Reproduction*. Raven Press, New York, pp. 753–836.

Setchell, B.P., Scott, T.W., Voglmayr, J.K. and Waites, G.M.H. (1969) Characterisation of testicular spermatozoa and the fluid which transports them into the epididymis. *Biology of Reproduction, Supplement* 1, 40–66.

Shannon, P. (1972) The effect of egg yolk level and dose rate on conception rate of semen diluted in caprogen. *Proceedings of the 7th International Congress of Animal Reproduction and AI*. Munich, pp. 279–280.

Sharma, R., Hogg, J. and Bromham, D. (1993) Is spermatozoan acrosin a predictor of fertilisation and embryo quality in the human? *Fertility and Sterility* 60(5), 881–887.

Sharman, O.P. (1976) Diurnal variations of plasma testosterone in stallions. *Biology of Reproduction* 15, 158–162.

Sharp, D.C. and Clever, B.D.I. (1993) Melatonin. In: McKinnon, A.O. and Voss, J.L. (eds) *Equine Reproduction*. Lea and Febiger, Philadelphia, pp. 100–108.

Shideler, R.K. (1993a) History. In: McKinnon, A.O. and Voss, E.L. (eds) *Equine Reproduction*. Lea and Febiger, Philadelphia, pp. 196–198.

Shideler, R.K. (1993b) External examination. In: McKinnon, A.O. and Voss, E.L. (eds) *Equine Reproduction*. Lea and Febiger, Philadelphia, pp. 199–203.

Shideler, R.K. (1993c) Rectal palpation. In: McKinnon, A.O. and Voss, E.L. (eds) *Equine Reproduction*. Lea and Febiger, Philadelphia, pp. 204–211.

Shideler, R.K., McChesney, A.E., Voss, J.L. and Squires, E.L. (1982) Relationship of endometrial biopsy and other management factors on fertility of broodmares. *Journal of Equine Veterinary Science* 2, 5–10.

Shin, S.J., Lein, D.H., Pattie, V.H. and Ruhnke, H. (1988) A new antibiotic combination for frozen bovine semen. 1. Control of mycoplasmas, *Campylobacter fetus* subspecies *venerealis* and *Haemophillus somnus*. *Theriogenology* 29, 557–591.

Shoemaker, C.F., Squires, E.L. and Shideler, R.K. (1989) Safety of altrenogest in pregnant mares and on health and development of offspring. *Journal of Equine Veterinary Science* 9, 67–72.

Sigler, D.H. and Kiracofe, G.H. (1988) Seminal characteristics of two–three year old Quarter Horse stallions. *Journal of Equine Veterinary Science* 8(2), 160–164.

Silva, J.F.S., Meinert, C., Ahlemeyer, B., Hueck, C. and Klug, E. (1990) Improvement to the artificial vagina for collecting semen from stallions. *Praktische Tierrarzt* 71(7), 69–70.

Silva, J.F.S., Mattos, R.C., Cruz, G.M., Carbalho, O.W.F. and Klug, E. (1996) Use of the Kenney testis index for the prediction of daily sperm output of horses. *Revista Brasileira de Reproducao Animal* 20(1), 18–22.

Silva Filho, J.M., Palares, M.S., Fonseca, F.A., Wanderley, A.T. and Oliveira, H.N. (1991a) Fertilising ability of transported stallion semen. 3. Use of a new procedure. *Anais, IX Congresso Brasileiro de Reproduao Animal, Belo Horizonte, Brazil, 22–26 July 1991*, Vol. II. Colegio Brasileiro de Reproducao Animal, Belo Horizonte, Brazil, p. 370.

Silva Filho, J.M., Fonseca, F.A., Carvalho, G.R., Mola, V.A.F., Palhares, M.S. and Oliveira, H.N. (1991b) Effect of insemination frequency on the fertility of mares inseminated with diluted fresh semen. *Anais, IX Congresso Brasileiro de Reproduao Animal, Belo Horizonte, Brazil, 22–26 July 1991*, Vol. II. Colegio Brasileiro de Reproducao Animal, Belo Horizonte, Brazil, p. 336.

Silva Filho, J.M., Santiago, M.L.D. and Palares, M.S. (1994) Fertility of mares inseminated with whole semen or semen diluted in minimal contamination extender. *Revista Brasileira de Reproducao Animal* 18(1–2), 69–80.

Simpson, R.B., Burns, S.J. and Snell, J.R. (1975) Microflora in stallion semen and their control with semen extender. *Proceedings of the 20th Annual Convention of the American Association of Equine Practitioners*, 225–260.

Singhvi, N.M. (1990) Studies on AI in equines. *Indian Journal of Animal Reproduction* 11(2), 99–104.

Sisson, S. and Grossman, J.D. (1975) Equine urogenital system. In: Getty, R. (ed.) *The Anatomy of Domestic Animals*, W.B. Saunders, Philadelphia, pp. 531–535.

Skatkin, P.N. (1952) Usoveršentvovanie metodov polučer semeni u žerebou (Improving the methods of collecting stallion semen). *Konevodstuo* 22, 25–30. *Animal Breeding Abstracts*, 20, 996.

Slaughter, G.R., Meistrich, M.L. and Means, A.R. (1989) Expression of RNAs for calmodulin, actins and tubulins in rat testis cells. *Biology of Reproduction* 40, 395–405.

Slonina, D., Okolski, A. and Baranska, K. (1995) Effect of capacitation method and calcium ionophore A23187 on stallion spermatozoa acrosome change. *Biology of Reproduction, Monograph Equine Reproduction VI* 1, 719–727.

Slusher, S.H. (1997) Infertility and diseases of the reproductive tract in stallions. In: Youngquist, R.S. (ed.) *Current Therapy in Large Animal Theriogenology*. W.B. Saunders, Philadelphia, pp. 16–23.

Smirnov, I.V. (1951) Storage of livestock semen at a temperature of −78°C to −183°C. *Socialist Zhivitn* 1, 94. *Animal Breeding Abstracts* 19, 155 (1951).

Smith, A.U. and Polge, C. (1950) Survival of spermatozoa at low temperatures. *Nature* 166, 668–669.

Smith, J.A. (1973) The occurrence of larvae of *Strongylus edentatus* in the testicles of stallions. *Veterinary Record* 93, 604–606.

Spallanzani, L. (1803) *Tracts on the Natural History of Animals and Vegetables*, 2nd edn. Creech and Constable, Edinburgh.

Spreckels, I. (1994) Investigations on the selection of spermatozoa by glass wool Sephadex filtration of stallion semen with special reference to cryopreservation. Thesis, Tierärztliche Hochschule Hannover, Germany.

Squires, E.L. (1993a) Progesterone. In: McKinnon, A.O. and Voss, J.L. (eds) *Equine Reproduction*. Lea and Febiger, Philadelphia, pp. 57–64.

Squires, E.L. (1993b) Progestin. In: McKinnon, A.O. and Voss, J.L. (eds) *Equine Reproduction*. Lea and Febiger, Philadelphia, pp. 311–318.

Squires, E.L., Amann, R.R., Hoyer, J.H., Nett, T.M. and Pickett, B.W. (1977) Effect of ejaculation on systemic levels of testosterone and LH in stallions. *Proceedings of the Annual Meeting (1977) of the American Society for Animal Science*. Madison, Wisconsin. Abstract 69, p. 210.

Squires, E.L., Pickett, B.W. and Amann, R.P. (1979a) Effect of successive ejaculation on stallion seminal characteristics. *Journal of Reproduction and Fertility, Supplement* 27, 7–12.

Squires, E.L., Stevens, W.B., McGlothin, D.E. and Pickett, V.W. (1979b) Effect of an oral progestagen on the oestrous cycle and fertility of mares. *Journal of Animal Science* 49, 729–735.

Squires, E.L., Berndtson, W.E., Hoyer, J.H., Pickett, B.W. and Wallach, S.J.R. (1981a) Restoration of reproductive capacity of stallions after suppression with exogenous testosterone. *Journal of Animal Science* 53, 1351–1359.

Squires, E.L., McGlothlin, D.E., Bowen, R.A., Berndtson, W.E. and Pickett, B.W. (1981b) The use of antibiotics in stallion semen for the control of *Klebsiella pneumoniae* and *Pseudomonas aeroginosa*. *Equine Veterinary Science* 1, 43–48.

Squires, E.L., Webel, S.K., Shideler, R.K. and Voss, J.L. (1981c) A review on the use of altrenogest for the broodmare. *Proceedings of the American Association of Equine Practitioners*, 221–231.

Squires, E.L., Wallace, R.A., Voss, J.L., Pickett, B.W. and Shideler, R.K. (1981d) The effectiveness of PGF2α and hCG and GnRH for appointment breeding of mares. *Journal of Equine Veterinary Science* 1, 5–9.

Squires, E.L., Todter, G.E., Berndtson, W.E. and Pickett, B.W. (1982) Effect of anabolic steroids on reproductive function of young stallions. *Journal of Animal Science* 54, 576–582.

Squires, E.L., Heesemann, C.P., Webel, S.K., Shideler, R.K. and Voss, J.L. (1983) Relationship of altrenogest to ovarian activity, hormone concentrations and fertility in mares. *Journal of Animal Science* 56, 901–910.

Squires, E.L., Voss, J.L., Villahoz, M.D. and Shideler, R.K. (1984) Use of ultrasound in broodmare reproduction. *Proceedings of the 29th Annual Convention of the American Association of Equine Practitioners*, Las Vegas, Nevada, pp. 27–43.

Squires, E.L., Amann, R.P., McKinnon, A.O. and Pickett, B.W. (1988) Fertility of equine spermatozoa cooled at 5°C or 20°C. *Proceedings of the International Congress of Animal Reproduction and AI*, Vol. 3, 297–299.

Squires, E.L., Badzinski, S.L., Amann, R.P., McCue, P.M. and Nett, T.M. (1997) Effects of altrenogest on total scrotal width, seminal characteristics, concentrations of LH and testosterone and sexual behaviour of stallions. *Theriogenology* 48(2), 313–328.

Stabenfeldt, G.H. and Hughes, J.P. (1977) Reproduction in horses. In: Cole, H. and Cupps, P. (eds) *Reproduction in Domestic Animals,* 3rd edn. Academic Press, London.

Steinmann, H. (1996) Use of trehalose as a cryoprotectant in the freezing of stallion semen. A laboratory study. Thesis, Tierärztliche Hochschule Hannover, Germany.

Stick, J.A. (1981) Surgical management of genital habronemiasis in a horse. *Veterinary Medicine Small Animal Clinician* 76, 410–414.

Stradaioli, G., Chiacchiarini, P., Monaci, M., Verini Supplizi, A., Martion, G. and Piermati, C. (1995) Reproductive characteristics and seminal plasma carnitine concentration in maiden mares and stallions. *Proceedings of the 46th Annual Meeting of the European Association for Animal Production,* Prague 4–7 September.

Strauss, S.S., Chen, C.L., Kalra, S.P. and Sharp, D.C. (1979) Localisation of gonadotrophin releasing hormone (GnRH) in the hypothalamus of ovarectomised pony mares by season. *Journal of Reproduction and Fertility, Supplement* 27, 123.

Strzemienski, P.J., Sertich, P.L., Varner, D.D. and Kenney, R.M. (1987) Evaluation of cellulose acetate/nitrate filters for the study of stallion sperm motility. *Journal of Reproduction and Fertility, Supplement* 35, 33–38.

Sugimoto, C., Isayama, Y., Sakazaki, R. and Kuramochi, S. (1983) Transfer of *Haemophilus equigenitalis* Taylor, *et al.* 1978 to the genus *Taylorella genera* known as *Taylorella equigenitalis* comb. nov. *Current Microbiology* 9, 155–162.

Sukalic, M., Herak, M. and Ljubesic, J. (1982) The first use of imported deep frozen stallion semen in Yugoslavia. *Veterinarski Glasnik* 36(12), 1027–1032.

Sullins, K.E., Bertone, J.J., Voss, J.L. and Pederson, S.J. (1988) Treatment of hemospermia in stallions: a discussion of 18 cases. *Compendium on Continuing Education for Practicing Veterinarians* 10(12), 1396–1400.

Sullivan, J.J. and Pickett, B.W. (1975) Influence of ejaculation frequency of stallions on characteristics of semen and output of spermatozoa. *Journal of Reproduction and Fertility, Supplement* 23, 29–34.

Sullivan, J.J., Parker, W.G. and Larson, L.L. (1973) Duration of oestrus and ovulation time in non lactating mares given hCG during three successive oestrus periods. *Journal of the American Veterinary Medical Association* 162, 895–898.

Sullivan, J.J., Turner, P.C., Self, L.C., Gutteridge, H.B. and Bartlett, D.E. (1975) Survey of reproductive efficiency in quarter-horse and thoroughbred. *Journal of Reproduction and Fertility, Supplement* 23, 315–318.

Swanson, E.W. and Bearden, H.J. (1951) An eosin–nigrosin stain for differentiating live and dead bovine spermatozoa. *Journal of Animal Science* 10, 981–987.

Swerczek, T.W. (1975) Immature germ cells in the semen of Thoroughbred stallions. *Journal of Reproduction and Fertility, Supplement* 23, 135–137.

Swierstra, E.E., Grebauer, M.R. and Pickett, B.W. (1974) Reproductive physiology of the stallion. I. Spermatogenesis and testes composition. *Journal of Reproduction and Fertility* 40, 113–123.

Swierstra, E.E., Pickett, B.W. and Gebauer, M.R. (1975) Spermatogenesis and duration of transit of spermatozoa through the excurrent ducts of stallions. *Journal of Reproduction and Fertility, Supplement* 23, 53–57.

Swire, P.W. (1962) Artificial insemination in the horse. In: Maule, J.P. (ed.) *The Semen of Animals and Artificial Insemination.* Technical communication No. 15, Commonwealth Agricultural Bureaux, Farnham Royal, UK, pp. 281–297.

Szollar, I., Magosi, Z., Totth-Polner, A. and Wekerle, L. (1993) Morphological examination of stallion spermatozoa. *Magyar Allatorvosok Lapja* 48(10), 603–607.

Szumowski, P. (1954) Essais de congélation du sperme de ceval. *Comptes Rendus de l'Académie d'Agriculture de France* 40, 156–162. *Animal Breeding Abstracts* 23, 523.

Tainturier, D. (1981) Bacteriology of *Hemophilus equigenitalis*. *Journal of Clinical Microbiology* 14, 355.

Tainturier, D. (1993) Évaluation de la gynécologie equine au cours des vingt dernières années. *Point Vétérinaria* 25(155), 52–537.

Tash, J.S. and Mann, T. (1972a) Adenosine 3′,5′-cyclic monophosphate in relation to motility in senescence of spermatozoa. *Proceedings of the Royal Society of Biology* 184, 109–114.

Tash, J.S. and Mann, T. (1973b) Relation of cyclic AMP to sperm motility. *Journal of Reproduction and Fertility* 35, 591.

Tearle, J.P., Smith, K.C., Boyle, M.S., Binns, M.M., Livesay, G.J. and Mumford, J.A. (1996) Replication of equid herpes virus-1 (EHV-1) in the testes and epididymides of ponies and venereal shedding of infectious virus. *Journal of Comparative Pathology* 115(4), 385–397.

Tekin, N., Wockener, A. and Klug, E. (1989) Effect of two different diluents and storage temperatures on the viability of stallion semen. *Deutsche Tierärztliche Wochenschrift* 96(5), 258–265.

Tessler, S. and Olds-Clarke, P. (1985) Linear and non linear mouse sperm motility patterns: a quantitative classification. *Journal of Andrology* 6, 35–44.

Teuscher, C., Kenney, R.M., Cummings, M.R. and Catten, M. (1994) Identification of two stallion sperm specific proteins and their autoantibody response. *Equine Veterinary Journal* 26(2), 148–151.

Thacker, D.L. and Almquist, J.O. (1953) Diluters for bovine serum. I. Fertility and motility of bovine spermatozoa in boiled milk. *Journal of Dairy Science* 36, 173–180.

Thibault, C., Leplaud, M. and Ortavant, R. (1948) L'électro-éjaculation chez le taureau: technique et résultats. (Electroejaculation in bulls: techniques and results.) *Comptes Rendus des Séances de la Société Biologie Acad. Sci.* (Paris) 226, 2006.

Thomas, P.G.A. and Ball, B.A. (1996) Cytofluorescent assay to quantify adhesion of equine spermatozoa to oviduct epithelial cells *in vitro*. *Molecular Reproduction and Development* 43(1), 55–61.

Thomas, P.G.A., Ball, B.A. and Brinsko, S.P. (1994a) Interaction of spermatozoa with oviduct epithelial cell explants is affected by estrous cycle and anatomic origin of explant. *Biology of Reproduction* 51(2), 222–228.

Thomas, P.G.A., Ball, B.A., Miller, P.G., Brinsko, S.P. and Southwood, L. (1994b) A subpopulation of morphologically normal, motile spermatozoa attach to equine oviductal epithelial cell monolayers. *Biology of Reproduction* 51(2), 303–309.

Thomas, P.G.A., Ball, B.A. and Brinsko, S.P. (1995a) Changes associated with induced capacitation influence the interaction between equine spermatozoa and oviduct epithelial cell monolayers. *Biology of Reproduction, Monograph Equine Reproduction VI* 1, 697–705.

Thomas, P.G.A, Ignotz, G.G., Ball, B.A., Brinsko, S.P. and Currie, W.B. (1995b) Effect of coculture with stallion spermatozoa on *de novo* protein synthesis and secretion by equine oviduct epithelial cells. *American Journal of Veterinary Research* 56(12), 1657–1662.

Thomas, P.G.A., Ball, B.A., Ignotz, G.G., Dobrinski, I., Parks, J.E. and Currie, W.B. (1997) Antibody directed against plasma membrane components of equine spermatozoa inhibits adhesion of spermatozoa to oviduct epithelial cells *in vitro*. *Biology of Reproduction* 56, 720–730.

Thomassen, R. (1988) The use of photometer in estimating the sperm concentration in horse semen. *Norsk Veterinaertidsskrift* 100(7–8), 553–555.

Thomassen, R. (1991) Use of frozen semen for artificial insemination in mares, results in 1990. *Norsk Veterinaertidsskrift* 103(3), 213–216.

Thomson, C.H., Thompson, D.L. Jr, Kincaid, L.A. and Nadal, M.R. (1996) Prolactin involvement with the increase in seminal volume after sexual stimulation in stallions. *Journal of Animal Science* 74(10), 2468–2472.

Thompson, D.L. Jr (1992) Reproductive physiology of stallions and jacks. In: Warren Evans, J. (ed.) *Horse Breeding and Management,* Elsevier, Amsterdam, pp. 237–261.

Thompson, D.L. Jr and Honey, P.G. (1984) Active immunisation of pre-pubertal colts against estrogens: hormonal and testicular responses after puberty. *Journal of Animal Science* 59, 189–196.

Thompson, D.L. Jr and Johnson, L. (1987) Effects of age, season and active immunization against estrogen, on serum prolactin concentrations in stallions. *Domestic Animal Endocrinology* 4, 17–22.

Thompson, D.L. Jr, Pickett, B.W., Squires, E.L. and Amann, R.P. (1979a) Testicular measurements and reproductive characteristics in stallions. *Journal of Reproduction and Fertility, Supplement* 27, 13–17.

Thompson, D.L. Jr, Pickett, B.W., Squires, E.L. and Nett, T.M. (1979b) Effect of testosterone and oestradiol 17β, alone and in combination, on LH and FSH concentrations in pituitary and blood serum of geldings and in serum after GnRH. *Biology of Reproduction* 21, 1231–1237.

Thompson, D.L. Jr, Pickett, B.W., Squires, E.L. and Nett, T.M. (1980) Sexual behaviour, seminal pH and accessory sex gland weights in geldings administered testosterone and/or estradiol 17β. *Journal of Animal Science* 51, 1358–1366.

Thompson, D.L. Jr, St George, R.L., Jones, L.S. and Garza, F.J. (1985) Patterns of secretion of LH, FSH and testosterone in stallions during summer and winter. *Journal of Animal Science* 60, 741–748.

Thompson, D.L. Jr, Johnson, L., St George, R.L. and Garza, F. Jr (1986) Concentrations of prolactin, LH, FSH in pituitary and serum of horses. Effect of sex, season and reproductive state. *Journal of Animal Science* 63, 854–860.

Threlfall, W.R. and Carleton, C.L. (1996) Mare's genital tract. In: Traub-Dargatz, J.L. and Brown, C.M. (eds) *Equine Endoscopy,* 2nd edn. The C.V. Mosby Company, Missouri, pp. 204–217.

Threlfall, W.R., Carelton, C.L., Robertson, J., Rosol, T. and Gabel, A. (1990) Recurrent torsion of the spermatic chord and scrotal testis in a stallion. *Journal of the American Veterinary Medical Association* 196, 1641–1643.

Timoney, P.J. and McCollum, W.H. (1985) The epidemiology of equine viral arteritis. *Journal of the American Veterinary Medical Association* 155, 318–322.

Timoney, P.J., McCollum, W.H., Murphy, T.W., Roberts, A.W., Willard, J.G. and Caswell, G.D. (1988a) The carrier state in equine arteritis virus infection in the stallion with specific emphasis on the venereal mode of virus transmission. *Journal of Reproduction and Fertility, Supplement* 35, 95–102.

Timoney, P.J., Umphenour, N.W. and McCollum, W.H. (1988b) Safety evaluation of a commercial modified live equine arteritis virus vaccine for use in stallions. In: Powell, D.G. (ed.) *Equine Infectious Diseases V: Proceedings of the Fifth International Conference.* University Press of Kentucky, Lexington, Kentucky, pp. 19–27.

Tischner, M. (1979) Evaluation of deep frozen semen in stallions. *Journal of Reproduction and Fertility, Supplement* 27, 53–59.

Tischner, M. (1992a) Artificial insemination in horses: a review of current status quo. *Equine Veterinary Education* 4(2), 89–92.

Tischner, M. (1992b) Equine artificial insemination in central and eastern Europe. *Acta Veterinaria Scandinavica, Supplement* 88, 111–115.

Tischner, M. and Kosiniak, K. (1986) Bacterial contamination of stallion semen collected by 'open' AV. *Vlaams Diergeneeskundig Tijdschrift* 55(2), 90–94.

Tischner, M., Kosiniak, K. and Bielanski, W. (1974) Analysis of the pattern of ejaculation in stallions. *Journal of Reproduction and Fertility* 41, 329–335.

Torres-Bogino, F., Sato, K., Oka, A., Kamo, Y., Hochi, S.I., Oguri, N. and Braun, J. (1995) Relationship among seminal characteristics, fertility and suitability for semen preservation in draft stallions. *Journal of Veterinary Medical Science* 57(2), 225–229.

Turner, A.S. and McIlwraith, C.W. (1982) Umbilical hernia therapy in the foal. In: Turner, A.S. and McIlwraith, C.W. (eds) *Techniques in Large Animal Surgery.* Lea and Febiger, Philadelphia, pp. 254–259.

Turner, R.M. (1998) Ultrasonography of the genital tract of the stallion. In: Reef, V.B. (ed.) *Equine Diagnostic Ultrasound.* W.B. Saunders, Philadelphia, pp. 446–479.

Turner, R.M., McDonnell, S.M. and Hawkins, J.F. (1995) Use of pharmacologically induced ejaculation to obtain semen from a stallion with a fractured radius. *Journal of the American Veterinary Medical Association* 206(12), 1906–1908.

Uphaus, H. and Kalm, E. (1994) Field and station tests for mares. 2. Use in the framework of a breeding programme. *Zuchtungskunde* 66(4), 268–280.

Vaillencourt, D., Fretz, P. and Orr, J.P. (1979) Seminoma in the horse. Report of 2 cases. *Journal of Equine Medical Surgery* 3, 213–218.

Vaillencourt, D., Gucy, P. and Higgins, R. (1993) The effectiveness of gentamicin or polymixin B for the control of bacterial growth in equine semen stored at 20°C or 5°C for up to forty eight hours. *Canadian Journal of Veterinary Research* 57(4), 277–280.

Vaissaire, J., Plateau, E. and Collobert-Laugier, C. (1987) Importance of bacteria in stallion semen. *Bulletin de l'Académie Vétérinaire de France* 60(2), 165–175.

Van Buiten, A., Zhang, J. and Boyle, M.S. (1989) Integrity of plasma membrane of stallion spermatozoa before and after freezing. *Journal of Reproduction and Fertility* 4, 18–22.

Van Denmark, N.L. (1952) Time and site of insemination in cattle. *Cornell Veterinarian* 42, 215–222.

Van der Holst, W. (1975) A study of the morphology of stallion semen during the breeding and non breeding season. *Journal of Reproduction and Fertility, Supplement* 23, 87–89.

Van der Holst, W. (1984) Stallion semen production in AI programs in the Netherlands. In: Courot, M. (ed.) *The Male in Farm Animal Production.* Martinus, Nijhoff, Boston, pp. 195–201.

Van der Veldon, M. (1988a) Ruptured inguinal hernia in new-born colt foals: a review of 14 cases. *Equine Veterinary Journal* 20, 178–181.

Van der Veldon, M. (1988b) Surgical treatment of acquired inguinal hernia in the horse: a review of 39 cases. *Equine Veterinary Journal* 20, 173–177.

Van Duijn, C. Jr and Hendrikse, J. (1968) Rational analysis of seminal characteristics of stallions in relation to fertility. *Instituut woor Veeteelkundig Onderzoek 'Schoonoord' Report B97*, Pretoria. (Cited in Pickett, 1993a).

Van Niekerk, C.H. and Van Heerden, J.S. (1972) Nutritional and ovarian activity of mares early in the breeding season. *Journal of the South African Veterinary Association* 43(4), 351-360.

Van Niekerk, C.H., Coughborough, R.I. and Doms, H.W.H. (1973) Progesterone treatment of mares with abnormal oestrus cycle early in the breeding season. *Journal of the South African Veterinary Association* 44(1), 37–45.

Varner, D.D. (1986) Collection and preservation of stallion spermatozoa. *Proceedings of the Annual Meeting (1986) of the Society of Theriogenology*, 13–33.

Varner, D.D. (1991) Composition of seminal extenders and its effect on motility of equine spermatozoa. *Proceedings of the Annual Meeting (1991) of the Society for Theriogenology*, 146–150.

Varner, D.D. (1992) Consideration regarding semen manipulation, extenders, cooling and freezing. *Proceedings of the Annual Meeting (1992) of the Society for Theriogenology*, 87–88.

Varner, D.D. and Schumacher, J. (1991) Diseases of the reproductive system: the stallion. In: Colahan, P.T., Mayhew, I.G., Merritt, A.M. and Moore, J.N. (eds) *Equine Medicine and Surgery*, Vol. 2, 4th edn. American Veterinary Publication Incorporated, Goleta, California, pp. 847–948.

Varner, D.D., Blanchard, T.L., Love, C.L., Garcia, M.C. and Kenney, R.M. (1987) Effects of sperm fractionation and dilution ratio on equine spermatozoal motility parameters. *Theriogenology* 28, 709–723.

Varner, D.D., Blanchard, T.L., Love, C.L., Garcia, M.C. and Kenney, R.M. (1988) Effects of cooling rate and storage temperature on equine spermatozoal motility parameters. *Theriogenology* 29, 1043–1054.

Varner, D.D., Blanchard, T.L., Meyers, P.J. and Meyers, S.A. (1989) Fertilising capacity of equine spermatozoa stored for 24 hours at 5°C or 20°C. *Theriogenology* 32, 515–525.

Varner, D.D., Vaughan, S.D. and Johnson, L. (1991a) Use of a computerized system for evaluation of equine spermatozoal motility. *American Journal of Veterinary Research* 52(2), 224–230.

Varner, D.D., Schumacher, J., Blanchard, T.L. and Johnson, L. (1991b) *Diseases and Management of Breeding Stallions.* American Veterinary Publication Incorporated, Goleta, California, 345 pp.

Varner, D.D., McIntosh, A.L., Forrest, D.W., Blanchard, T.L. and Johnson, L. (1992) Potassium penicillin G, amikacin sulphate or a combination in seminal extenders for stallions: effects on spermatozoal motility. *Proceedings of the International Congress of Animal Reproduction and AI* 12, 1496–1498.

Varner, D.D., Taylor, T.S. and Blanchard, T.L. (1993) Seminal vesiculitis. In: McKinnon, J.L. and Voss, J.L. (eds) *Equine Reproduction.* Lea and Febiger, Philadelphia, pp. 861–864.

Veeramachaneni, D.N.R., Moeller, C.L., Pickett, B.W., Shiner, K.A. and Sawyer, H.R. (1993) On processing and evaluation of equine seminal samples for cytopathology and fertility assessment: the utility of electron microscopy. *Journal of Equine Veterinary Science* 13(4), 207–215.

Venzke, W.G. (1975) Endocrinology. In: Getty, R. (ed.) *Sisson and Grossman's Anatomy of Domestic Animals,* Vol. 1, 5th edn. W.B. Saunders, Philadelphia, pp. 550–551.

Veselinovic, S., Perkucin, R., Miljkovic, V., Beslin, R., Salasi, I., Petrujkic, T. and Milojevic, Z. (1980) Results of the application of the artificial fertilising of mares by the use of deep frozen semen taken from stallions in Vojvodina. *Proceedings of the 9th International Congress on Animal Reproduction and Artificial Insemination,* pp. 347–351.

Veznik, Z., Svecova, D., Pospisil, L. and Diblikova, I. (1996) Detection of *Chlamydia* spp. in animal and human semen by direct immunoflorescence. *Veterinarni Medicina* 41(7), 201–206.

Vidament, M., Dupere, A.M., Julienne, P., Evain, A., Noue, P. and Palmer, E. (1997) Equine frozen semen: freezability and fertility field results. *Theriogenology* 48(6), 907–917.

Vieira, R.C., Klug, E. and Rath, D. (1980) Determination of acrosin activity using autoradiographic film plates as a control parameter for the different steps of deep freezing stallion semen. *Proceedings of the International Congress on Animal Reproduction and Artificial Insemination* 5, 141–142.

Vlachos, K. (1960) Die kunstliche beasmung der stuten griechenland durch versand von konserviertem und gekuhltem sperma. (Artificial insemination of mares in Greece with semen that has been cooled and transported.) *Berliner und Münchener Tierärztliche Wochenschrift* 73, 424–427. *Animal Breeding Abstracts* 29, 669.

Vlachos, K. and Paschaleri, E. (1969) Research on some factors influencing fertility in solipeds. *Animal Breeding Abstracts* 37, 1227.

Voglmayr, J.K. (1975) Endocrinology. In: Hamilton, D.W. and Greep, R.O. (eds) *Handbook of Physiology, Volume 5, Male Reproductive System.* American Physiological Society, Washington DC, pp. 437–451.

Voglmayr, J.K., White, I.G. and Parks, R.P. (1978) The fertilising capacity of ram testicular spermatozoa, freshly collected and after storage in cauda epididymal fluid *Theriogenology* 10(4), 313–321.

Volkmann, D.H. (1987) A modified phantom (dummy mount) for stallion semen collection. *Equine Veterinary Journal* 19(4), 339–341.

Volkmann, D.H. and Van Zyl, D. (1987) Fertility of stallion semen frozen in 0.5ml straws. *Journal of Reproduction and Fertility, Supplement* 35, 143–148.

Voss, J.L. (1993) Human chorionic gonadotrophin. In: McKinnon, A.O. and Voss, J.L. (eds) *Equine Reproduction*. Lea and Febiger, Philadelphia, pp. 325–328.

Voss, J.L. and McKinnon, A.O. (1993) Hemospermia and urospermia. In: McKinnon, A.O. and Voss, J.L. (eds) *Equine Reproduction*. Lea and Febiger, Philadelphia, pp. 864–870.

Voss, J.L. and Pickett, B.W. (1973) The effect of nutritional supplement on conception rates in mares. *Proceedings of the 19th Annual Convention of the American Association of Equine Practitioners*. Atlanta, 49–54.

Voss, J.L. and Pickett, B.W. (1975) Diagnosis and treatment of haemospermia in the stallion. *Journal of Reproduction and Fertility, Supplement* 23, 151–154.

Voss, J.L. and Pickett, B.W. (1976) Reproductive management of the broodmare. *Animal Reproduction Laboratory General Series Bulletin No. 961*, Colorado State University, Fort Collins.

Voss, J.L., Sullivan, J.J., Pickett, B.W., Parker, W.G., Burwash, L.D. and Larson, L.L. (1975) The effect of hCG on the duration of oestrus, ovulation time and fertility in mares. *Journal of Reproduction and Fertility, Supplement* 23, 297–301.

Voss, J.L., Pickett, B.W. and Shideler, R.K. (1976) The effect of hemospermia on fertility in horses. *Proceedings of the 8th International Congress on Animal Reproduction and AI*. Krakow, Vol. 4, pp. 1093–1095.

Voss, J.L., Wallace, R.A., Squires, E.L., Pickett, B.W. and Shideler, R.K. (1979) Effects of synchronisation and frequency of insemination on fertility. *Journal of Reproduction and Fertility, Supplement* 27, 257–261.

Voss, J.L., Squires, E.L., Pickett, B.W. and Amman, R.P. (1980) Factors affecting reproductive performance of the stallion. *Proceedings of the Annual Convention of the American Association of Equine Practitioners*, 25, 33–49.

Voss, J.L., Pickett, B.W. and Squires, E.L. (1981) Stallion spermatozoal morphology and motility and their relationship to fertility. *Journal of the American Veterinary Medical Association* 178, 287–289.

Voss, J.L., Pickett, B.W. and Loomis, P.R. (1982) The relationship between seminal characteristics and fertility in Thoroughbred stallions. *Journal of Reproduction and Fertility, Supplement* 32, 635–636.

Voss, J.L., Pickett, B.W., Bowen, R.A., Squires, E.L. and McKinnon, A.O. (1988) Effects of breed, month and age on stallion seminal characteristics. *Proceedings of the 11th International Congress on Animal Reproduction and AI*, University College, Dublin, Republic of Ireland, pp. 119–210.

Wagenaar, G. and Grootenhuis, G. (1953) Sterility and sperm examination in stallions. Abstract in *Veterinary Bulletin* 1954, 24, No. 282. *Proceedings of the 15th International Veterinary Congress*, 1953, 2, Part 1, pp. 726–730.

Waide, Y. and Niwa, T. (1961) Storage and insemination trials with frozen semen in goats. *Proceedings of 8th International Congress on Animal Production*. Hamburg, p. 170.

Waites, G.M.H. (1968) Temperature and fertility in mammals. *Proceedings of the 6th International Congress on Animal Reproduction and Artificial Insemination*. Paris, Vol. 1, pp. 235–256.

Waites, G.M.H. and Ortavant, R. (1968) Effets précoces d'une brève élévation de la température testiculaire sur la spermatogenèse du Bélier. (Early effects of brief elevation of testicular temperature on spermatiogenesis in the ram.) *Annals of Biology, Animal Biochemistry and Biophysics* 8, 323–331.

Walker, D.F. and Vaughan, J.T. (1980) *Bovine and Equine Urogenital Surgery*. Lea and Febiger, Philadelphia, pp. 105–182.

Walton, A. (1936) *Notes on Artificial Insemination of Sheep, Cattle and Horses.* Hoborn Surgical Instruments Co. Ltd, London, 19 pp.

Walton, A. and Rawochenski, M. (1936) An experiment in eutelegenesis. *Journal of Heredity* 27, 341–344.

Ward, C.R., Varner, D.D., Storey, B.T. and Kenney, R.M. (1988) Chlorotetracycline fluorescent probe analysis of equine spermatozoa. *Proceedings of the 11th International Congress on Animal Production and Artificial Insemination, University College Dublin, Ireland,* Vol. 3, Brief communications, Paper 309. University College Dublin, Dublin, Republic of Ireland.

Watamabe, M. and Sugimori, Y. (1957) Studies on the artificial insemination in ducks. *Zootec. E. Vet. Milan* 12, 119. *Animal Breeding Abstracts* 25, 318 (1957).

Watson, E.D. (1988) Uterine defence mechanisms in mares resistant and susceptible to persistent endometritis. *Proceedings of the 34th American Association of Equine Practitioners,* 279–387.

Watson, E. (1995) Artificial insemination in horses. *In Practice* February 1995, 54–58.

Watson, E.D. and Nikolakopoulos, E. (1996) Sperm longevity in the mare's uterus. *Journal of Equine Veterinary Science* 16(9), 390–392.

Watson, E.D., Clarke, C.J., Else, R.W. and Dixon, P.M. (1994) Testicular degeneration in three stallions. *Equine Veterinary Journal* 26(6), 507–510.

Watson, P.F. (1976) The protection of ram and bull spermatozoa by the low density lipoprotein fraction of egg yolk during storage at 5°C and deep freezing. *Journal of Thermological Biology* 1, 137–141.

Watson, P.F. (1979) The preservation of semen in mammals. In: Finn, C.A. (ed.) *Oxford Reviews in Biology,* Vol. 1. Clarendon Press, Oxford, pp. 283–350.

Watson, P.F. (1990) AI and the preservation of semen. In: Lamming, G.E. (ed.) *Marshall's Physiology of Reproduction, Volume 2, Male Reproduction.* Churchill, Livingstone, London, pp. 747–869.

Watson, P.F. and Duncan, A.G. (1988) Effect of salt concentration and unfrozen water fraction on the viability of slowly frozen ram spermatozoa. *Cryobiology* 25, 131–142.

Watson, P.F. and Morris, G.J. (1987) Cold shock injury in animal cells. In: Bowler, K. and Fuller, B.J. (eds) *Temperature and Animal Cells.* Company of Biologists Ltd, Cambridge, UK, pp. 311–340.

Watson, P.F. and Plummer, J.M. (1985) The responses of boar and sperm membranes to cold shock and cooling. *Proceedings of the 1st International Conference on Deep Freezing of Boar Semen.* Swedish University of Agricultural Science, Uppsala, pp. 113–127.

Watson, P.F., Plummer, J.M. and Allen, W.E. (1987) Quantitative assessment of membrane damage in cold-shocked spermatozoa of stallions. *Journal of Reproduction and Fertility, Supplement* 35, 651–653.

Webb, R.L., Evans, J.W., Arns, M.J., Webb, G.W., Taylor, T.S. and Potter, G.D. (1990) Effects of vesiculectomy on stallion spermatozoa. *Journal of Equine Veterinary Science* 10(3), 218–223.

Webb, G.W., Arns, M.J. and Pool, K.C. (1993) Spermatozoa concentration influences the recovery of progressively motile spermatozoa and the number of inseminates shipped in conventional containers. *Journal of Equine Veterinary Science* 13, 486–489.

Weber, J.A. and Woods, G.L. (1992) Transrectal ultrasonography for the evaluation of stallion accessory sex glands. *Veterinary Clinics of North America, Equine Practice* 8(1), 183–190.

Weber, J.A. and Woods, G.L. (1993) Ultrasonic measurements of stallion accessory glands and excurrent ducts during seminal emission and ejaculation. *Biology of Reproduction* 49(2), 267–273.

Weber, J.A., Geary, R.T. and Woods, G.L. (1990) Changes in accessory sex glands of stallions after sexual preparation and ejaculation. *Journal of the American Veterinary Medical Association* 186(7), 1084–1089.

Wesson, J.A., Orr, E.L., Quay, W.B. and Ginther, O.J. (1979) Seasonal relationship between pineal hyd-oxindale-O-methyl-transferase (HIOMT) activity and reproductive activity in the pony. *General Compendium of Endocrinology* 38, 46–52.

Wetheridge, S. (1996) The use of equine AI and ET in the United Kingdom. BSc Thesis, University of Aberystwyth, Wales.

White, I.G. (1980) Accessory sex organs and fluids of the male reproductive tract. In: *Animal Models for Research on Contraception and Fertility.* Harper and Row, Maryland, pp. 105–123.

White, I.G. and Wales, R.G. (1960) The susceptibility of spermatozoa to cold shock. *International Journal of Fertility* 5, 195–201.

White, I.G., Blackshaw, A.W. and Emmens, C.W. (1954) Metabolic and motility studies relating to low temperature storage of ram and bull spermatozoa. *Australian Veterinary Journal* 30, 85–94.

Wierzbowski, S. (1958) Pzebieg odruchu ejakulacji u ogierow przy pobudzeniu naturalnym oraz przy uzyciu sztucznej pochwy. (Ejaculatory reflexes in stallions following natural stimulation and the use of the artificial vagina.) *Ptodnosc I nieptodnosc zwierzat domowych. Zesz. Probl. Postep. Nauk roln. Polsk. Akad. Nauk* 11, 153–156. *Animal Breeding Abstracts* 26, 1775.

Wilhelm, K.M., Graham, J.K. and Squires, E.L. (1996) Comparison of the fertility of cryopreserved stallion spermatozoa with sperm motion analyses, flow cytometry evaluation and zona-free hamster oocyte penetration. *Theriogenology* 46(4), 559–578.

Williams, J. and Foster, P.M.D. (1988) The production of lactate and pyruvate as sensitive indices of altered rat sertoli cell function *in vitro* following the addition of various testicular toxicants. *Toxicology and Applied Pharmacology* 9, 160–170.

Wilson, C.G., Downie, C.R., Hughes, J.P. and Roser, J.F. (1990) Effects of repeated hCG injection on reproductive efficiency in mares. *Journal of Equine Veterinary Science* 10, 301–308.

Windsor, D.P., Evans, G. and White, I.G. (1993) Sex predetermination by separation of X and Y chromosome-bearing sperm: a review. *Reproduction, Fertility and Development* 5, 155–171.

Winzler, J.R. (1970) Carbohydrates in cell surfaces. *International Review of Cytology* 29, 77–125.

Witte, A. (1989) Investigations on the preservation of chilled horse semen using different methods of dilution. Laboratory and field studies. Thesis, Tierärztliche Hochschule Hannover, Germany.

Wockener, A. and Colenbrander, B. (1993) Liquid storage and freezing of semen from New Forest and Welsh pony stallions. *Deutsche Tierärztliche Wochenschrift* 100(3), 125–126.

Wockener, A. and Schuberth, H.J. (1993) Freezing of maiden stallion semen – motility and morphology findings in sperm cells assessed by various staining methods including a monoclonal antibody with reactivity against an antigen in the acrosomal ground substance. *Reproduction in Domestic Animals* 28(6), 265–272.

Wockener, A., Papa, F.O., Sieme, H. and Bader, H. (1990) Untersuchungen zur Flussigkonservierung von Hengstsperma. *Pferdeheilkunde* 6, 129–135.

Woods, G.L. (1989) Pregnancy loss. A major cause of infertility in the mare. *Equine Practice* 11, 29–32.

Woods, J., Berfelt, D.R. and Ginther, O.J. (1990) Effects of time of insemination relative to ovulation on pregnancy rate and embryonic loss rate in mares. *Equine Veterinary Journal* 22, 410–415.

Woodward, R.A. (1987) The current Australian breed policies governing the use of artificial insemination and embryo transfer in the horse. *Proceedings of the Post-graduate Committee in Veterinary Science, Stallion artificial breeding and embryo transfer* 94, 327–333.

Wright, J.G. (1963) The surgery of the inguinal canal in animals. *Veterinary Record* 75, 1352–1363.

Wright, P.J. (1980) Serum sperm agglutinins and semen quality in the bull. *Australian Veterinary Journal* 56(1), 10–13.

Yanagimachi, R. (1990) Capacitation and the acrosome reaction. In: Asch, R.H., Balmaceda, J.P. and Johnston, I. (eds) *Gamete Physiology*. Norwell, MA Serono Symposium, pp. 31–42.

Yanagimachi, R., Yanagimachi, H. and Roger, B.J. (1976) The use of a zona-free animal ova as a test system for the assessment of the fertilising capacity of human spermatozoa. *Biology of Reproduction* 15, 471–472.

Yates, D.J. and Whitacre, M.D. (1988) Equine artificial insemination. *Veterinary Clinics of North America, Equine Practice* 4(2), 291–304.

Yi, L.H., Li, Y.J. and Hao, B.Z. (1983) A study on the conception rate of frozen horse semen. *Chinese Journal of Animal Science* 3, 2–4.

Yu Xueli, Pang YouZhi, Li YingHua and Wang ZhanBin (1997) Study on freezing canine pellet semen by the new method of frying in liquid nitrogen. *Chinese Journal of Veterinary Science* 17(3), 289–291.

Yurdaydin, N., Sevinc, A. and Wlader, W. (1985) Studies on freezing semen of stallions of different breeds. *Ankara Universitesi Veteriner Fakultesi Dergisi* 32(3), 446–455.

Yurdaydin, N., Tekin, N., Gulyuz, F., Aksu, A. and Klug, E. (1993) Field trials of oestrus synchronisation and artificial insemination results in Arab broodmare herd in the National Stud Hasire/Eskisehir (Turkey). *Deutsche Tierärztliche Wochenschrift* 100(11), 432–434.

Zafracas, A.M. (1975) *Candida* infection of the genital tract in Thoroughbred mares. *Journal of Reproduction and Fertility, Supplement* 23, 349.

Zafracas, A.M. (1994) The equines in Greece nowadays. *Bulletin of the Veterinary Medical Society* 45(4), 333–338.

Zall, R.R. (1990) Control and destruction of micro-organisms, microbiology of milk. In: Robinson, R.R. (ed.) *Dairy Microbiology*, 2nd edn. Elsevier Science, Amsterdam, pp. 126–127.

Zavos, P.M. and Gregory, G.W. (1987) Employment of the hyposmotic swelling (HOS) test to assess the integrity of the equine sperm membrane. *Journal of Andrology* 8, 25.

Zemjanis, R. (1970) Collection and evaluation of semen. In: *Diagnostic and Therapeutic Techniques in Animal Reproduction*, Williams and Wilkins, Baltimore, pp. 139–155.

Zgorniak-Nowosielska, I., Bielanski, W. and Kosiniak, K. (1984) Mycoplasmas in stallion semen. *Animal Reproduction Science* 7(4), 343–350.

Zhang, J., Boyle, M.S., Smith, C.A. and Moore, H.D.M. (1990a) Acrosome reaction of stallion spermatozoa evaluated with monoclonal antibody and zona-free hamster eggs. *Molecular Reproduction and Development* 27(2), 152–158.

Zhang, J., Rickett, S.J. and Tanner, S.J. (1990b) Antisperm antibodies in the semen of a stallion following testicular trauma. *Equine Veterinary Journal* 22, 138–141.

Zhang, J., Muzs, L.Z. and Boyle, M.S. (1991) Variations in structural and functional changes of stallion spermatozoa in response to calcium ionophore A23187. *Journal of Reproduction and Fertility, Supplement* 44, 199–205.

Zidane, N., Vaillancourt, D., Guay, P., Poitras, P. and Bigras-Poulin, M. (1991) Fertility of fresh equine semen preserved for up to 48 hours. *Journal of Reproduction and Fertility, Supplement* 44, 644.

Ziegler, R. (1991) Computerised videomicrographic evaluation of stallion semen before and after freezing in a diluent containing egg yolk. Thesis, Tierarztliche Fakulat der Ludwig.

Zivotkov, H.I. (1939) Fistuljyl metod polucenija spermy ot zerebcov I ego primenenie na praktike. (The fistula method of collecting sperm from stallions and its practical application.) *Sovstsk. Zooteh.* 1939, 4, 44–52. *Animal Breeding Abstracts* 8, 112.

Zmurin, L. (1959) K voprosu o sohranenii v zamorozenuom sostojanii semeni zerebca. (The storage of stallion semen by freezing.) *Konevodstvo* 29(4), 24–27. *Animal Abstracts* 27, 1223.

Zwain, I., Gaillard, J.-L., Dintinger, T. and Silberzahn, P. (1989) Down-regulation of testicular aromatization in the horse. *Biology of Reproduction* 40, 503–510.

Index

accessory glands 45, 46–50
    bacterial infection 222
acetylcarnitine 226
β-*N*-acetylglucosaminidase 87
acid phosphatase 226
acrosin 91, 144, 226, 227
acrosome 90–91, 112
acrosome reaction 144
acrosome vesicle 112–113
Actinomyces bovis 28
activin 59, 74, 77
adenosine triphosphate 226, 227
advantages of AI 1–5
Aerobacter 222, 249
age
    effect on semen characteristics 80
    effect on seminal volume 195
    effect on spermatogenesis 120–121
alkaline phosphatase 226
altrenogest 130, 173, 318, 319, 320
ampullae 49–50, 79, 80
anabolic steroids 130
anaphase 111
anaphase 2 112
anthelmintics 130–131
antibiotics in extenders 225, 248–250
arabinose 248
arginine 258
artificial vagina (AV) 151, 152
    assembly 161–162, 163
    Cambridge model 15, 151–153
    cleaning 160–161
    Colorado/Lane model 151, 152, 154–155
    CSU model 151–152, 154
    development 15–17
    in dummy mare 176
    filters 162, 164
    Hannover model 158, 159
    liner 161
    lubrication 165–166
    Missouri model 16, 156–158
    Nishikawa model 16, 155–156
    open-ended/Polish model 159, 160, 257
    preparation and maintenance 160–166
    Roanoke model 158–159
    semen collection procedure 181–185
    water jacket 162–165
ascorbic acid in extenders 250–251
aspartate aminotransferase 226
ATP 226, 227
AV *see* artificial vagina
axenome 95, 97, 99, 100, 101, 113

Bacillus 222
bacterial classification 222
bacteriology of semen 221–225
bacteriospermia 221
betaine 288
biochemical analysis of semen 226–227
blindfolding of stallions 188
blood–testis barrier 59, 138
blood typing 6
bovine serum albumin 247–248
breed
    and spermatogenesis 125–126
    and volume of semen produced 80, 81
breed societies
    acceptance of AI 22–24, 337, 339
    role in control of AI 25–26
breeding season
    effect on conception rate with frozen semen 298–300
    effect on seminal volume 195

breeding season *continued*
    effect on spermatogenesis 118–120
    extension 5
    stallions 63–67
Brucella abortus 28
BSA 247–248
bulbospongiosus muscle 37, 41, 45
bulbourethral glands 46, 48, 79
bute 131

calcium in seminal plasma 87
Candida 29
caprogen in extenders 251
Caslick index 31–32, 306, 309
Caslick operation 32, 306, 309, 310, 332, 337–338
Celle 274
CEM 223
centrifugation of semen 253–257
ceruloplasmin 59
cervical mucus penetration test 230
cervix 304, 305, 307
Chlamydia 222
cholesterol 91, 103, 144
chromatin analysis 232
CIDRs 318
citric acid 18, 86, 247
clomipramine 180
cobalt supplements 128
cold shock 262–266
competitive fertilization 231
concanavalin A 288
conception rates 335
    chilled semen 276
    frozen semen 296–300
    raw/fresh semen 260–261
condom semen collector 14, 166–167
contagious equine metritis 223
controlled internal drug-releasing dispensers 318
copper supplements 128
corona glandis 39, 41
corpus cavernosus penis 39, 41, 45
corpus cavernosus urethra 41, 45
cost of AI 7
Cowpers glands *see* bulbourethral glands
creatine 258
cremaster muscle 51, 53, 54, 55
cryoprotectants 277, 278–280, 340
cryptorchidism 55, 131–134
cytology of semen 221

diakinesis 111
diazepam 188
dihydrotestosterone 74, 75, 76

diplotene 111
disadvantages of AI 5–8
disease
    control 4
    transfer risks 7
dourine 29
drugs affecting spermatogenesis 129–131
Druschia 199
ductus deferens 49, 50–51
dummy mare 16–17, 171, 174–177
dynein 99, 146, 147

eCG 70
ejaculation 45–46, 80
    encouragement of 170–177
    frequency
        and seminal volume 195
        and spermatogenesis 123–125
    pharmacologically induced 180
    retrograde 198
electroejaculation 15, 78, 181
embryo sexing 338–339
emission 45
endometritis 311
    causative bacteria 222
    post-coital 3, 31
    prediction 306
endoscopy of mares reproductive tract 311–312
environmental influences on spermatogenesis 126–131
eosin–nigrosin stain 213, 214, 219–220
epididymis 45, 46, 51–52, 53, 55, 59
equilenin 74
equilin 74
equine chorionic gonadotrophin 70
equine coital exanthema 28–29
Equine Express 274
equine herpes viruses 29, 225
equine influenza 28
equine viral arteritis 28–29, 225–226
Equitainer 19, 28, 269, 271–273, 274
erection 41, 45
ergothionine 18, 86
Escherichia coli 28, 199, 222, 249, 311
ethics 6, 337–338
EVA 28–29, 225–226
exercise, effect on spermatogenesis 131
Expecta Foal 274
extenders 18–19, 202–203, 225, 235, 250–252
    antibiotics in 225, 248–250
    bovine serum albumin 247–248
    in centrifugation 254–255
    chilled semen 261–262, 268, 269–270, 275–276
    citric acid 247

cream and gel 242–244
current use 237–238
egg yolk 245–246
energy sources 248
fresh/raw semen 259–260
frozen semen 280–288, 295
historical development 235–237
illini variable temperature (IVT) 268
milk and milk-product 238–242
osmolarity 258
pH 258
thawing 325
TRIS 244–245

fertility
　effects of sperm abnormalities on 216–219
　and sperm motility 265
fertilization 88, 145
　competitive 231
filtering of semen 193–194, 257, 261
filtration assay 228–229
fixed-time insemination 4, 321
Flehman 76
flow cytometry 228, 232
follicle-stimulating hormone, stallions 62, 69, 70–71, 75, 76–77
France, control of equine AI 24–25
fraud 6
fructose
　in extenders 248
　in seminal plasma 85
FSH, stallions 62, 69, 70–71, 75, 76–77
funding 340

galactose 248
β-galactosidase 87
α-1, 4-galactosyltransferase 103–104
GAP 70
gene banks 3, 30, 338
genetic improvement of stock 1–2
genetic pool reduction 5
Germany, control of equine AI 25–26
glanders 28
glandula bulbourethralis *see* bulbourethral glands
glans penis 38–41, 42, 43
glucose
　in extenders 248
　in seminal plasma 85
　in sperm metabolism 141–142
glutamic oxaloacetic transaminase 226
glycerol
　as cryoprotectant 277, 279, 280, 285–288, 340
　in extenders 252

glycerylphosphorylcholine 86–87
glycine in extenders 252
glycocalyx 104, 264–265
glycosidases 87
GnRH-associated protein 70
gonadotrophin-releasing hormone (GnRH) 68, 70, 75, 76, 77
　in manipulation of oestrus 317
　pharmacological administration 130
gonocytes 62
GPC 86–87
granulosa cell tumour, ovaries 174

Habronema 29, 199, 200
haemocytometry 209–210, 211, 212
Haemophilus equigenitalis 311
haemospermia 196, 198–201
hCG 21, 130, 173–174, 315–317, 320, 321
hemizona assay 232
heparin-binding proteins 84
hernia 134–135
heterospermic insemination 231
history 1, 9–12
hobbles 174, 175
hormones
　effects of abnormalities on spermatogenesis 131
　hypothalamic–pituitary–gonadal axis 69–71
　in regulation and control of spermatogenesis 75–77
　testicular 72–75
horse population, decline in 12–13
Horse Race Betting Levy Board, Codes of Practice 223
HSP-1 84
human chorionic gonadotrophin 21, 130, 173–174, 315–317, 320, 321
hyaluronidase 91, 226
hydrogen peroxide 142
hymen 305
hypo-osmotic stress test 229–230
hypothalamic–pituitary–gonadal axis, stallions 67–69, 69–71, 75–77
hypothyroidism 131

imipramine 180
immunological infertility 138
implantation fossa 113
infection
　mares 31, 303, 311
　reduction of risk of transfer 28–29
　stallions 33–34, 139, 221–226
inguinal canal 51

inhibin 59, 74, 77
injury
    mares 31
    removal of risk 4–5
    stallions 5, 33
inositol 86
insemination 302, 335–336
    development of techniques 20–21
    factors affecting success 332–335
    fixed-time 4, 321
    laparoscopic 331
    methods 329–332
    per rectum 21, 330–331
    per vagina 329–330, 331, 332
    post-ovulatory 334–335
    preparation of mares 312–322
    preparation of semen 322–329
    repeat 7
    selection of mares 302–312
    timing and frequency 333–335
iron deficiency 128
ischiocavernosus muscles 41, 45
isolation 4

jump mare 171, 172–174, 175, 186

keep fees 6
Kenney index 122–123
Klebsiella 28, 29, 222, 223–224, 225, 249, 311

labour costs 35
lactate dehydrogenase 226
lactic acid
    secretion by Sertoli cells 59
    in seminal plasma 85
lactose 248
lamina propria 58
laminin 58
lectins 227
Leidig cells 60–61, 72, 129
    effects of toxins on 129
leptotene 109
luteinizing hormone (LH), stallions 62, 69, 70–71, 75, 76, 77, 129

magnesium 87
manchette 113–114
manganese supplements 128
mannose 248
mannosidases 87
mares

    age 304
    condition 304
    difficult 30–32
    history 303
    infection 31, 303, 311
    injury 31
    insemination 302–326
    ovariectomized 174
    perineal conformation 31–32
    preparation 312–322, 323
    psychological problems 32
    rectal palpation 308, 322
    reproductive tract
        bacterial culturing 311
        competence 304–312
        endoscopy 311–312
        external examination 304–308, 309, 310
        internal examination 5, 308, 310–312
        ultrasonography 310, 322
    selection 302–312
    temperament 303–304
    veterinary examination 322
meiosis in spermatogenesis 109–112
melatonin 67, 69–70, 75
MEM 251
membrane integrity tests 227–228
metaphase 111
metaphase 2 112
methylxanthinines in extenders 251
Micrococcus 222
minerals in seminal plasma 87
minimum essential medium 251
mycoplasma, semen 226

native stock improvement 29–30
Neisseria 222
Netherlands, control of equine AI 26
neurophysin 74
nexin 99
nutritional effect on spermatogenesis 128–129

oestradiol
    in manipulation of oestrus 320–321
    stallions 74, 75
oestrogens, stallions 74–75, 76, 77
oestrone, stallions 74, 75
oestrous cycle
    manipulation 20–21, 172–174, 312–321
    natural 312–313
orchitis 139
osmometer 201
ovariectomy 174
ovaries

cystic 174
    granulosa cell tumour 174
oviductal epithelial cell explant test 230
ovulation, manipulation of timing 20–21, 172–174, 312–321
ovum, duration of viability 333
oxytocin, stallions 74, 77

pachytene 111
packaging
    chilled semen 271
    frozen semen 289–292
pampiniform plexus 54
paraventricular nucleus 67
pelvic floor, ideal height 305, 308
penis
    anatomy 37–41, 42–44
    body 41, 44
    erection 41, 45
    roots 41
performance index 25
perineal conformation 31–32
phantom mare 16–17, 171, 174–177
phenylbutazone 131
phosphatidylcholine 263
phosphatidylethanolamine 263
phospholipids 103, 140, 263–264
phosphorus 87
pineal gland 67, 69, 75
pituitary 69, 70–71
plasma membrane, spermatozoa 101–104
    effects of cold shock 262–265
pneumovagina 305
polyvinyl alcohol in extenders 251
post-coital immunological response 2
potassium 87
Pouret operation 306
preoptic nucleus 67, 68
prepuce 37, 38
prevalence of equine AI 21–24
PRIDs 318, 320–321
primordial cells 62
proacrosin 91
progestagens 130
    in manipulation of oestrus 172–173, 318
progesterone, in manipulation of oestrus 172–173, 318–321
progesterone-releasing intervaginal devices 318, 320–321
prolactin, stallions 70, 71, 77
prophase 109–111
prophase 2 112
prostaglandins, in manipulation of oestrus 172–174, 313–317, 319–320, 321

prostate gland 48–49, 79
protein
    deficiency 128
    in plasma membrane 264
    in seminal plasma 84–85
proteinase inhibitors in extenders 251
Proteus 28, 222, 249
Pseudomonas
    mare's reproductive tract 311
    reduction of risk of transfer 28, 29
    in semen 221, 222, 223, 224, 225, 249
    in urethritis 199
psychological problems
    mares 32
    stallions 34–35
puberty 62
pyospermia 196, 201

quality control 341

radiation, effects on spermatogenesis 128
rare breed preservation 4
regulations
    for equine AI 24–27, 339
    for semen transport 29
reproductive activity, stallions
    behavioural control 75, 76
    commencement 62–63
    decline with age 63
    hormonal control 67–75
    seasonality 63–67
reproductive anatomy, stallions 37–61
reproductive control, stallions 62–77
research
    decline in 12–13
    funding 340
responsibility for conception 7–8
rete testis 51, 52, 59
retractor penis muscle 37, 41, 45
ribose 248
rigs 55, 131
rose penis *see* glans penis
rostral nucleus 67
rubber bag semen collector 14

scrotum
    anatomy 53–56
    skin 56
seeding 278
selenium deficiency 128
semen 78–82
    appearance 196–201

semen *continued*
   bacteriology 221–225
   biochemical analysis 226–227
   centrifugation 253–257
   characteristics
      factors affecting 80–82
      normal range of values 78
   chilled 10, 18, 261–276
      conception rates 276
      cooling rate 266–267
      dilution rates 275–276
      length of storage 268–269
      packaging 271
      preparation for use 326
      storage temperature 267–268
      transport 27–28, 271–275
   collection 14, 151–189
      by manual stimulation 178
      collecting area 170
      development of techniques 14–17
      frequency 185–186
      minimal restraint during 178–179
      preparation of stallion 167–170
      procedure 181–185
      risks involved 8
      sexual stimulation of stallion 170–177
      training the stallion 186–189
      without mounting 179–181
   concentration 34
   cytology 221
   evaluation 190–233
      development of techniques 17–18
      functional tests 226–232
      gross 193–202
      microscopic 202–226
   filtering 193–194, 257, 261
   fresh/raw 258–261
   frozen 13, 276–301, 340–341
      conception rates 296–300
      cooling rates 292–294
      development of techniques 19–20
      dilution rates 295–296
      packaging 289–292
      preparation for use 326, 327, 328
      thawing 294–295, 323–325
   handling 18–20, 191–193
   hypertonicity 201
   hypotonicity 201
   inadequacy 34
   insemination dose 326–327
   insemination volume 328–329
   mycoplasma 226
   osmolarity 201
   ownership problems 6
   pH 201–202
   post-coital fraction 78, 80
   post-sperm/gel fraction 46, 78, 79–80
   preparation for insemination 322–329
   pre-sperm fraction 46, 78, 79
   routine evaluation and monitoring 2
   sperm-rich fraction 46, 78, 79
   storage 234–301
      development of techniques 18–20
      long-term 3
   transport 27–28, 29, 271–275
   variations in quality 6–7
   virology 225–226
   volume 17
      determination 194
      factors affecting 195–196
seminal plasma 46, 78
   abnormalities 87–88
   characteristics 18
   composition 82–87
   functions 252
   pH 18
   removal 252–257, 260, 271, 288–289
seminal vesicles 49, 79
seminal vesiculitis 199
seminiferous tubules 58–60, 105
   epithelium cycle 115–118
seminoma 137
Sertoli cells 59–60, 72, 105, 108, 117
   effects of toxins on 129
   seasonal changes in numbers 118
smegma 38–39
sodium 87
sodium chloride 252
sorbitol 85
spectrophotometry 18, 210–213
sperm counter 212–213
spermatids 59, 62, 108, 109, 112–115
spermatocytes 62, 108, 109, 112
spermatocytogenesis 106–109
spermatogenesis 62, 88, 104–115
   factors affecting 118–139
   hormonal regulation and control 75–77
spermatogonia 62, 106–108
   seasonal changes in numbers 118–119
spermatozoa 78, 88, 149–150
   abnormalities 17, 215–219
   acrosome region 90–91, 112
   anatomy 88–104
   annulus 98, 100
   average daily production 105–106
   axenome 95, 97, 99, 100, 101, 113
   capacitation 143–145
   capitulum 93–94
   cold shock 262–266
   concentration 17–18, 209–213

deposition 41, 45–46
effects of bacterial contamination on 224–225
end-piece 101, 102
epididymal storage 52, 121
filtration assay 228–229
flagellum *see* spermatozoa, principal piece
fragility 19
function 145
glycocalyx 104, 264–265
head 88–92, 93
hemizona assay 232
hypo-osmotic stress test 229–230
live:dead ratio 219–221
longevity 208–209, 333
maturation 52
membrane integrity tests 227–228
metabolism 140–143
mid-piece 95, 96–100
morphology 213–219
motility 203–208, 234, 265, 325
movement 146–149, 150
neck 92–96
normal measurements 17
nucleus 89–90
oviductal epithelial cell explant test 230
plasma membrane 101–104
    effects of cold shock 262–265
post-acrosomal region 91–92
principal piece 100–101, 113, 114–115, 145–148
proximal centriole 94
sexing 232, 339
staining 213–214, 215
zona-free hamster ova penetration assay 231
spermiation 115
spermiodensimeter 211–212
spermiogenesis 112–115
sphingomyelin 263
staining, spermatozoa 213–214, 215
stallion rings 199
stallions
    blindfolding 188
    classification 186–187
    concurrent careers 3–4
    difficult 33–35
    infection 33–34, 139, 221–226
    injury 5, 33
    manual stimulation 178
    minimal restraint 178–179
    preparation for semen collection 167–170
    psychological problems 34–35
    reproductive anatomy 37–61
    reproductive control 62–77
    sexual stimulation 170–177
    training for semen collection 186–189
    use after death 338
    washing 168–170, 224
Staphylococcus 28, 222, 249, 311
Starstedt 274
starvation regime 21
state control 24–26, 340
strangles 28
Streptococcus 28, 199, 222, 224, 225, 249, 311
Strongylus endenatus 188
Stud SCORE Kit 207
sub-fertility of stallions 2
sucrose in extenders 248
summer sores 29, 199
supraoptic nucleus 67
Sweden, control of equine AI 26
Switzerland, control of equine AI 26
synapsis 109–110

taurine in extenders 251
Taylorella equigenitalis 28, 29, 222–223, 224, 225
teasing 168, 171, 172–174, 175, 186
    effect on seminal volume 196
teasing board 180
telophase 111
telophase 2 112
temperament, mares 303–304
temperature, effect on spermatogenesis 126–128
teratoma 137, 138
testes 51
    anatomy 53–61
    degeneration 136–137
    disease and infection 139
    herniation 134–135
    hormone secretion 72–75
    hypoplasia 136
    internal structure 56–58
    interstitial tissue 60–61
    neoplasms 137, 138
    parenchyma 57–58
    size 55–56
        and spermatogenesis 121–123
    torsion 137
    undescended 55, 131–134
testosterone 60–61, 72–73, 75, 76
    low levels 131
    pharmacological use 130
thyroid-stimulating hormone 70
toxic chemicals, effect on spermatogenesis 129–131
transferrin 59
transport of semen 27–28, 29, 271–275
tris(hydroxymethyl)-aminomethane (TRIS) 244–245
Trypanosoma 29
TSH 70
tubulin 99, 146

tunica albuginea 39, 41, 57–58, 123
tunica dartos 56
tunica vaginalis 56–57
Tyrode's solution 251

ultrasonography of mare's reproductive tract 310, 322
United Kingdom, control of equine AI 26–27
urethra 38, 48
    ulceration 199
urethral fistula 14
urethral fossa 38
urethralis muscle 45
urethritis 199, 200–201
urethrostomy, subischial 200
urospermia 196–198
uterine infection 303

vagina 304, 305, 307
vaginal sponge 14
vas deferens 49, 50–51
vesicular glands 49, 79

virology, semen 225–226
vitamin C in extenders 250–251
vitamin deficiencies 128
vitamin E in extenders 250–251
vulva 304, 305, 306, 307
vulva constrictor muscle 305

washing of stallions 168–170, 224
workload
    and seminal volume 195
    and spermatogenesis 123–125

xylazine 180
xylose 248

zinc
    deficiency 128–129
    in seminal plasma 87
    supplements 128
zona-free hamster ova penetration assay 231
zygotene 109–110